Lecture Notes in Physics

The Lecture Notes in Physics

The series Lecture Notes in Physics (LNP), founded in 1969, reports new developments in physics research and teaching—quickly and informally, but with a high quality and the explicit aim to summarize and communicate current knowledge in an accessible way. Books published in this series are conceived as bridging material between advanced graduate textbooks and the forefront of research and to serve three purposes:

- to be a compact and modern up-to-date source of reference on a well-defined topic
- to serve as an accessible introduction to the field to postgraduate students and nonspecialist researchers from related areas
- to be a source of advanced teaching material for specialized seminars, courses and schools

Both monographs and multi-author volumes will be considered for publication. Edited volumes should, however, consist of a very limited number of contributions only. Proceedings will not be considered for LNP.

Volumes published in LNP are disseminated both in print and in electronic formats, the electronic archive being available at springerlink.com. The series content is indexed, abstracted and referenced by many abstracting and information services, bibliographic networks, subscription agencies, library networks, and consortia.

Proposals should be sent to a member of the Editorial Board, or directly to the managing editor at Springer:

Christian Caron
Springer Heidelberg
Physics Editorial Department I
Tiergartenstrasse 17
69121 Heidelberg/Germany
christian.caron@springer.com

For other titles published in this series, go to
www.springer.com/series/5304

Jonas Fransson

Non-Equilibrium Nano-Physics

A Many-Body Approach

Springer

Prof. Jonas Fransson
Dept. Physics
Uppsala University
Uppsala
Sweden
jonas.fransson@fysik.uu.se

Fransson, J., *Non-Equilibrium Nano-Physics: A Many-Body Approach*, Lect. Notes Phys. 809 (Springer, Dordrecht 2010), DOI 10.1007/978-90-481-9210-6

ISSN 0075-8450 e-ISSN 1616-6361
ISBN 978-90-481-9209-0 e-ISBN 978-90-481-9210-6
DOI 10.1007/978-90-481-9210-6
Springer Dordrecht Heidelberg London New York

Library of Congress Control Number: 2010930713

Cover design: eStudio Calamar S.L.

Printed on acid-free paper

Springer is part of Springer Science+Business Media (www.springer.com)

I want to climb this ladder up to the stars
I feel no fear, from this height
Though there's no rescue in sight
A hollow yearning, and nothing learning
An ancient look at the watch on my wrist
My life has just been dismissed

I'm about to die, and I think I will
Have nothing left to loose, as I never had any skill

Preface

At some point earlier I would probably have named this book *f-electron methods applied to nanoscale systems*, which is a formally correct statement, however, it feels less relevant nowadays. Why? Because, although it is appealing to make the connection to the tradition of strongly correlated electron system and related issues, there are many questions traditionally considered in a textbook on strong electron correlations, that will not be covered within the present text. The intention with this book is not give an account of strongly correlated systems as such. Rather, the intention is to present a formulation of the non-equilibrium physics in nanoscale systems in terms of many-body states and operators and, in addition, discuss a diagrammatic approach to Green functions expressed by many-body states. Thus, the issues focused on in this book are results of typical questions that arise when addressing nanoscale systems from a practical point of view, e.g. current-voltage asymmetries, negative differential conductance, spin-dependent tunneling, local vibrations, and coupling to superconducting leads.

The use of many-body states and operators constructed of such states was previously introduced by, e.g. Hubbard 1963, but others have preceded him and many more will doubtlessly pick up ides along those lines. It is my aim to give a reasonable introduction to a formalism of non-equilibrium Green functions (NEGFs) expressed in terms of many-body operators. It is, however, more interesting to provide meaningful reasons for considering and using many-body states and many-body operator Green functions (MBGFs) in studies of localized electrons interacting with a de-localized environment. The strengths of any method based on many-body states becomes best visualized in systems where the localized electrons interact via e.g. Coulomb repulsion, hopping/tunneling, and exchange, and where the energy scales of these interactions are comparable. In other words, in systems where it does not make sense to pick out a single energy scale and consider it large in comparison with the others, it is often preferable to transform the localized subsystem into, e.g. its many-body eigenstates. Such a formulation gives a freedom in varying the energy scales of the localized subsystem without worrying about their mutual relationship. In short, the focus will be on nanoscale systems constituted of complexes of subsystems interacting with one another, under non-equilibrium conditions, in which the local properties of the subsystems are preferably being described in terms of its (many-body) eigenstates.

Although the content of this book is discussed from the perspective of the physics, the book may still be considered as a book on a technique, or combinations of techniques. The discussion above mentions the many-body operator Green functions, which will be introduced and discussed at length. The discussion will, however, be focused on non-equilibrium conditions, which means that only little space will be spent on special techniques that may be used under strict equilibrium conditions. For those interested in strongly correlated electrons in the equilibrium case I refer to the excellent book by Ovchinnikov and Val'kov. Here, the technique will, thus, be set in the framework of non-equilibrium Green functions (NEGFs) and the formalism developed by Kadanoff and Baym, and Keldysh.

In order to develop a systematic approach that we can both apply to non-equilibrium conditions while still being efficient in the treatment of correlated states, one should be working with imaginary time contour ordered averages of operators. This enables a systematic diagrammatic expansion of our averages, expansions that usually are necessary to conduct in the class of systems under considerations. The diagrammatic expansions discussed here, are outlined by means of functional differentiations of averages.

My aim is that this book can be read by graduate students that have some experience in quantum mechanical field theory, Dirac formalism, second quantization, and quantum statistical methods. I certainly hope that experienced researcher will take up this book as well. Much of the content will be presented in a basic language, such as equation of motion and expansions, and I will not go into the deeper aspects given in a path integral approach. With this said, I thus hope that the present text will be accessible to many more readers that only to those who have a very deep fundamental understanding of the intricate world of quantum field theory.

The organization of this book is thought of as a bit evolutionary, in the sense that it begins with a class of problems where one encounters problems when working with conventional field theoretical methods. Then, the concept of many-body states and many-body operators is introduced and the Green functions are constructed and discussed. Only after this, the systems are being simplified in order to better illustrate the technique itself. This is meant to turn focus on the technique rather than on the complexity of the physical system. As the concepts are becoming familiar we can again add complexity

Finally, it is with a great pleasure I thank I. Sandalov for teaching me about non-equilibrium, Green functions, many-body operators, and strongly correlated electron systems. I would also like to thank A.V. Balatsky and J.-X. Zhu for introducing me into STM techniques and spin dynamics in non-equilibrium. Further, I am grateful to M. Galperin for sharing his views on extensions of the Hubbard operator scheme to include electron-vibron coupled systems. My Ph.D. student P. Berggren has done a good job in proof reading parts of the text, for which I thank him. I am indebted to O. Eriksson and L. Nordström for being understanding and patient with my questions, discussions, and ideas concerning correlated electron systems and the use of Hubbard operators in various possible and impossible instances. Last but not least, I want to express my gratitude towards my wife Johanna, and my children Eugenia, Elmer, Wilbur, and Werner, which have been and still are tremendously patient with me.

Uppsala Jonas Fransson

Contents

Chapter 1
Many-Body Representation of Physical Systems

Abstract We begin the discussion with a consideration of the type of nanoscale systems that we are interested in. Using those we introduce the concept of many-body operators through and discuss some reasons for taking this route.

1.1 Many-Body States

In a nanoscale system, e.g. quantum dot or molecule coupled to leads, or atomic cluster on surface, one often has to simultaneously deal with different energy and length scales. That these different scales meet at the same playground can be regarded as one of the hallmarks of nanoscale science, since it gives the opportunity to, atom by atom, engineer structures with properties that would never occur naturally. It is, however, one of the obstacles with nanoscale systems, which also leads to the great challenges and possibilities. Here, we will not dwell on the possibilities with nanoscience, but rather discuss a technique and ways to theoretically treat some problems that arise in non-equilibrium nanoscale systems from a physical point of view.

Suppose that we are interested in the electronic and magnetic properties of a cluster of Mn atoms lying on a Cu surface, and suppose that we want to make a systematic study with respect to the number of Mn atoms in the cluster. Such issues were addressed experimentally by, e.g. Hirjibehedin et al. [1]. One of the major challenges in giving a theoretical description of such a system is the modeling of the cluster of Mn atoms itself. But also, this cluster is interacting with the substrate surface, in the present case a metallic Cu surface. In the experiment [1], there was also a layer of insulating CuN in between the cluster and surface. These additional interactions provide further complications to the description, nevertheless, the question still remains, i.e. how do we model the cluster on the surface?

Another example is found by considering quantum dots being coupled to one another by electrostatic forces, tunneling, and spin interactions. Here, we have already given a view point of how the model will appear. However, we need to go to more

J. Fransson, *Non-Equilibrium Nano-Physics*,
Lecture Notes in Physics 809,
DOI 10.1007/978-90-481-9210-6_1,

detail when actually writing down the model used for calculations. We assume, for instance, that the quantum dot creates a quantum well which carries a number of electron levels. Will the electrons in the different levels interact through Coulomb forces, tunneling and exchange? Which interactions between the quantum dots shall we consider. In the end, all comes down to which properties we are interested in describing.

Therefore, let us do it more tangible. Suppose that the quantum dots each carry a set of localized levels, labeled $\varepsilon_{\nu i\sigma}$, where ν is the quantum dot label, i some state label, and σ a spin label. Also, assume that we include the Coulomb and tunneling interactions within the quantum dots. We can thus model one of the quantum dots as

$$\begin{aligned}\mathcal{H}_\nu = &\sum_{i\sigma} \varepsilon_{\nu i\sigma} d^\dagger_{\nu i\sigma} d_{\nu i\sigma} + \sum_i U_{\nu i} n_{\nu i\uparrow} n_{\nu i\downarrow} - 2\sum_{ij} J_{\nu ij} \mathbf{s}_{\nu i} \cdot \mathbf{s}_{\nu j} \\ &+ \sum_{ij} (U_{\nu ij} - J_{\nu ij}/2)(n_{\nu i\uparrow} + n_{\nu i\downarrow})(n_{\nu j\uparrow} + n_{\nu j\downarrow}) \\ &+ \sum_{\nu ij\sigma} (t_{\nu ij\sigma} d^\dagger_{\nu i\sigma} d_{\nu j\sigma} + H.c.). \end{aligned} \tag{1.1}$$

Here, $d^\dagger_{\nu i\sigma}$ ($d_{\nu i\sigma}$) creates (annihilates) an electron at the energy level $\varepsilon_{\nu i\sigma}$, $n_{\nu i\sigma} = d^\dagger_{\nu i\sigma} d_{\nu i\sigma}$ is the number operator, $U_{\nu i}$ ($U_{\nu ij}$) is the intra-level (inter-level) Coulomb interaction, and $J_{\nu ij}$ and $t_{\nu ij\sigma}$ is the exchange and tunneling interaction, respectively. The spin operators $\mathbf{s}_{\nu i} = \sum_{\sigma\sigma'} d^\dagger_{\nu i\sigma} \hat{\tau}_{\sigma\sigma'} d_{\nu i\sigma'}$, where $\hat{\tau}$ is the vector of Pauli matrices.

One can see that already at this level, where we only consider electronic interactions within a single quantum dot, that the model becomes quite involved. How about joining the quantum dots together? Which interactions will be relevant to include between the quantum dots? Then, in the end we would also like to connect the ensemble of quantum dots to some external bath, perhaps through leads, in order to establish a non-equilibrium situation. In a realistic system, we should also have to worry about electrons interacting with local vibrational modes, or other sources of inelastic scattering. In short, the systems we want to study grow in complexity each time we add another (fermionic or bosonic) degree of freedom to the system.

The way that will be pursued in this text is through diagonalization, i.e. finding the many-body eigenstates, of the subsystem under main interest. The diagonalization of the subsystem leads to a many-body representation of the localized states in which all internal parameters become implicit in the model, and only the interactions with the environment appear explicitly. Thus, one benefit of this new representation is that we can think of the subsystem as an effective unit in presence of external fields.

Explicitly, the model given in (1.1) has an associated system of eigenenergies $E_{\nu Nn}$ and eigenstates $|\nu Nn\rangle$ with N electron in the nth state. Thus, the intricate

Hamiltonian can be replaced by the diagonal form

$$\mathcal{H}_\nu = \sum_{Nn} E_{\nu Nn} |\nu Nn\rangle\langle \nu Nn|. \tag{1.2}$$

One has to be aware that this is merely a change in representation and that there is no simplification made going from the models in (1.1) and (1.2).

The system of eigenstates satisfy the closure relation $\sum_{Nn} |\nu Nn\rangle\langle \nu Nn| = 1$. Thus, the transformation between the two representations can expanded as

$$d_{\nu i\sigma} = \sum_{NN'nn'} |\nu Nn\rangle\langle \nu Nn| d_{\nu i\sigma} |\nu N'n'\rangle\langle \nu N'n'|. \tag{1.3}$$

The operator $d_{\nu i\sigma}$ removes one electron in the ith level with spin σ from the n'th state in the N' electron configuration. Hence, the above relation will only make sense if $N' = N + 1$ which admits us to write

$$d_{\nu i\sigma} = \sum_{Nnn'} \langle \nu Nn| d_{\nu i\sigma} |\nu N+1n'\rangle |\nu Nn\rangle\langle \nu N+1n'|. \tag{1.4}$$

Note that the single annihilation Fermi operator $d_{\nu i\sigma}$ is represented by the sum of removing an electron from all states within the local subsystem. This is natural, partly since we do not know *a priori* from which state the electron is removed, hence we need to include all possibilities, but also since this is exactly one of the strengths with the Fermi operator representation, one does not have to keep track of the individual states.

In the diagonal representation, which henceforth will be referred to as the *many-body representation*, the operators are the outer products (projection operators) $|\nu Nn\rangle\langle \nu N'n'|$. In this representation we must keep track of the individual states, which is a price we have to pay for the benefits of other advantages. Nevertheless, sometimes it is desirable to keep track of the individual states and to have an explicit record of what is going on between the states. Therefore, the transition matrix elements $\langle \nu Nn| d_{\nu i\sigma} |\nu N+1n'\rangle$ play a crucial role in the many-body operator representation, which will become clear in Sect. 1.4.

The matrix element $\langle \nu Nn| d_{\nu i\sigma} |\nu N+1n'\rangle$ describes the single electron transition between the states $|\nu Nn\rangle$ and $|\nu N+1n'\rangle$ when removing an electron in the ith level with spin σ. This need not be possible at all, of course, but then this matrix element vanishes. For instance, the logics that led to (1.4), i.e. trowing away all the matrix elements that do not represent a single electron transition from an $N+1$ to an N electron state, is the result of not writing lots of vanishing matrix elements, which vanish due to impossible transitions. Any matrix element that is non-zero, however, represents a physical single electron transition that *can and will* occur within the subsystem. The probability for this process to take place may be large or small, a real number between 0 and 1, and is given by the square of the probability amplitude $|\langle \nu Nn| d_{\nu i\sigma} |\nu N+1n'\rangle|^2$.

In this section, we have used the quantum dot term in a rather non-precise manner and we will continue in this way. Henceforth, we shall, thus, discuss correlated (sub-) systems in terms of a general abstract quantum dot concept when we want to avoid any reference to specific types and classes of systems. This is a quite common approach in literature, here we might bend the concept a little further.

1.2 Two-Level System

For the sake of being explicit, we consider a concrete example constituted of a double quantum dot system coupled to external leads. We assume that the level separation in the quantum dots is sufficiently large in order to neglect influences from all levels but one in each quantum dot. Thus, we have a two level system where the levels are spatially separated and located in different quantum dots. We assume that electrons in the two levels are interact through tunneling and charging interactions. We may also assume that the electrons interact through direct spin-spin interactions.

We realize that the above physics is very well captured by the model given in (1.1), however, reduced to

$$\begin{aligned}\mathcal{H} = &\sum_\sigma \varepsilon_{A\sigma} d^\dagger_{A\sigma} d_{A\sigma} + U_A n_{A\uparrow} n_{A\downarrow} + \sum_\sigma \varepsilon_{B\sigma} d^\dagger_{B\sigma} d_{B\sigma} + U_B n_{B\uparrow} n_{B\downarrow} \\ &+ (U' - J/2)(n_{A\uparrow} + n_{A\downarrow})(n_{B\uparrow} + n_{B\downarrow}) - 2J \mathbf{s}_A \cdot \mathbf{s}_B \\ &+ t \sum_\sigma (d^\dagger_{A\sigma} d_{B\sigma} + H.c.). \end{aligned} \tag{1.5}$$

Our goal is to transform this Hamiltonian model into a diagonal form $\mathcal{H} = \sum_{Nn} E_{Nn} |Nn\rangle\langle Nn|$, where the states $|Nn\rangle$ are eigenstates of the Hamiltonian. By including spin into the treatment, we find that there are 16 eigenstates of the model; one state with no electrons ($N = 0$, $n = 1$), four states with one electron ($N = 1$, $n = 1, \ldots, 4$), and so on, i.e. ($N = 2$, $n = 1, \ldots, 6$), ($N = 3$, $n = 1, \ldots, 4$), ($N = 4$, $n = 1$).

We can find the eigenstates through different means. The present model is sufficiently small to be reasonable for analytical calculations. The empty and the four electron states are trivial to find, and we can write them as $|01\rangle = |0\rangle_A|0\rangle_B$ and $|41\rangle = |\uparrow\downarrow\rangle_A|\uparrow\downarrow\rangle_B$ in terms of the Fock states $|p\rangle_A|q\rangle_B$, $p, q = 0, \uparrow, \downarrow, \uparrow\downarrow$, of the quantum dot.

Throughout this book we will use the following conventions for the Fock states:

$$|p\rangle_A|q\rangle_B = d^\dagger_{Bq} d^\dagger_{Ap}|0\rangle, \tag{1.6a}$$

$$|\uparrow\downarrow\rangle_{A(B)} = d^\dagger_{A(B)\downarrow} d^\dagger_{A(B)\uparrow}|0\rangle, \tag{1.6b}$$

where $|0\rangle$ denotes the vacuum state. Hence, $|\uparrow\downarrow\rangle_A|\uparrow\downarrow\rangle_B = d^\dagger_{B\downarrow} d^\dagger_{B\uparrow} d^\dagger_{A\downarrow} d^\dagger_{A\uparrow}|0\rangle$ etc.

The energies for these states are $E_{01} = 0$ and $E_{41} = \sum_\sigma(\varepsilon_{A\sigma} + \varepsilon_{B\sigma}) + U_A + U_B + 4(U' - J/2)$, respectively.

The one-electron states $|1n\rangle$ are found by acting with the one-electron Fock states onto the Hamiltonian, e.g.

$$\mathcal{H}|\uparrow\rangle_A|0\rangle_B = \varepsilon_{A\uparrow}|\uparrow\rangle_A|0\rangle_B + t|0\rangle_A|\uparrow\rangle_B,$$

$$\mathcal{H}|0\rangle_A|\uparrow\rangle_B = \varepsilon_{B\uparrow}|0\rangle_A|\uparrow\rangle_B + t|\uparrow\rangle_A|0\rangle_B,$$

and analogously for the spin down states. These equations lead to the one-electron states

$$|1n\rangle = \alpha_n|\uparrow\rangle_A|0\rangle_B + \beta_n|0\rangle_A|\uparrow\rangle_B, \quad n = 1, 3, \tag{1.7}$$

$$|1n\rangle = \alpha_n|\downarrow\rangle_A|0\rangle_B + \beta_n|0\rangle_A|\downarrow\rangle_B, \quad n = 2, 4. \tag{1.8}$$

These states are clearly seen to be superpositions of the available states for each spin projection. Such states are also known as coherent states, although such a terminology is not quite correct in Fermi systems. By also allowing for spin-flip in the tunneling between the levels, these superpositions would consist of all four one-electron Fock states.

The expansion coefficients are subject to normalization of the eigenstates, and here given by

$$\left.\begin{aligned}\alpha_n^2 &= \frac{1}{2}\frac{\xi_n^2}{1+\xi_n^2+\sqrt{1+\xi_n^2}\,\mathrm{sign}\,\Delta\varepsilon},\\ \beta_n^2 &= \frac{1}{2}\frac{(1+\sqrt{1+\xi_n^2}\,\mathrm{sign}\,\Delta\varepsilon)^2}{1+\xi_n^2+\sqrt{1+\xi_n^2}\,\mathrm{sign}\,\Delta\varepsilon},\end{aligned}\right\} \quad n = 1, 2, \tag{1.9a}$$

$$\left.\begin{aligned}\alpha_n^2 &= \frac{1}{2}\frac{\xi_n^2}{1+\xi_n^2-\sqrt{1+\xi_n^2}\,\mathrm{sign}\,\Delta\varepsilon},\\ \beta_n^2 &= \frac{1}{2}\frac{(1-\sqrt{1+\xi_n^2}\,\mathrm{sign}\,\Delta\varepsilon)^2}{1+\xi_n^2-\sqrt{1+\xi_n^2}\,\mathrm{sign}\,\Delta\varepsilon},\end{aligned}\right\} \quad n = 3, 4, \tag{1.9b}$$

where $\xi_n = 2t/\Delta\varepsilon$, whereas $\Delta\varepsilon = \varepsilon_{A\sigma} - \varepsilon_{B\sigma}$, which difference is assumed to be spin-independent. The corresponding eigenenergies are given by

$$E_{11} = \frac{1}{2}\left(\varepsilon_{A\uparrow} + \varepsilon_{B\uparrow} - \sqrt{(\varepsilon_{A\uparrow} - \varepsilon_{B\uparrow})^2 + 4t^2}\right), \tag{1.10a}$$

$$E_{13} = \frac{1}{2}\left(\varepsilon_{A\uparrow} + \varepsilon_{B\uparrow} + \sqrt{(\varepsilon_{A\uparrow} - \varepsilon_{B\uparrow})^2 + 4t^2}\right), \tag{1.10b}$$

and analogously for $n = 2, 4$ (spin-$\downarrow$).

Before continuing, it is worth to briefly discuss the physical interpretation of the eigenstates. In the case when the dimensionless parameters $|\xi_n| \to \infty$, the energy levels $|\varepsilon_{A\sigma} - \varepsilon_{B\sigma}| \to 0$, or $t \to \infty$. Both situations lead to that α_n, $\beta_n \to 1/2$, which is a mathematical representation of resonant conditions of the quantum dot, although the first may be more reasonable to think of in realistic nanoscale systems. Physically, resonance of the quantum dot means that the electron (wavefunction) is more or less equally distributed throughout the whole structure, i.e. the eigenstates are being equally weighted in both quantum dots (equally weighted on both Fock states constituting the eigenstate). In the other limit, i.e. $|\xi_n| \to 0$ and $\Delta\varepsilon > 0$, we find that $\alpha_{1(2)} \to 0$ and $\beta_{1(2)} \to 1$, while $\alpha_{3(4)} \to 1$ and $\beta_{3(4)} \to 0$. The ground state of the quantum dot becomes localized on the level $\varepsilon_{B\sigma}$, whereas the first excited state is localized on $\varepsilon_{A\sigma}$. Reversing the conditions to $\Delta\varepsilon < 0$ switches the order of the ground state and the first excited state.

The three electron states are found to have analogous characteristics as the one-electron states, which is expected by symmetry. We thus write the three electron states as $|3n\rangle = \gamma_n|\uparrow\rangle_A|\uparrow\downarrow\rangle_B + \kappa_n|\uparrow\downarrow\rangle_A|\uparrow\rangle_B$, $n = 1, 3$ and similarly for $n = 2, 4$ (spin-$\downarrow$). The three-electron energy eigenvalues are given by

$$E_{31} = \frac{1}{2}\biggl(\sum_{i=A,B}(2\varepsilon_{i\uparrow} + \varepsilon_{i\downarrow} + U_i) + 4(U' - J/2) \\ - \sqrt{(\varepsilon_{A\downarrow} + U_A - \varepsilon_{B\downarrow} - U_B)^2 + 4t^2}\biggr), \tag{1.11a}$$

$$E_{33} = \frac{1}{2}\biggl(\sum_{i=A,B}(2\varepsilon_{i\uparrow} + \varepsilon_{i\downarrow} + U_i) + 4(U' - J/2) \\ + \sqrt{(\varepsilon_{A\downarrow} + U_A - \varepsilon_{B\downarrow} - U_B)^2 + 4t^2}\biggr), \tag{1.11b}$$

and similarly for $n = 2, 4$.

The two electron states, finally are separated into three states with spin $S = 0$ and three states with spin $S = 1$. The $S = 1$ states are in the spin-degenerate case, i.e. $\varepsilon_{A\sigma} = \varepsilon_A$ and $\varepsilon_{B\sigma} = \varepsilon_B$, easily found as

$$|21\rangle = |\uparrow\rangle_A|\uparrow\rangle_B, \tag{1.12a}$$

$$|22\rangle = |\downarrow\rangle_A|\downarrow\rangle_B, \tag{1.12b}$$

$$|23\rangle = [|\uparrow\rangle_A|\downarrow\rangle_B + |\downarrow\rangle_A|\uparrow\rangle_B]/\sqrt{2}. \tag{1.12c}$$

These states are known as the triplet states since their energies coincide, and are given by

$$E_{21} = E_{22} = E_{23} = \varepsilon_A + \varepsilon_B + U' - J. \tag{1.13}$$

This triplet configuration will, on the other hand, separate as soon as the quantum dot is subject to, e.g. magnetic fields which lift the spin degeneracy of the single electron levels. It may be noted that the triplet states are always uniformly distributed between the two quantum dots, independently of the tunneling. This may be regarded as a consequence of the Pauli exclusion principle, since for example two electrons with spin $\uparrow$ cannot simultaneously rest in one of the quantum dots, but have to be located at different sites. This is to say, that a single level cannot acquire the total spin $S = 1$.

The $S = 0$ states must, in the spin-degenerate case, be written as a superposition of the three Fock states $|\Phi^{AB}_{S=0}\rangle = [|\uparrow\rangle_A|\downarrow\rangle_B - |\downarrow\rangle_A|\uparrow\rangle_B]/\sqrt{2}$, $|\Phi^{A}_{S=0}\rangle = |\uparrow\downarrow\rangle_A|0\rangle_B$, and $|\Phi^{B}_{S=0}\rangle = |0\rangle_A|\uparrow\downarrow\rangle_B$. The eigenenergy equation becomes cubic, but although it is analytically solvable, the solutions are non-transparent and are therefore not displayed here. It is, nonetheless, worth mentioning that the state with the lowest energy is often referred to as the singlet state. The energy of this state is in the present model lower than the triplet state energy under the assumption that the spin-spin interaction parameter $J \leq 0$, giving rise to anti-ferromagnetic coupling between the quantum dots. Whenever the charging energy $U_{A(B)} > \varepsilon_{A(B)\uparrow} + \varepsilon_{A(B)\downarrow}$, the singlet state is mainly weighted on the combination $|\Phi^{AB}_{S=0}\rangle = [|\uparrow\rangle_A|\downarrow\rangle_B - |\downarrow\rangle_A|\uparrow\rangle_B]\sqrt{2}$, which extends evenly on both quantum dots.

The energies E_{21}, E_{22} can be found analytically within the present model, as well as the corresponding states. The other two-electron energies and their corresponding eigenstates are generally given by solving the system

$$\begin{pmatrix} E-\gamma_1 & J/2 & -t & t \\ J/2 & E-\gamma_2 & -t & t \\ -t & t & E-\gamma_3 & 0 \\ -t & t & 0 & E-\gamma_4 \end{pmatrix} \begin{pmatrix} |\uparrow\rangle_A|\downarrow\rangle_B \\ |\downarrow\rangle_A|\uparrow\rangle_B \\ |\uparrow\downarrow\rangle_A|0\rangle_B \\ |0\rangle_A|\uparrow\downarrow\rangle_B \end{pmatrix} = 0, \tag{1.14}$$

where $\gamma_1 = \varepsilon_{A\uparrow} - \varepsilon_{B\downarrow} - U'$, $\gamma_2 = \varepsilon_{A\downarrow} - \varepsilon_{B\uparrow} - U'$, $\gamma_3 = \varepsilon_{A\uparrow} - \varepsilon_{A\downarrow} - U_A$, and $\gamma_4 = \varepsilon_{B\uparrow} - \varepsilon_{B\downarrow} - U_B$. Here, we allow for spin-dependent levels achieved for instance by magnetic fields.

We now have a full many-body description of the two level system represented by the quantum dot, and we can write the resulting Hamiltonian as

$$\mathcal{H} = \sum_{Nn} E_{Nn}|Nn\rangle\langle Nn|. \tag{1.15}$$

We have also seen that there are many lessons to learn about the system just by extracting the electronic structure of the quantum dot through the procedure of finding the many-body states. This is, of course, not surprising since the problem of finding the electronic structure for a given system, is perhaps one the most challenging problems in materials science, which has led to the invention of e.g. density functional theory.

1.3 Many-Body Operators

The notation of the many-body states in terms of kets is sometimes inconvenient, e.g. when making algebraic manipulations, or when one introduces Green functions in terms of the many-body states. It is therefore motivated to introduce another notation of the projection operators in, e.g. (1.4). Following the Hubbard [2, 3], we introduce the X-operators

$$X_{NN'}^{nn'} = |Nn\rangle\langle N'n'|, \tag{1.16}$$

also known as *Hubbard operators*. We will, however, introduce some more notation. The usage of the X-operators will mainly be restricted to represent transitions between states differing by and odd number of electrons. We shall refer to such transitions as *Fermi-like*. The expansion in (1.4) is made strictly out of Fermionic transitions. Transitions between states differing by an even number of electrons will be referred to as *Bose-like* and denoted by Z, e.g. $Z_{NN\pm2}^{nn'} = |Nn\rangle\langle N\pm 2n'|$. Finally, one often encounters $|Nn\rangle\langle Nn|$, which can be regarded as a transition from one state to the state itself. Such *diagonal* transitions are certainly Bose-like, but since they are frequently occurring we reserve the notation $h_N^n = |Nn\rangle\langle Nn| (= Z_{NN}^{nn})$ for the diagonal cases. In cases when the change in the number of electrons is not known, we will only use the X-operators.

There is an algebra emerging around the introduced operators and, while everything resides on the properties and consequences of the eigenstates used for the construction of the projection operators, we summarize a few important properties here for convenience.

First, we have multiplication and commutator relations, i.e.

$$X_{NN'}^{nn'} X_{MM'}^{mm'} = \delta_{n'm}\delta_{N'M} X_{NM'}^{nm'}, \tag{1.17a}$$

$$[X_{NN'}^{nn'}, X_{MM'}^{mm'}]_{\pm} = \delta_{n'm}\delta_{N'M} X_{NM'}^{nm'} \pm \delta_{nm'}\delta_{NM'} X_{MN'}^{mn'}, \tag{1.17b}$$

and the closure relation

$$\sum_{Nn} h_N^n = 1. \tag{1.17c}$$

As a consequence, one finds that multiplication of a Fermi-like and Bose-like transitions always results in a Fermi-like transition. Likewise, multiplication of two Fermi-like, or two Bose-like, transitions results in a Bose-like transition. We shall take as convention to always have anti-commutation between two Fermi-like operators while using commutation for all other possible combinations, that is, between two Bose-like operators and between one Fermi- and one Bose-like operator.

1.4 Transition Matrix Elements

From the expansion (1.4) we see that the Fermi operator, e.g. $d_{A\sigma}$ is given in terms of the Hubbard operators $X_{NN+1}^{nn'}$ with expansion coefficients $\langle Nn|d_{A\sigma}|N+1n'\rangle$. These matrix elements are important quantities that have to be treated with some care. For example, we cannot assume that they can be replaced by a single value, since some matrix elements are identically zero, while all the others are real numbers between 0 and 1. On the other hand, if we gain knowledge about all matrix elements, we will be able to provide a closer description of the physics in that we can decide which transitions that contribute to the processes of interest.

What we need is a systematic way to calculate the matrix elements. We can certainly always make direct calculations. Borrowing the states and notation from Sect. 1.2 we have, for example

$$\langle 11|d_{A\uparrow}|21\rangle = \Big[\alpha_1\, {}_B\langle 0|_A\langle\uparrow| + \beta_1\, {}_B\langle\uparrow|_A\langle 0|\Big] d_{A\uparrow}|\uparrow\rangle_A|\uparrow\rangle_B = -\beta_1. \tag{1.18}$$

This procedure works well whenever the eigenstates are sufficiently simple to handle. A more involved example is $\langle 11|d_{A\downarrow}|24\rangle$, where $|24\rangle = A_4|\Phi^{AB}_{S=0}\rangle + B_4|\Phi^{A}_{S=0}\rangle + C_4|\Phi^{B}_{S=0}\rangle$. The operator $d_{A\downarrow}$ acting on $|\Phi^{B}_{S=0}\rangle$ gives zero contribution. The non-zero contribution is calculated through

$$\begin{aligned}\langle 11|d_{A\downarrow}|21\rangle =& \big[\alpha_1\, {}_B\langle 0|_A\langle\uparrow| + \beta_1\, {}_B\langle\uparrow|_A\langle 0|\big] d_{A\downarrow}\big[A_4|\Phi^{AB}_{S=0}\rangle + B_4|\Phi^{A}_{S=0}\rangle\big]\\ =& \beta_1 A_4\, {}_B\langle\uparrow|_A\langle 0|d_{A\downarrow}|\Phi^{AB}_{S=0}\rangle + \alpha_1 B_4\, {}_B\langle 0|_A\langle\uparrow|d_{A\downarrow}|\Phi^{A}_{S=0}\rangle\\ =& \beta_1 A_4 + \alpha_1 B_4.\end{aligned} \tag{1.19}$$

Obviously, we have to consider the transition matrix element between each of the Fock states in order to acquire information about the matrix elements between the eigenstates. We want to avoid unnecessary repetition of our calculations, hence, we should make the calculations more systematic.

Thus, consider the set of eigenstates $\{|Nn\rangle\}$ expressed in terms of the Fock states $\{|a\rangle\}$, where n denotes the state in the N electron configuration, whereas a are different states possible on the level ε_A. Any given matrix element $\langle Nn|d_a|N'n'\rangle$ can be expressed through

$$\langle Nn|d_a|N'n'\rangle = \sum_{a'a''} \langle Nn|a'\rangle\langle a'|d_a|a''\rangle\,\langle a''|N'n'\rangle\,. \tag{1.20}$$

This expression provides a systematic approach to calculating the matrix elements between the eigenstates, since the matrix elements $\langle a'|d_a|a''\rangle$ between the Fock states can be calculated once and for all, since we do not change the Fock basis that has been introduced. The overlap integrals $\langle a|Nn\rangle$ establish projections of the eigenstates onto the Fock basis, and these are also needed to be calculated only once. Notice that the overlap integrals $\langle a|Nn\rangle$ equal the weight, or expansion, coefficient of the eigenstate $|Nn\rangle$ in terms of the Fock state $|a\rangle$. To see this, let $|Nn\rangle = \sum_a C_{an}|a\rangle$, for some coefficients C_{an} in the Fock basis of N-electron states. Hence,

$$\langle a|Nn\rangle = \langle a|\sum_{a'} C_{na'}|a'\rangle = \sum_{a'} C_{na'}\,\langle a|a'\rangle = \sum_{a'} C_{na'}\delta_{aa'} = C_{na}. \tag{1.21}$$

As an example, we have the transitions between the one-electron and two-electrons states when an electron in quantum dot B is removed. Using the one and two electron bases previously defined, we have (omitting the subscripts A, B)

$$
\begin{aligned}
&\langle 1n|d_{B\sigma}|2m\rangle \\
&= \langle 1n| \begin{pmatrix} |\uparrow\rangle|0\rangle \\ |\downarrow\rangle|0\rangle \\ |0\rangle|\uparrow\rangle \\ |0\rangle|\downarrow\rangle \end{pmatrix}^T \begin{pmatrix} \langle 0|\langle\uparrow| \\ \langle 0|\langle\downarrow| \\ \langle\uparrow|\langle 0| \\ \langle\downarrow|\langle 0| \end{pmatrix} d_{B\sigma} \begin{pmatrix} |\uparrow\rangle|\uparrow\rangle \\ |\downarrow\rangle|\downarrow\rangle \\ \frac{[|\uparrow\rangle|\downarrow\rangle+|\downarrow\rangle|\uparrow\rangle]}{\sqrt{2}} \\ \frac{[|\uparrow\rangle|\downarrow\rangle-|\downarrow\rangle|\uparrow\rangle]}{\sqrt{2}} \\ |\uparrow\downarrow\rangle|0\rangle \\ |0\rangle|\uparrow\downarrow\rangle \end{pmatrix}^T \begin{pmatrix} \langle\uparrow|\langle\uparrow| \\ \langle\downarrow|\langle\downarrow| \\ \frac{\langle\downarrow|\langle\uparrow|+\langle\uparrow|\langle\downarrow|}{\sqrt{2}} \\ \frac{\langle\downarrow|\langle\uparrow|-\langle\uparrow|\langle\downarrow|}{\sqrt{2}} \\ \langle 0|\langle\uparrow\downarrow| \\ \langle\uparrow\downarrow|\langle 0| \end{pmatrix} |2m\rangle \\
&= \begin{pmatrix} \alpha_{n\uparrow} & \alpha_{n\downarrow} & \beta_{n\uparrow} & \beta_{n\downarrow} \end{pmatrix} \begin{pmatrix} \delta_{\sigma\uparrow} & 0 & \frac{\delta_{\sigma\downarrow}}{\sqrt{2}} & \frac{\delta_{\sigma\downarrow}}{\sqrt{2}} & 0 & 0 \\ 0 & \delta_{\sigma\downarrow} & \frac{\delta_{\sigma\uparrow}}{\sqrt{2}} & -\frac{\delta_{\sigma\uparrow}}{\sqrt{2}} & 0 & 0 \\ 0 & 0 & 0 & 0 & 0 & \delta_{\sigma\downarrow} \\ 0 & 0 & 0 & 0 & 0 & -\delta_{\sigma\uparrow} \end{pmatrix} \\
&\quad \times \begin{pmatrix} \delta_{m1} \\ \delta_{m2} \\ \delta_{m3} \\ \sum_{n=4,5,6} \delta_{nm} A_m \\ \sum_{n=4,5,6} \delta_{nm} B_m \\ \sum_{n=4,5,6} \delta_{nm} C_m \end{pmatrix} \\
&= \alpha_{n\uparrow}\delta_{\sigma\uparrow}\delta_{m1} + \alpha_{n\downarrow}\delta_{\sigma\downarrow}\delta_{m2} + \frac{\alpha_{n\uparrow}\delta_{\sigma\uparrow} + \alpha_{n\downarrow}\delta_{\sigma\downarrow}}{\sqrt{2}}\delta_{m3} \\
&\quad + \sum_{m'=4,5,6} \delta_{mm'} \left(\frac{\alpha_{n\uparrow}\delta_{\sigma\uparrow} - \alpha_{n\downarrow}\delta_{\sigma\downarrow}}{\sqrt{2}} A_m + [\beta_{n\uparrow}\delta_{\sigma\downarrow} - \beta n \downarrow \delta_{\sigma\uparrow}] C_m \right).
\end{aligned}
\tag{1.22}
$$

The calculation may look somewhat elaborate, however, all these steps are in principle necessary for the calculations of the matrix elements.

More generally, we can calculate the matrix element e.g. $\langle Nn|d_{A\sigma}|N+1m\rangle$ according to the following consideration. Let $\{|a\rangle\}$ and $\{|b\rangle\}$ be Fock bases for the N- and $N+1$-electron configurations such that $|Nn\rangle = \sum_a C_{na}|a\rangle$ and $|N+1m\rangle = \sum_b D_{mb}|b\rangle$, for some coefficients C_{na} and D_{mb}, which we allow to be complex for the sake of generality. The matrix element $\langle Nn|d_{A\sigma}|N+1m\rangle$ can, thus, be expanded according to

$$
\langle Nn|d_{A\sigma}|N+1m\rangle = \sum_{ab} C^*_{na} D_{mb} \langle a|d_{A\sigma}|b\rangle, \tag{1.23}
$$

where $\langle a|d_{A\sigma}|b\rangle$ is the $a \times b$-matrix of the transitions between the $N+1$- and N-electron Fock states that occur by removing a spin σ electron in quantum dot A.

We have, thus, seen how we, after defining the properties of the quantum dot system, can evaluate the eigenstates and eigenenergies and transition matrix elements. Those are calculated once and for all, under the given conditions, and can now be considered as being know through the remainder of this text.

References

1. Hirjibehedin, C., Lutz, C.P., Heinrich, A.J.: Spin coupling in engineered atomic structures. Science **312**, 1021–1024 (2006)
2. Hubbard, J.: Proc. R. Soc. A **276**, 238 (1963)
3. Hubbard, J.: Proc. R. Soc. A **277**, 237 (1963)

Chapter 2
Occupation Number Formalism

Abstract We introduce our systems into a perturbing environment and study the equations for properties of the local subsystems. An occupation number formalism for the localized states is discussed.

2.1 Perturbing the Local Levels

The perturbations we have in mind here, are perturbations that enable transport of e.g. charge carriers, spin, thermal energy, etc., through a single quantum dot or a network or quantum dots. Typically, we will consider charge transport. The perturbations may then be called leads, however, they would have the character of thermal baths coupled to the quantum dots. In either case, the important effects on the transport that we are interested in arise in the quantum dots when coupled to the leads. Hence, the leads typically have non-interesting features and are described in a free electron-like model like

$$\mathcal{H}_\chi = \sum_{k\sigma} \varepsilon_{k\sigma} c^\dagger_{k\sigma} c_{k\sigma}, \quad k\sigma \in \chi, \tag{2.1}$$

where χ labels the lead, however any degree of complicated structure may be ascribed to the leads. The operators $c^\dagger_{k\sigma}$ $(c_{k\sigma})$ denotes creation (destruction) of an electron in the lead at the energy $\varepsilon_{k\sigma}$ with spin σ. In what follows and when appropriate, we will usually discuss in terms of left and right leads and use the symbols L and R to denote the left and right lead, respectively, with chemical potentials μ_L and μ_R. The electron energy $\varepsilon_{k\sigma}$ is given relative to the chemical potential in the corresponding lead.

The leads are meant to be coupled to the quantum dots such that a transport may be conducted between the leads. While the coupling is taken to be rather simple minded here, there is no limit to the possible complexity that may be present. Assuming that we have a network of quantum dots, each of which may be described by the model in (1.1) or (1.2). The coupling between the quantum dot ν and a number

J. Fransson, *Non-Equilibrium Nano-Physics*,
Lecture Notes in Physics 809,
DOI 10.1007/978-90-481-9210-6_2,

of χ leads may be modeled by

$$\mathcal{H}_T^{\nu}=\sum_{i\sigma}\sum_{r=1}^{\chi}\sum_{k\in\chi_r}v_{ki\sigma}c_{k\sigma}^{\dagger}d_{\nu i\sigma}+H.c. \tag{2.2}$$

Converting the d operators in the tunneling term into Hubbard operators, using (1.4), the tunneling Hamiltonian becomes

$$\mathcal{H}_T^{\nu}=\sum_{Nnm}\left[\sum_{i\sigma}\sum_{r=1}^{\chi}\sum_{k\in\chi_r}v_{ki\sigma}c_{k\sigma}^{\dagger}(d_{\nu i\sigma})_{NN+1}^{nm}\right]X_{NN+1}^{nm}+H.c., \tag{2.3}$$

where $(d_{\nu i\sigma})_{NN+1}^{nm}\equiv\langle Nn|d_{\nu i\sigma}|N+1m\rangle$. Again, we see that the Hubbard operator does not care about from which level in the quantum dot the electron is removed, or added, and that all such information is taken care of in the transition matrix elements. Also, notice that each Hubbard operator couples to all leads, which illustrates the fact that the many-body states are extended throughout the quantum dot and hence are influenced by the perturbations of all levels, irrespective of whether some of the levels are not connected to all leads.

2.2 Occupations Numbers

The occupation and occupation numbers of electron levels, or particle levels in general, is an important notion within quantum physics. Through the occupation numbers we can determine the state of the system. The state of the system can be expressed in terms of the eigenstates, which in the present formulation is the most convenient. Using this information about the quantum dot, we can for instance calculate the charge current through the quantum dot, when the quantum dot is coupled to leads, or to deduce the spin state of the quantum dot.

The occupation numbers $N_{Nn}=\langle h_N^n\rangle$ is the average of the diagonal Hubbard operators. The occupation numbers can be determined through the equation of motion with respect to the Hamiltonian system we consider. For instance, suppose that the quantum dot is coupled to leads as discussed above, but to no other external perturbations. Let the system be described by the Hamiltonian $\mathcal{H}=\sum_{r=1}^{\chi}\mathcal{H}_{\chi_r}+\sum_{Nn}E_{Nn}h_N^n+\sum_{\nu}\mathcal{H}_T^{\nu}$, with $\mathcal{H}_T^{\nu}$ as in (2.3). Then, the occupation number N_{Nm} satisfies the equation

$$i\hbar\frac{\partial}{\partial t}N_{Nn}=\langle[h_N^n,\mathcal{H}]\rangle=\sum_{\nu}\langle[h_N^n,\mathcal{H}_T]\rangle, \tag{2.4}$$

since $[h_N^n,h_M^m]=0$ for all Nn, Mm. The dynamics of the occupation numbers is determined by the tunneling between the quantum dot and the leads. This equation expresses the dynamics of the occupation under influence of the perturbative baths in the leads, and since an electron being in the state $|Nn\rangle$ may tunnel in to the leads, we realize that this state is coupled to all the states $|N-1n'\rangle$. On the other hand, an electron may certainly also tunnel into the quantum dot from the baths, thus also coupling the state $|Nn\rangle$ to all the states $|N+1n'\rangle$. Hence, all the configurations with N electrons couple to all $N\pm 1$-electron configurations through the baths, and we find that the dynamics of the occupation numbers is solved through a coupled system which dimensions scale as 2^{2N}, where N is the number of levels in the quantum dot (considering the possibility of 2 electrons per level, spin $\uparrow$ or spin $\downarrow$). In order to be concrete and to start build our intuition, we consider two simple examples.

2.3 Single-Level System

Assume a single quantum dot with a single level. This can be described with $\mathcal{H}_{QD} = \sum_\sigma \varepsilon_\sigma d_\sigma^\dagger d_\sigma + Un_\uparrow n_\downarrow$, where ε_σ is the energy for the spin-projection σ, whereas U is the charging energy. We rewrite the quantum dot in terms of Hubbard operators, i.e. $\mathcal{H}_{QD} = \sum_{p=0,\sigma,2} E_p h^p$, where $E_0 = 0$, $E_\sigma = \varepsilon_\sigma$, and $E_2 = \varepsilon_\uparrow + \varepsilon_\downarrow + U$. Coupling the quantum dot to left and right leads, we can write the total system as

$$\mathcal{H} = \sum_{k\sigma\in L\cup R} \varepsilon_{k\sigma} c_{k\sigma}^\dagger c_{k\sigma} + \sum_p E_p h^p + \sum_{k\sigma} [v_{k\sigma} c_{k\sigma}^\dagger (X^{0\sigma} + \sigma X^{\bar{\sigma}2}) + H.c.], \quad (2.5)$$

since $d_\sigma = X^{0\sigma} + \sigma X^{\bar{\sigma}2}$, where the factor $\sigma = \pm 1$. This model is the so-called Anderson model [1], here extended to include two electron baths (the left and right leads).

The equation for the occupation number N_0 is, thus, given by

$$\begin{aligned} i\hbar\frac{\partial}{\partial t} N_0(t) &= \sum_{k\sigma} [v_{k\sigma} \langle c_{k\sigma}^\dagger [h^0, X^{0\sigma}](t)\rangle + v_{k\sigma}^* \langle [h^0, X^{\sigma 0}] c_{k\sigma}(t)\rangle] \\ &= \sum_{k\sigma} v_{k\sigma} \langle c_{k\sigma}^\dagger(t) X^{0\sigma}(t)\rangle - c.c. = i2\,\mathrm{Im} \sum_{k\sigma} v_{k\sigma} \langle c_{k\sigma}^\dagger(t) X^{0\sigma}(t)\rangle, \end{aligned} \quad (2.6a)$$

where $c.c.$ denotes complex conjugate. Similarly, we have the equations for the numbers N_σ and N_2 given by

$$i\hbar\frac{\partial}{\partial t} N_\sigma(t) = -2i\,\mathrm{Im} \sum_k [v_{k\sigma} \langle c_{k\sigma}^\dagger(t) X^{0\sigma}(t)\rangle - \bar{\sigma} v_{k\bar{\sigma}} \langle c_{k\bar{\sigma}}^\dagger(t) X^{\sigma 2}(t)\rangle], \quad (2.6b)$$

$$i\hbar\frac{\partial}{\partial t} N_2(t) = -i2\,\mathrm{Im} \sum_{k\sigma} \sigma v_{k\sigma} \langle c_{k\sigma}^\dagger(t) X^{0\bar{\sigma}}(t)\rangle. \quad (2.6c)$$

As we expected from the previous general discussion, the dynamics of the occupation number N_p depends on all transitions to and from the state $|p\rangle$. The next step is to derive suitable equations for the averages on the right hand sides of the equations. Using the equation of motion for the averages to linear order in the tunneling, along with the rate of change of the Hubbard operator $i\partial_t X^{pq} =$

$[X^{pq}, \mathcal{H}] = \Delta_{qp} X^{pq} + [X^{pq}, \mathcal{H}_T]$, where $\Delta_{qp} = E_q - E_p$ is the energy for the transition $|q\rangle \to |p\rangle$, we find

$$\left(i\hbar\frac{\partial}{\partial t} - \Delta_{\sigma 0} + \varepsilon_{k\sigma}\right)\langle c^{\dagger}_{k\sigma} X^{0\sigma}\rangle = -v^{*}_{k\sigma}\langle h^0\rangle + \sum_{k'\sigma'}[-v_{k\bar{\sigma}}\langle c^{\dagger}_{k\sigma} c^{\dagger}_{k\bar{\sigma}} Z^{02}\rangle$$
$$+ v^{*}_{k'\sigma'}\langle c^{\dagger}_{k\sigma\sigma}(h^0\delta_{\sigma'\sigma} + Z^{\sigma'\sigma}) c_{k'\sigma'}\rangle]. \quad (2.7)$$

The second term on the right hand side of this equation contains the two electron process describing tunneling of two electrons from the quantum dot to the leads. Such processes are typically important when the leads are superconducting, however, considering normal metallic leads these processes are negligible. The last term accounts for spin-flip tunneling ($\sigma' = \bar{\sigma}$) and since such processes are not contained in the Hamiltonian, this contribution vanishes exactly in the linear order approximation. Hence, the equation for the average becomes

$$\left(i\hbar\frac{\partial}{\partial t} - \Delta_{\sigma 0} + \varepsilon_{k\sigma}\right)\langle c^{\dagger}_{k\sigma} X^{0\sigma}\rangle = -v^{*}_{k\sigma}\langle h^{\sigma}\rangle + \sum_{k} v^{*}_{k\sigma}\langle c^{\dagger}_{k\sigma}(h^0 + h^{\sigma}) c_{k\sigma}\rangle. \quad (2.8)$$

Approximating the last average by $(N_0 + N_\sigma)\langle n_{k\sigma}\rangle = (N_0 + N_\sigma) f(\varepsilon_{k\sigma})$, where $f(\varepsilon_{k\sigma}) = [e^{\beta(\varepsilon_{k\sigma} - \mu_\chi)} + 1]^{-1}$ is the Fermi distribution function with $\beta^{-1} = k_B T$ (see Sect. 3.5 for a derivation), we find that

$$\langle c^{\dagger}_{k\sigma} X^{0\sigma}\rangle(t) = -i v^{*}_{k\sigma}\int_{-\infty}^{t} [f(\varepsilon_{k\sigma}) N_0 - \{1 - f(\varepsilon_{k\sigma})\} N_\sigma] e^{-i(\Delta_{\sigma 0} - \varepsilon_{k\sigma})(t-t')} dt'. \quad (2.9)$$

Assuming, further, that we are considering stationary conditions only, we can apply the Markovian approximation (neglecting memory effects) giving

$$\langle c^{\dagger}_{k\sigma} X^{0\sigma}\rangle(t) = v^{*}_{k\sigma}\frac{f(\varepsilon_{k\sigma}) N_0 - \{1 - f(\varepsilon_{k\sigma})\} N_\sigma}{\varepsilon_{k\sigma} - \Delta_{\sigma 0} + i\delta}, \quad (2.10)$$

where $\delta > 0$ has been added for convergence. Putting this expression into (2.6a), we obtain

$$\frac{\partial}{\partial t} N_0 = -\frac{2\pi}{\hbar}\sum_{k\sigma} |v_{k\sigma}|^2 \big(f(\varepsilon_{k\sigma}) N_0 - [1 - f(\varepsilon_{k\sigma})] N_\sigma\big)\delta(\varepsilon_{k\sigma} - \Delta_{\sigma 0}). \quad (2.11)$$

Making the analogous derivations for the other averages in (2.6a)–(2.6c), defining $\Gamma^{\chi}_{0\sigma} = 2\pi \sum_k |v_{k\sigma}|^2 \delta(\varepsilon_{k\sigma} - \Delta_{\sigma 0})/\hbar$ and $\Gamma^{\chi}_{\sigma 2} = 2\pi \sum_k |v_{k\bar{\sigma}}|^2 \delta(\varepsilon_{k\bar{\sigma}} - \Delta_{2\sigma})/\hbar$, $f^{+}_{\chi}(\omega) = f(\omega - \mu_\chi)$ and $f^{-}_{\chi}(\omega) = 1 - f^{+}_{\chi}(\omega)$, and using that $\partial_t N_p = 0$ under stationary conditions, we write the equation for the occupation numbers as

$$\frac{\partial}{\partial t}N_0 = 0 = \sum_{\chi\sigma} \Gamma^{\chi}_{0\sigma}[N_0 f^{+}_{\chi}(\Delta_{\sigma 0}) - N_\sigma f^{-}_{\chi}(\Delta_{\sigma 0})], \tag{2.12a}$$

$$\frac{\partial}{\partial t}N_\sigma = 0 = \sum_{\chi}[\Gamma^{\chi}_{0\sigma} N_0 f^{+}_{\chi}(\Delta_{\sigma 0}) - \{\Gamma^{\chi}_{0\sigma} f^{-}_{\chi}(\Delta_{\sigma 0}) + \Gamma^{\chi}_{\sigma 2} f^{+}_{\chi}(\Delta_{2\sigma})\}N_\sigma$$
$$+ N_2 \Gamma^{\chi}_{\sigma 2} f^{-}_{\chi}(\Delta_{2\sigma})], \tag{2.12b}$$

$$\frac{\partial}{\partial t}N_2 = 0 = \sum_{\chi\sigma} \Gamma^{\chi}_{\sigma 2}[N_\sigma f^{+}_{\chi}(\Delta_{2\sigma}) + N_2 f^{-}_{\chi}(\Delta_{2\sigma})]. \tag{2.12c}$$

Within the given approximation, those four equations completely describe the dynamics of the electron occupation in the quantum dot in terms of the electron flow between the leads and the quantum dot. The solution can be written in analytical form (left to the reader) which can be used for calculation of e.g. the charge current through the system under non-equilibrium conditions.

Both N_p and the Fermi function describe the occupation in the corresponding subsystem, whereas the couplings $\Gamma^{\chi}_{0\sigma}$ and $\Gamma^{\chi}_{\sigma 2}$ account for the transfer of electrons between the subsystems. Hence, physically the above rate equations describes the electron (im-)balance between the leads as being coupled through the quantum dot. By increasing the chemical potential in, say, the left lead relative to the right (which corresponds to a bias voltage applied between the leads), electrons begin to flow from the left to the right. The presence of the quantum dot makes this flow more complicated than it would be in e.g. a simple metal-insulator-metal junction, since not only the occupation in the leads but also the occupation in the quantum dot has to be calculated. In particular only one electron at the time can be transferred between the left lead and the quantum dot, and between the quantum dot and the right lead. By varying the bias voltage we obtain a variation of the quantum dot occupation. The quantum dot occupation also depends on the position of the level relative to the chemical potentials of the leads. Shifting the position of the level or the chemical potentials of the leads correspond to experimentally applying a gate voltage over the quantum dot.

2.4 Two-Level System in the Pauli Spin Blockade

We take the procedure from the previous discussion to the two-level system introduced in Sect. 1.2, that is, two quantum dots coupled in series, where each of the quantum dots is coupled to one lead. The coupling may be modeled by

$$\mathcal{H}_T = \sum_{p\sigma} v_{p\sigma} c^{\dagger}_{p\sigma} d_{A\sigma} + \sum_{q\sigma} v_{q\sigma} c^{\dagger}_{q\sigma} d_{B\sigma} + H.c., \tag{2.13}$$

where $v_{p(q)\sigma}$ is the tunneling (or hybridization, or hopping) between the left (right) lead, and where we have taken indices p (q) in the left (right) lead.

Converting the d operators in the tunneling term into Hubbard operators, using (1.4), the tunneling Hamiltonian becomes

$$\mathcal{H}_T = \sum_{Nnm}\Big[\sum_{p\sigma} v_{p\sigma} c^\dagger_{p\sigma} (d_{A\sigma})^{nm}_{NN+1} + \sum_{q\sigma} v_{q\sigma} c^\dagger_{q\sigma} (d_{B\sigma})^{nm}_{NN+1}\Big] X^{nm}_{NN+1} + H.c., \tag{2.14}$$

where $(d_{A(B)\sigma})^{nm}_{NN+1} \equiv \langle Nn|d_{A(B)\sigma}|N+1m\rangle$. We again see that the Hubbard operator does not care about in which quantum dot the electron is removed, or added, and that all such information is taken care of by the transition matrix elements.

Considering the occupation numbers N_{Nn} to the same level of approximation as in the single-level system, we obtain the following set of equations (owing to the condition $\partial_t N_{Nn} = 0$)

$$\hbar\partial_t N_{01} = -\sum_{\chi n} \Gamma^\chi_{01,1n}[f^+_\chi(\Delta_{1n,01})N_{01} - f^-_\chi(\Delta_{1n,01})N_{1n}], \tag{2.15a}$$

$$\begin{aligned}\hbar\partial_t N_{Nn} = \sum_{\chi n'}\big(&\Gamma^\chi_{N-1n',Nn}[f^+_\chi(\Delta_{Nn,N-1n'})N_{N-1n'} - f^-_\chi(\Delta_{Nn,N-1n'})N_{Nn}]\\ &- \Gamma^\chi_{Nn,N+1n'},[f^+_\chi(\Delta_{N+1n',Nn})N_{Nn} - f^-_\chi(\Delta_{N+1n',Nn})N_{N+1n'}]\big),\\ &N = 1,2,3,\end{aligned} \tag{2.15b}$$

$$\hbar\partial_t N_{41} = \sum_{\chi n} \Gamma^\chi_{3n,41}[f^+_\chi(\Delta_{41,3n})N_{3n} - f^-_\chi(\Delta_{41,3n})N_{41}], \tag{2.15c}$$

where $\Gamma^\chi_{Nn,N-1n'} = 2\pi \sum_{k\sigma} |v_{k\sigma}(d_{A(B)\sigma})^{n'n}_{N-1N}|^2 \delta(\Delta_{Nn,N-1n'} - \varepsilon_{k\sigma})$.

We make a brief analysis of the occupation numbers under conditions studied experimentally by Ono et al. [2], for a particularly interesting case from a physical point of view, namely the Pauli spin-blockade phenomenon.

Following the experimental estimates we take $\varepsilon_{A(B)\sigma} = \varepsilon_{A(B)}$ and $U_A = U_B = U = 2U' = 2\Delta\varepsilon$, where $\Delta\varepsilon = \varepsilon_A - \varepsilon_B > 0$, assume weakly coupled quantum dots, i.e. $2t \ll \Delta\varepsilon$, and that the direct Heisenberg exchange $J \approx 0$. We further assume that $\varepsilon_A - \mu = -U/2$, where μ is the equilibrium chemical potential in the system. The one-electron transition energies for this set-up are depicted in Fig. 2.1. Under the given conditions we find that only $|\Delta_{2n,1n'} - \mu| \ll U/2$ for $n = 1,\dots,5$, and $n' = 1,2$, and $|\Delta_{3n,26} - \mu| \ll U/2$ for $n = 3,4$. For all other transitions we have $|\Delta_{Nn,N-1n'} - \mu| \gg U/2$. Those estimates allow for a reasonably simple analytical analysis of the occupation numbers.

By letting the bias voltage $eV = \mu_L - \mu_R \in [0.1, 1]U$ adjust the left and right chemical potentials symmetrically around μ, it is easily found that

$$f^+_\chi(\Delta_{1n',01}) = 1, \quad \forall n', \tag{2.16a}$$

$$f^+_L(\Delta_{2n,1n'}) = 1 \quad \text{and} \quad f^+_R(\Delta_{2n,1n'}) = 0, \quad n = 1,\dots,5,\ n' = 1,2, \tag{2.16b}$$

$$f^+_\chi(\Delta_{2n,1n'}) = 1, \quad n = 1,\dots,5,\ n' = 3,4, \tag{2.16c}$$

$$f_\chi^+(\Delta_{26,1n'}) = 0, \quad \forall n', \tag{2.16d}$$

$$f_\chi^+(\Delta_{3n',2n}) = 0, \quad n' = 1,\dots,4,\ n = 1,\dots,5, \tag{2.16e}$$

$$f_\chi^+(\Delta_{3n',26}) = 1, \quad n' = 1,2, \tag{2.16f}$$

$$f_L^+(\Delta_{3n',26}) = f_R^-(\Delta_{3n',26}) = 1, \quad n' = 3,4, \tag{2.16g}$$

$$f_\chi^+(\Delta_{41,3n'}) = 0, \quad \forall n'. \tag{2.16h}$$

Using these observations we directly deduce that $N_{01} = N_{41} = 0$, as expected since the transition energies $\Delta_{1n',01}$ and $\Delta_{41,3n'}$ lie far from resonance, that is, outside the window between the left and right chemical potentials opened by the bias voltage. It also follows that N_{26} vanishes, since $\Delta_{26,1n'} - \mu \gtrsim U/2$. Therefore, there cannot occur any transition between the state $|2,6\rangle$ and any of the three-electron states, although the transition energies $\Delta_{3n',26} \in [\mu_R, \mu_L]$, $n' = 3,4$. Taking this observation along with the fact that $\Delta_{3n',2n} \gtrsim U/2$, $n = 1,\dots,5$, $n' = 1,\dots,4$, it is clear that the occupation numbers $N_{3n'} = 0$ for all n'. This, further, implies that the one-electron occupation numbers $N_{1n'} = 0$, $n' = 3,4$. Hence, the only non-vanishing occupation numbers are $N_{1n'}$, $n' = 1,2$, and N_{2n}, $n = 1,\dots,5$, which leads to the equations

$$N_{1n'} = \frac{\sum_{n=1}^{5} \Gamma^R_{1n',2n} N_{2n}}{\sum_{m=1}^{5} \Gamma^L_{1n',2m}}, \quad n' = 1,2, \tag{2.17a}$$

$$0 = \sum_{n'=1,2} [\Gamma^L_{1n',2n} N_{1n'} - \Gamma^R_{1n',2n} N_{2n}], \quad n = 1,\dots,5. \tag{2.17b}$$

Under spin-degenerate conditions the triplet configurations $|2,n\rangle$, $n = 1,2,3$, have to have equal probability, hence, $N_{2n} = N_T/3$, for $n = 1,2,3$. For the same

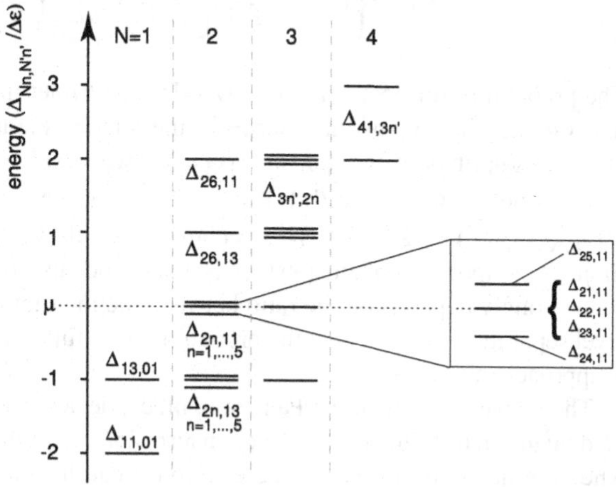

Fig. 2.1 Equilibrium distribution of the one-electron transition energies $\Delta_{Nn,N-1n'}$ in terms of $\Delta\varepsilon$, for $\varepsilon_A = -U/2$, $\Delta\varepsilon = U/2 = U'$, $2t/\Delta\varepsilon \ll 1$, and $J = 0$. In absence of magnetic fields we have degenerate transitions $\Delta_{12,01} = \Delta_{11,01}$, $\Delta_{14,01} = \Delta_{13,01}$, $\Delta_{26,12} = \Delta_{26,11}$, and $\Delta_{26,14} = \Delta_{26,13}$. [From [3]]

Table 2.1 Matrix elements for the transitions $X_{12}^{n'n}$ given in terms of the eigenstates of the model, where $\alpha_n = \alpha$ and $\beta_n = \beta$, $n = 1, 2$, and where $|2, n\rangle = A_n|\Phi_{S=0}^{AB}\rangle + B_n|\Phi_{S=0}^{A}\rangle + C_n|\Phi_{S=0}^{B}\rangle$

$\lvert(d_{A\uparrow})_{12}^{11}\rvert^2$	$\lvert(d_{A\downarrow})_{12}^{22}\rvert^2$	β^2	
$\lvert(d_{A\uparrow})_{12}^{12}\rvert^2$	$\lvert(d_{A\downarrow})_{12}^{21}\rvert^2$	0	
$\lvert(d_{A\uparrow})_{12}^{13}\rvert^2$	$\lvert(d_{A\downarrow})_{12}^{23}\rvert^2$	$\beta^2/2$	
$\lvert(d_{A\uparrow})_{12}^{1n}\rvert^2$	$\lvert(d_{A\downarrow})_{12}^{2n}\rvert^2$	$(\beta A_n/\sqrt{2}+\alpha B_n)^2$,	$n=4,5$
$\lvert(d_{B\uparrow})_{12}^{11}\rvert^2$	$\lvert(d_{B\downarrow})_{12}^{22}\rvert^2$	α^2	
$\lvert(d_{B\uparrow})_{12}^{12}\rvert^2$	$\lvert(d_{B\downarrow})_{12}^{21}\rvert^2$	0	
$\lvert(d_{B\uparrow})_{12}^{13}\rvert^2$	$\lvert(d_{B\downarrow})_{12}^{23}\rvert^2$	$\alpha^2/2$	
$\lvert(d_{B\uparrow})_{12}^{1n}\rvert^2$	$\lvert(d_{B\downarrow})_{12}^{2n}\rvert^2$	$(\alpha A_n/\sqrt{2}+\beta C_n)^2$,	$n=4,5$

reason, the non-vanishing one-electron occupation numbers have to be equal as well, i.e. $N_{11} = N_{12} = N_1/2$, which can also be deduced from the matrix elements $(d_{A(B)\sigma})_{12}^{n'n}$, see Table 2.1, when putting $\alpha_1 = \alpha_2 = \alpha$ and $\beta_1 = \beta_2 = \beta$. We thus find that N_1 and N_{2n}, $n = 4, 5$ are related to N_T through the equations

$$N_1 = \frac{2}{3}\frac{\Gamma^R}{\Gamma^L}\left(\frac{\alpha}{\beta}\right)^2 N_T, \tag{2.18a}$$

$$N_{2n} = \frac{1}{3}\left(\frac{L_n}{R_n}\cdot\frac{\alpha}{\beta}\right)^2 N_T, \quad n = 4, 5, \tag{2.18b}$$

where $\Gamma^\chi = 2\pi\sum_{k\sigma}|v_k|^2\delta(\omega - \varepsilon_{k\sigma})$, while $L_n = \beta A_n/\sqrt{2} + \alpha B_n$ and $R_n = \alpha A_n/\sqrt{2}+\beta C_n$ which are bounded and finite for all $\xi = 2t/\Delta\varepsilon$, and slowly varying functions of ξ. Thus, by the normalization condition (closure relation), here reduced to $1 = N_1 + N_T + N_{24} + N_{25}$, we obtain

$$N_T = \left\{1 + \frac{1}{3}\left(\frac{\alpha}{\beta}\right)^2\left[2\frac{\Gamma^R}{\Gamma^L} + \sum_{n=4,5}\left(\frac{L_n}{R_n}\right)^2\right]\right\}^{-1}. \tag{2.19}$$

The probability for occupying the two-electron triplet in the restricted bias voltage interval can, thus, be given in terms of the internal parameters of the quantum dot and the couplings to the leads. Especially, we observe that N_T approaches 1 for $\xi \ll 1$, since then $\alpha^2 = \xi^2/[(1+\sqrt{1+\xi^2})^2+\xi^2] \approx 0$ and $\beta^2 = (1+\sqrt{1+\xi^2})^2/[(1+\sqrt{1+\xi^2})^2+\xi^2] \approx 1$. This is consistent with $N_{2n} \to 0$, $n = 4, 5$, in the same limit. The important conclusion then is that the occupation of the two-electron triplet state strongly depends on the ratio between the interdot tunneling t and the relative level separation $\Delta\varepsilon$, and that the probability of a fully occupied two-electron triplet is approached as $\xi \to 0$.

The above analysis of the Pauli spin blockade achieved in the serial weakly coupled quantum dot can be interpreted in terms of transition matrix elements. Namely, when the quantum dot has a one electron it can host a second electron in several different configurations, and in the considered bias regime it is most likely that the second electron enters the quantum dot from the left lead. The quantum dot may make a transition into any of the triplet or singlet states, all transitions occur with a rather high probability. Transitions between the singlet and the one-electron states may, however, occur with finite, and reasonably large, probability by letting one electron out to the right lead. The occupation of the singlet states therefore become small, since once any such state is occupied, it immediately leaks electron density to the right lead. In contrast, transitions between the triplet and one-electron states by letting one electron out to the right lead has a small probability. Therefore, the electron density tends to accumulate in the triplet states, for which the occupation eventually approach unity. As the triplet has become fully occupied, no more transport of electrons from or to either the left or right lead can occur—the system has entered the Pauli spin blockade.

We learn an important lesson from this analysis. In the quantum dot, the lowest two-electron singlet state has an energy which is lower than the triplet state energy. Using equilibrium arguments, one would then have concluded that this singlet state had become occupied since its energy is the lowest among the two-electron energies. Under non-equilibrium conditions, however, one cannot use simple ground state arguments. The reasons is that, if there are states that can mediate electron density from one state to another, these states will be, at least fractionally, occupied by electrons.

References

1. Anderson, P.W.: Phys. Rev. **124**, 41 (1961)
2. Ono, K., Austing, D.G., Tokura, Y., Taurcha, S.: Science **297**, 1313–1317 (2002)
3. Fransson, J., Råsander, M.: Phys. Rev. B **73**, 205333 (2006)

Chapter 3
Many-Body Operator Green Functions

Abstract The formalism is extended to propagators, Green functions, in equilibrium and non-equilibrium. The contoured ordered Green functions, used in non-equilibrium, are introduced and we discuss the analytical continuation to real time correlation functions.

3.1 Basic Definitions

In general we shall be interested in describing the behavior of many-electron systems at finite temperatures. For an open system in thermodynamic equilibrium the expectation value of an operator ψ is computed using the grand canonical ensemble of statistical mechanics. We have

$$\langle\psi\rangle = \frac{\sum_n \langle n|\psi|n\rangle e^{-\beta(E_n-\mu N)}}{\sum_n e^{-\beta(E_n-\mu N_n)}} = \frac{\mathrm{tr}\, e^{-\beta(\mathcal{H}-\mu N)}\psi}{\mathrm{tr}\, e^{-\beta(\mathcal{H}-\mu N)}}, \tag{3.1}$$

where $\{|n\rangle, E_n\}$ represents an ortho-normal eigensystem of the many-electron system, and N_n the number of particles in state $|n\rangle$, whereas μ is the chemical potential. In the expression to the far right, $\mathcal{H} = \sum_n E_n h^n$ is the Hamiltonian of the system with N electrons.

The Green functions, which from now on shall form the basic tool in our discussions of many-electron systems, are thermodynamic averages of products of operators. Usually, the one-electron Green function is defined as the propagation of an electron created at position $\mathbf{r}'$ and time t', denoted $\psi^\dagger(\mathbf{r}', t')$, and annihilated at $\mathbf{r}$ and t, denoted $\psi(\mathbf{r}, t)$, e.g. $G(\mathbf{r}, \mathbf{r}'; t, t') = (-i)\langle \mathrm{T}\psi(\mathbf{r}, t)\psi^\dagger(\mathbf{r}', t')\rangle$, where T is the time-ordering operator. The time-ordering operator acts on the other operators in the average such that

$$G(\mathbf{r}, \mathbf{r}'; t, t') = \begin{cases} (-i)\langle\psi(\mathbf{r}, t)\psi^\dagger(\mathbf{r}', t')\rangle, & t > t', \\ (\mp i)\langle\psi^\dagger(\mathbf{r}', t')\psi(\mathbf{r}, t)\rangle, & t < t', \end{cases} \tag{3.2}$$

J. Fransson, *Non-Equilibrium Nano-Physics*,
Lecture Notes in Physics 809,
DOI 10.1007/978-90-481-9210-6_3,

where the plus (minus) sign in the second row refers to Fermionic (Bosonic) operators.

When working in terms of Hubbard operators, we will have to regard the creation of an electron as a transition $|N+1,n'\rangle\langle N,n|$ to a state $|N+1,n'\rangle$ with one more electron than the initial state $|N,n\rangle$. The annihilation is thus regarded as the transition to a state with one electron less. Moreover, due to the identity in (1.4), any one-electron Green function written in terms of Fermi operators, e.g. $G_{\alpha\alpha'}(t,t') = (-i)\langle \mathrm{T} d_\alpha(t) d^\dagger_{\alpha'}(t')\rangle$, has to be expanded in the whole set of Hubbard operator, i.e.

$$G_{\alpha\alpha'}(t,t') = (-i)\langle \mathrm{T} d_\alpha(t) d^\dagger_{\alpha'}(t')\rangle$$

$$= \sum_{pp'qq'} (d_\alpha)^{pq} (d^\dagger_{\alpha'})^{q'p'} (-i)\langle \mathrm{T} X^{pq}(t) X^{q'p'}(t')\rangle. \tag{3.3}$$

As we know from the earlier sections, the transition matrix elements $(d_\alpha)^{pq}$ take care of the amplitude/importance of each of the propagators $(-i)\langle \mathrm{T} X^{pq}(t) X^{q'p'}(t')\rangle$.

From time to time it will be inconvenient to use the four index notation. We therefore define the Latin multi-indices $a, b, c, \ldots$, such that $X^a = X^{pq}$, and the conjugate indices $\bar{a}, \bar{b}, \bar{c}, \ldots$, such that $X^{\bar{a}} = X^{qp} = (X^{pq})^\dagger$. Then, we can write the Hubbard operator Green functions as $G_{a\bar{b}}(t,t') = (-i)\langle \mathrm{T} X^a(t) X^{\bar{b}}(t')\rangle$, such that the above identity becomes

$$G_{\alpha\alpha'}(t,t') = \sum_{ab} (d_\alpha)^a (d^\dagger_{\alpha'})^{\bar{b}} G_{a\bar{b}}(t,t'). \tag{3.4}$$

We shall, however, not be so strict with the notation, but rather use any notation that is convenient for the moment. Nonetheless, whenever we are to discuss generalities or whenever there is no desire to refer to specific states, we will resort to this auxiliary notation.

We have omitted all explicit $\mathbf{r}$ dependence in the Green functions above. We have done this in order to keep notation to its minimal. Whenever we will be interested in spatial dependence of some quantity we will include this into the description, however, the main interest lie at the moment on the time dependence of the propagators.

We finally notice that the time-ordering acts on the Hubbard operators according to the scheme

$$(-i)\langle \mathrm{T} X^a(t) X^{\bar{b}}(t')\rangle = \begin{cases} (-i)\langle X^a(t) X^{\bar{b}}(t')\rangle, & t > t', \\ (\mp i)\langle X^{\bar{b}}(t') X^a(t)\rangle, & t < t', \end{cases} \tag{3.5}$$

where, again, the plus (minus) sign in the second row refers to Fermi-like (Bose-like) transitions.

3.2 Equilibrium

Although most of the content in this book is devoted to non-equilibrium conditions, it will be useful to briefly consider the equilibrium case. For concreteness, we take a single localized level interacting with a conduction channel, e.g.

$$\mathcal{H}=\sum_{\mathbf{k}\sigma}\varepsilon_{\mathbf{k}\sigma}c_{\mathbf{k}\sigma}^{\dagger}c_{\mathbf{k}\sigma}+\sum_{p=0\sigma 2}E_p h^p+\sum_{\mathbf{k}\sigma}[v_{\mathbf{k}\sigma}c_{\mathbf{k}\sigma}^{\dagger}(X^{0\sigma}+\sigma X^{\bar{\sigma}2})+H.c.]. \quad (3.6)$$

The conduction channel is described by the Hamiltonian $\sum_{\mathbf{k}\sigma}\varepsilon_{\mathbf{k}\sigma}c_{\mathbf{k}\sigma}^{\dagger}c_{\mathbf{k}\sigma}$, where the energy $\varepsilon_{\mathbf{k}\sigma}$ is related to the momentum $\mathbf{k}$ by some dispersion relation.

We are to study the evolution of the localized electron, and therefore we are interested in the Green functions $G_{a\bar{b}}(t,t')$, with $a,b=\{0\sigma,\bar{\sigma}2\}$. The equation of motion for $G_{0\sigma\bar{a}}(t,t')$ becomes

$$i\partial_t G_{0\sigma\bar{a}}(t,t')=\delta(t-t')\langle\{X^{0\sigma},X^{\bar{a}}\}(t)\rangle+(-i)\langle\mathrm{T}[X^{0\sigma},\mathcal{H}](t)X^{\bar{a}}(t')\rangle. \quad (3.7)$$

Here and henceforth, we will use units such that $\hbar=1$. We introduce the notation $P_{0\sigma\bar{a}}(t)=\langle\{X^{0\sigma},X^{\bar{a}}\}(t)\rangle$, the end-factor, which provides the amplitude of the Green function. In case of $\bar{a}=\sigma 0$, it simply provides the sum of the occupation numbers in the states $|0\rangle$ and $|\sigma\rangle$. This interpretation becomes slightly awkward in all other cases, since it might be strange to talk about transitions, like e.g. $Z^{\bar{\sigma}\sigma}$, as states. In the present case, averages of all such off-diagonal transitions, or coherences, vanish. There may, however, be situations when the averages of those transitions are finite, e.g. when such transitions are explicitly present in the Hamiltonian.

By performing the commutator $[X^{0\sigma},\mathcal{H}]$ the resulting equation of motion becomes

$$\begin{aligned}(i\partial_t-E_\sigma+E_0)&G_{0\sigma\bar{a}}(t,t')\\
&=\delta(t-t')P_{0\sigma\bar{a}}(t)-\sum_{\mathbf{k}}\bar{\sigma}v_{\mathbf{k}\bar{\sigma}}(-i)\langle\mathrm{T}(c_{\mathbf{k}\bar{\sigma}}^{\dagger}Z^{02})(t)X^{\bar{a}}(t')\rangle\\
&\quad+\sum_{\mathbf{k}}[v_{\mathbf{k}\sigma}(-i)\langle\mathrm{T}(h^0+h^\sigma)(t)c_{\mathbf{k}\sigma}(t)X^{\bar{a}}(t')\rangle\\
&\quad+v_{\mathbf{k}\bar{\sigma}}(-i)\langle\mathrm{T}(Z^{\bar{\sigma}\sigma}c_{\mathbf{k}\bar{\sigma}})(t)X^{\bar{a}}(t')\rangle.\end{aligned} \quad (3.8)$$

Here, the second and third terms on the left hand side of the above equation provides the transition energy between the states $|0\rangle$ and $|\sigma\rangle$. The second term on the left hand side is negligible unless the conduction channel is in the superconducting state. The third term, i.e.

$$\sum_{\mathbf{k}}v_{\mathbf{k}\sigma}(-i)\langle\mathrm{T}(h^0+h^\sigma)(t)c_{\mathbf{k}\sigma}(t)X^{\bar{a}}(t')\rangle\approx\sum_{\mathbf{k}}v_{\mathbf{k}\sigma}P_{0\sigma\sigma 0}(t)F_{\mathbf{k}\sigma\bar{a}}(t,t'), \quad (3.9)$$

by simple de-coupling. Here, $F_{\mathbf{k}\sigma\bar{a}}(t,t') = (-i)\langle \mathrm{T} c_{\mathbf{k}\sigma}(t) X^{\bar{a}}(t')\rangle$ describes the transfer, or tunneling, of an electron between the localized state and the conduction channel. The last term in the equation of motion vanishes to this order of approximation, i.e. $(-i)\langle \mathrm{T}(Z^{\bar{\sigma}\sigma} c_{\mathbf{k}\bar{\sigma}})(t) X^{\bar{a}}(t')\rangle \approx P_{0\sigma\bar{\sigma}0}(t) F_{\mathbf{k}\bar{\sigma}\bar{a}}(t,t') = 0$. In terms of this approximation, we thus have the two equations

$$(i\partial_t - \Delta_{\sigma 0}) G_{0\sigma\bar{a}}(t,t') = \delta(t-t') P_{0\sigma\bar{a}}(t) + P_{0\sigma\sigma 0} \sum_{\mathbf{k}} v_{\mathbf{k}\sigma} F_{\mathbf{k}\sigma\bar{a}}(t,t'), \tag{3.10a}$$

$$(i\partial_t - \Delta_{2\bar{\sigma}}) G_{\bar{\sigma}2\bar{a}}(t,t') = \delta(t-t') P_{\bar{\sigma}2\bar{a}}(t) + \sigma P_{\bar{\sigma}22\bar{\sigma}} \sum_{\mathbf{k}} v_{\mathbf{k}\sigma} F_{\mathbf{k}\sigma\bar{a}}(t,t'), \tag{3.10b}$$

where $\Delta_{qp} = E_q - E_p$ is the transition energy for the process $|q\rangle \to |p\rangle$, and where we used the same approach for the Green function $G_{\bar{\sigma}2\bar{a}}(t,t')$. The equation of motion for the transfer Green function $F_{\mathbf{k}\sigma\bar{a}}(t,t')$ is given by

$$(i\partial_t - \varepsilon_{\mathbf{k}\sigma}) F_{\mathbf{k}\sigma\bar{a}}(t,t') = v_{\mathbf{k}\sigma}[G_{0\sigma\bar{a}}(t,t') + \sigma G_{\bar{\sigma}2\bar{a}}(t,t')]. \tag{3.11}$$

We Fourier transform the above equation of motions, i.e. we use

$$G(i\omega_n) = \int_0^{-i\beta} G(t,t') e^{i\omega_n(t-t')} dt, \qquad \omega_n = \begin{cases} i\frac{2n}{\beta}\pi + \mu, & \text{Bose-like}, \\ i\frac{2n+1}{\beta}\pi + \mu, & \text{Fermi-like}, n \in \mathbb{Z}, \end{cases} \tag{3.12a}$$

$$G(t,t') = \frac{i}{\beta} \sum_n G(i\omega_n) e^{-i\omega_n(t-t')}, \qquad \begin{cases} 0 \le it \le \beta, \\ 0 \le it' \le \beta, \end{cases} \tag{3.12b}$$

which is allowed since the propagators only depend on the time difference $t - t'$ in equilibrium. It should be noticed that the time-integration is taken along the contour that begins and terminates at 0 and $-i\beta$, respectively, and circumventing the positive real axis. We, thus, obtain

$$F_{\mathbf{k}\sigma\bar{a}}(i\omega) = v_{\mathbf{k}\sigma}[G_{0\sigma\bar{a}} + \sigma G_{\bar{\sigma}2\bar{a}}]/(i\omega - \varepsilon_{\mathbf{k}\sigma}). \tag{3.13}$$

Keeping in mind that the Fourier frequency ω_n here is an odd integer of $i\pi/\beta$, we shall omit the subscript n. Using this result in the aligns for the localized states, we arrive at the solvable set of aligns

$$\big(i\omega - \Delta_{\sigma 0} - P_{0\sigma\sigma 0} V_\sigma(i\omega)\big) G_{0\sigma\bar{a}}(i\omega) = P_{0\sigma\bar{a}} + \sigma P_{0\sigma\sigma 0} V_\sigma(i\omega) G_{\bar{\sigma}2\bar{a}}(i\omega), \tag{3.14a}$$

$$\big(i\omega - \Delta_{2\bar{\sigma}} - P_{\bar{\sigma}22\bar{\sigma}} V_\sigma(i\omega)\big) G_{\bar{\sigma}2\bar{a}}(i\omega) = P_{\bar{\sigma}2\bar{a}} + \sigma P_{\bar{\sigma}22\bar{\sigma}} V_\sigma(i\omega) G_{0\sigma\bar{a}}(i\omega). \tag{3.14b}$$

Here, the propagator $V_\sigma(i\omega) = \sum_{\mathbf{k}} |v_{\mathbf{k}\sigma}|^2/(i\omega - \varepsilon_{\mathbf{k}\sigma})$ provides a broadening to the localized state due to the interaction between electrons in the localized state and the conduction channel. In order to make the notation shorter, we put $G_{pq} = G_{pqqp}$ and $P_{pq} = P_{pqqp}$.

We are now in position to solve for the diagonal Green functions. It is evident, in the equations above, that we also need to find the equations for the off-diagonal Green functions $G_{0\sigma 2\bar{\sigma}}$ and $G_{\bar{\sigma}2\sigma 0}$. Although they describe mysterious sequences of transitions, these propagators are non-zero, and are given by

$$G_{0\sigma 2\bar{\sigma}}(i\omega) = \sigma \frac{P_{0\sigma} V_\sigma(i\omega)}{i\omega - \Delta_{\sigma 0} - P_{0\sigma} V_\sigma(i\omega)} G_{\bar{\sigma}2}(i\omega), \tag{3.15a}$$

$$G_{\bar{\sigma}2\sigma 0}(i\omega) = \sigma \frac{P_{\bar{\sigma}2} V_\sigma(i\omega)}{i\omega - \Delta_{2\bar{\sigma}} - P_{\bar{\sigma}2} V_\sigma(i\omega)} G_{0\sigma}(i\omega). \tag{3.15b}$$

We thus have

$$G_{0\sigma}(i\omega) = \frac{P_{0\sigma}}{i\omega - \Delta_{\sigma 0} - P_{0\sigma} V_\sigma(i\omega) - \frac{P_{0\sigma} V_\sigma(i\omega) P_{\bar{\sigma}2}}{i\omega - \Delta_{2\bar{\sigma}} - P_{\bar{\sigma}2} V_\sigma(i\omega)} V_\sigma(i\omega)}, \tag{3.16a}$$

$$G_{\bar{\sigma}2}(i\omega) = \frac{P_{\bar{\sigma}2}}{i\omega - \Delta_{2\bar{\sigma}} - P_{\bar{\sigma}2} V_\sigma(i\omega) - \frac{P_{\bar{\sigma}2} V_\sigma(i\omega) P_{0\sigma}}{i\omega - \Delta_{\sigma 0} - P_{0\sigma} V_\sigma(i\omega)} V_\sigma(i\omega)}. \tag{3.16b}$$

If we were to solve for the localized level using Fermi operator notation, we would have been interested in the Green function $\mathcal{G}_\sigma(t,t') = (-i)\langle \mathrm{T} d_\sigma(t) d_\sigma^\dagger(t')\rangle$. Expanding this Green function in terms of Hubbard operators, i.e. $\mathcal{G}_\sigma = G_{0\sigma} + \sigma[G_{0\sigma 2\bar{\sigma}} + G_{\bar{\sigma}2\sigma 0}] + G_{\bar{\sigma}2}$, we find that the Green function for the localized level is given by

$$\mathcal{G}_\sigma(i\omega) = \frac{[i\omega - \Delta_{2\bar{\sigma}}]P_{0\sigma} + [i\omega - \Delta_{\sigma 0}]P_{\bar{\sigma}2}}{[i\omega - \Delta_{\sigma 0} - P_{0\sigma} V_\sigma(i\omega)][i\omega - \Delta_{2\bar{\sigma}}] - [i\omega - \Delta_{\sigma 0}]P_{\bar{\sigma}2} V_\sigma(i\omega)}. \tag{3.17}$$

This expression is complicated and not very appealing. However, we shall remember that, cf. Sect. 2.3, $\Delta_{\sigma 0} = E_\sigma - E_0 = \varepsilon_\sigma$ and $\Delta_{2\bar{\sigma}} = E_2 - E_{\bar{\sigma}} = \sum_\sigma \varepsilon_\sigma + U - \varepsilon_{\bar{\sigma}} = \varepsilon_\sigma + U$. Furthermore, the end-factors $P_{0\sigma} = \langle h^0 + h^\sigma \rangle = N_0 + N_\sigma$ and, analogously, $P_{\bar{\sigma}2} = N_{\bar{\sigma}} + N_2$. By the closure relation, (1.17c), we have $1 = \sum_p \langle h^p \rangle = \sum_p N_p$, hence, $P_{0\sigma} + P_{\bar{\sigma}2} = 1$. The expansion $n_\sigma = d_\sigma^\dagger d_\sigma = h^\sigma + h^2$ also suggests that $P_{\bar{\sigma}2} = \langle n_{\bar{\sigma}} \rangle$, giving $P_{0\sigma} = 1 - \langle n_{\bar{\sigma}} \rangle$. We can thus rewrite the expression above as

$$\begin{aligned}
\mathcal{G}_\sigma(i\omega) &= \frac{i\omega - \Delta_{\sigma 0} - U P_{0\sigma}}{[i\omega - \Delta_{\sigma 0} - V_\sigma(i\omega)][i\omega - \Delta_{2\bar{\sigma}}] - P_{\bar{\sigma}2} V_\sigma(i\omega) U} \\
&= \frac{i\omega - \varepsilon_\sigma - (1 - \langle n_{\bar{\sigma}} \rangle)U}{[i\omega - \varepsilon_\sigma - V_\sigma(i\omega)][i\omega - \varepsilon_\sigma - U] - \langle n_{\bar{\sigma}} \rangle V_\sigma(i\omega) U},
\end{aligned} \tag{3.18}$$

where the expression on the second line is the usual self-consistent Hartree-Fock approximation found in literature. Finally, the end-factors may be calculated using the spectral theorem, noting that

$$N_0 = \langle h^0 \rangle = \sum_\sigma \langle X^{0\sigma} X^{\sigma 0} \rangle = -\frac{1}{\pi} \int [1 - f(\omega)] \operatorname{Im} \sum_\sigma G^r_{0\sigma}(\omega) d\omega, \quad (3.19a)$$

$$N_\sigma = \langle h^\sigma \rangle = \langle X^{\sigma 0} X^{0\sigma} \rangle + \langle X^{\sigma 2} X^{2\sigma} \rangle$$
$$= -\frac{1}{\pi} \int [f(\omega) \operatorname{Im} G^r_{0\sigma}(\omega) + [1 - f(\omega)] \operatorname{Im} G^r_{\sigma 2}(\omega)] d\omega, \quad (3.19b)$$

$$N_2 = \langle h^2 \rangle = \sum_\sigma \langle X^{2\sigma} X^{\sigma 2} \rangle = -\frac{1}{\pi} \int f(\omega) \sum_\sigma \operatorname{Im} G^r_{\sigma 2}(\omega) d\omega. \quad (3.19c)$$

In this way we obtain the self-consistent aligns for the localized level. The summations appear since the occupation numbers are coupled to the other states in the system through the transitions in the Green functions. We must simply account for all density of electrons that couple to each occupation number.

Using Hubbard operators for this model up to the mean field approximation may seem as making the mathematics only more complicated. On the other hand, the derivation was performed in order to illustrate some important questions that may arise when working in terms of Hubbard operators, namely interpretation of the transition energies $\Delta_{qp} = E_q - E_p$ and the end-factors $P_{pqq'p'}$.

3.3 Contour Ordered Green Functions

We are now entering the non-equilibrium formalism of the many-body operator Green functions. We generalize the single-electron Green function to the complex contour $[t_0, t_0 - i\beta]$, where t_0 is arbitrary, through

$$G(\mathbf{r}, \mathbf{r}'; t, t'; U) = (-i) \langle \mathrm{T} \psi(\mathbf{r}, t) \psi^\dagger(\mathbf{r}', t') \rangle_U \equiv (-i) \frac{\langle \mathrm{T} S \psi(\mathbf{r}, t) \psi^\dagger(\mathbf{r}', t') \rangle}{\langle \mathrm{T} S \rangle}, \quad (3.20)$$

where T means imaginary time, or contour, ordering and the action operator S is given by

$$S = \exp\left(-i \int_{t_0}^{t_0 - i\beta} \mathcal{H}'(t') dt' \right), \quad (3.21)$$

where $\mathcal{H}'(t')$ is a disturbance potential defined on the space and times on the contour $[t_0, t_0 - i\beta]$.

There is a major difference between the non-definitions of the propagators, actions, Hamiltonians, etc. in equilibrium and non-equilibrium. In the former case, there is always a ground state defined with respect to which we can perform all our calculations. This ground state is defined for all times t, which implies that we may define the complex time contour $[t_0, t_0 - i\beta]$ arbitrarily. Especially, we may put $t_0 = 0$.

The situation is very different under non-equilibrium conditions, however, since there is no ground state defined with respect to which we can undertake our calculations. It is, therefore, important that we include some arbitrary time t_0 in the definition of the complex time contour $[t_0, t_0 - i\beta]$, in order not to restrict our treatment to equilibrium. On the other hand, in all our considerations of non-equilibrium, we shall assume that there was a ground state defined at $t_0 \to -\infty$. By taking this mathematical construction for granted, we can refer all our calculations to the ground state at $t_0 \to -\infty$, simply by extending the complex time contour to both begin and terminate at $-\infty$ and circumventing the positive real axis.

The exact form of the disturbance potential $\mathcal{H}'(t')$ may vary from case to case. Generally, this potential will be of the form $\mathcal{H}'(t') = \sum_\xi U_\xi(t') Z^\xi$, where ξ is a Bose-like transition index, and where $U_\xi(t')$ is an external source field acting on the system. In the following, we shall denote the integration over the imaginary time interval $[t_0, t_0 - i\beta]$, or some equally appropriately defined imaginary time contour, by the subscript C, e.g. $S = \exp\{-i \int_C \mathcal{H}'(t')dt'\}$.

The single-electron Green functions defined for equilibrium and non-equilibrium conditions satisfy the boundary conditions $G(t,t';U)_{|t=t_0} = -e^{\beta\mu} G(t,t';U)_{|t=t_0-i\beta}$ [1]. This is easily derived under equilibrium conditions using the definition of the average, since

$$
\begin{aligned}
G(t,t';U)_{|t=0} &= i\langle \psi^\dagger(t')\psi(0)\rangle \\
&= i\frac{\operatorname{tr} e^{-\beta(\mathcal{H}-\mu N)}\psi^\dagger(t')\psi(0)}{\operatorname{tr} e^{\beta(\mathcal{H}-\mu N)}} = i\frac{\operatorname{tr} \psi(0) e^{-\beta(\mathcal{H}-\mu N)}\psi^\dagger(t')}{\operatorname{tr} e^{\beta(\mathcal{H}-\mu N)}} \\
&= i\frac{\operatorname{tr} e^{-\beta(\mathcal{H}-\mu N)}[e^{\beta(\mathcal{H}-\mu N)}\psi(0)e^{-\beta(\mathcal{H}-\mu N)}]\psi^\dagger(t')}{\operatorname{tr} e^{\beta(\mathcal{H}-\mu N)}},
\end{aligned}
\tag{3.22}
$$

where we have used that the trace is invariant under even permutations of the operators. The number operator $N = \psi^\dagger\psi$ commutes with the Hamiltonian, hence, $e^{\beta(\mathcal{H}-\mu N)} = e^{\beta\mathcal{H}} e^{-\beta\mu N}$. Moreover, due to the fact that $[\psi, N] = \psi$, it is a straight forward exercise to show that $e^{-\beta\mu N}\psi(0)e^{\beta\mu N} = e^{\beta\mu}\psi(0)$. By finally using the property that $\psi(t) = e^{i\mathcal{H}t}\psi(0)e^{-i\mathcal{H}t}$, we arrive at the relation

$$
G(t,t')_{|t=0} = -e^{\beta\mu}(-i)\frac{\operatorname{tr} e^{-\beta(\mathcal{H}-\mu N)}\psi(-i\beta)\psi^\dagger(t')}{\operatorname{tr} e^{\beta(\mathcal{H}-\mu N)}} = -e^{\beta\mu} G(t,t')_{|t=-i\beta}.
\tag{3.23}
$$

This derivation can be straight forwardly generalized to non-equilibrium, where we have

$$
G(t,t';U)_{|t=t_0} = -e^{\beta\mu} G(t,t';U)_{|t=t_0-i\beta}.
\tag{3.24}
$$

In terms of Hubbard operators, however, these boundary conditions become slightly altered in that the exponent $e^{\beta\mu}$ is replaced by $e^{0\cdot\beta\mu} = 1$. The transition $X^a = X^{qq'}$ commutes with the number operator $N = \sum_p h^p$ according to $[X^a, N] = \sum_p [X^{qq'}, h^p] = \sum_p [\delta_{pq'} X^{qp} - \delta_{pq} X^{pq'}] = 0$, which leads to that

$e^{-\beta\mu N} X^a e^{\beta\mu N} = X^a$. It is then easy to derive

$$G_{a\bar{b}}(t,t';U)_{t=t_0} = i\langle X^{\bar{b}}(t')X^a(0)\rangle = -(-i)\langle X^a(-i\beta)X^{\bar{b}}(t')\rangle$$
$$= -G_{a\bar{b}}(t,t';U)_{t=t_0-i\beta}. \tag{3.25}$$

Our generalized non-equilibrium Green functions satisfy similar equations of motion as the corresponding equilibrium Green functions. Taking the single level system, for instance, we have the equations of motion (dropping the reference U to the disturbance potential in the Green functions $G_{pqq'p'}$)

$$\big(i\partial_t - \Delta_{\sigma 0} - \Delta U_{\sigma 0}(t)\big)G_{0\sigma\bar{a}}(t,t') - U_{\sigma\bar{\sigma}}(t)G_{0\bar{\sigma}\bar{a}}(t,t')$$
$$= \delta(t-t')P_{0\sigma\bar{a}}(t) + \sum_{\mathbf{k}} v_{\mathbf{k}\sigma}(-i)\langle \mathrm{T}([h^0+h^\sigma]c_{\mathbf{k}\sigma})(t)X^{\bar{a}}(t')\rangle_U$$
$$+ \sum_{\mathbf{k}} v_{\mathbf{k}\bar{\sigma}}(-i)\langle \mathrm{T}(Z^{\bar{\sigma}\sigma}c_{\mathbf{k}\bar{\sigma}})(t)X^{\bar{a}}(t')\rangle_U, \tag{3.26a}$$

$$\big(i\partial_t - \Delta_{2\bar{\sigma}} - \Delta U_{2\bar{\sigma}}(t)\big)G_{\bar{\sigma}2\bar{a}}(t,t') + U_{\sigma\bar{\sigma}}(t)G_{\sigma 2\bar{a}}(t,t')$$
$$= \delta(t-t')P_{\bar{\sigma}2\bar{a}}(t) + \sigma\sum_{\mathbf{k}} v_{\mathbf{k}\sigma}(-i)\langle \mathrm{T}([h^{\bar{\sigma}}+h^2]c_{\mathbf{k}\sigma})(t)X^{\bar{a}}(t')\rangle_U$$
$$+ \bar{\sigma}\sum_{\mathbf{k}} v_{\mathbf{k}\bar{\sigma}}(-i)\langle \mathrm{T}(Z^{\bar{\sigma}\sigma}c_{\mathbf{k}\bar{\sigma}})(t)X^{\bar{a}}(t')\rangle_U, \tag{3.26b}$$

where $\Delta U_{qp}(t) = U_q(t) - U_p(t)$. We derive those equation exactly in the same way as in the previous discussion. The new features that appear are the source fields $U_\xi(t)$, which are generated by the time-derivative of the Green functions, and here using the disturbance potential $\mathcal{H}'(t') = U_0(t')h^0 + \sum_\sigma [U_\sigma(t')h^\sigma + U_{\sigma\bar{\sigma}}(t')Z^{\sigma\bar{\sigma}}]$. This is most easily seen on the time contour $[t_0, t_0 - i\beta]$ by noting that

$$\mathrm{T}SX^a(t) = \mathrm{T}\{e^{-i\int_t^{t_0-i\beta}\mathcal{H}'(t')dt'}X^a(t)e^{-i\int_{t_0}^{t}\mathcal{H}'(t')dt'}\}, \tag{3.27}$$

which, upon time differentiation, gives

$$i\frac{\partial}{\partial t}\mathrm{T}SX^a(t) = \mathrm{T}\Big\{e^{-i\int_t^{t_0-i\beta}\mathcal{H}'(t')dt'}[-\mathcal{H}'(t)X^a(t)$$
$$+ i\frac{\partial}{\partial t}X^a(t) + X^a(t)\mathcal{H}'(t)]e^{-i\int_{t_0}^{t}\mathcal{H}'(t')dt'}\Big\}$$
$$= \mathrm{T}S[X^a(t), \mathcal{H} + \mathcal{H}'(t)]. \tag{3.28}$$

Then, for instance, the commutator $[X^{0\sigma}, \mathcal{H}'] = (U_\sigma - U_0)X^{0\sigma} + U_{\sigma\bar{\sigma}}X^{0\bar{\sigma}}$, which explains the presence of the Green function $G_{0\bar{\sigma}\bar{a}}$ in the equation for $G_{0\sigma\bar{a}}$.

We did introduce the auxiliary source fields $U_\xi(t)$ for a reason, which we shall discuss in more detail below. Further consequences of introducing these fields will also be discussed in Chap. 5. We notice in (3.26a), (3.26b) the presence of three-operator Green functions, e.g. $(-i)\langle \mathrm{T}([h^0 + h^\sigma]c_{\mathbf{k}\sigma})(t)X^{\bar{a}}(t')\rangle_U$. We may deal with those according to the decoupling procedure applied in Sect. 3.2, or by the Wick's theorem following the approach described in detail in [2]. The latter of those procedures, however, is inappropriate to use in non-equilibrium situations. Therefore, we will follow another approach that provides a diagrammatic expansion of the propagators.

Consider a general propagator with two Hubbard operators, e.g.

$$G_{a\bar{b}}(t,t') = (-i)\frac{\langle \mathrm{T}SX^a(t)X^{\bar{b}}(t')\rangle}{\langle \mathrm{T}S\rangle}, \tag{3.29}$$

where the transitions a, b are arbitrary. Varying this Green function with respect to the sources $U_\xi(t)$ in the action $S = \exp\{-i\int_C \sum_\xi U_\xi(t)Z^\xi dt'\}$, we obtain

$$\begin{aligned}\delta G_{a\bar{b}}(t,t') = -i\int_C \sum_\xi \delta U_\xi(t'')\Bigg\{&(-i)\frac{\langle \mathrm{T}SZ^\xi(t'')X^a(t)X^{\bar{b}}(t')\rangle}{\langle \mathrm{T}S\rangle} \\ &-(-i)\frac{\langle \mathrm{T}SX^a(t)X^{\bar{b}}(t')\rangle}{\langle \mathrm{T}S\rangle}\frac{\langle \mathrm{T}SZ^\xi(t'')\rangle}{\langle \mathrm{T}S\rangle}\Bigg\}dt''.\end{aligned} \tag{3.30}$$

Here, we recognize the first factor in the second term as $G_{a\bar{b}}(t,t')$. By rearranging the above expression, we, thus, find that the propagator containing three Hubbard operators can be expressed in terms of the one-electron Green function and functional derivatives thereof, according to

$$(-i)\langle \mathrm{T}Z^\xi(t'')X^a(t)X^{\bar{b}}(t')\rangle_U = \left(\langle \mathrm{T}Z^\xi(t'')\rangle_U + i\frac{\delta}{\delta U_\xi(t'')}\right)G_{a\bar{b}}(t,t'). \tag{3.31}$$

Keeping in mind that all averages $\langle \mathrm{T}Z^\xi(t)\rangle_U$ naturally appearing in the aligns of motion for the Green functions $G_{a\bar{b}}(t,t')$, are results of anti-commutators as e.g. $P_{a\bar{b}}(t) = \langle \mathrm{T}\{X^a, X^{\bar{b}}\}(t)\rangle_U = \langle \mathrm{T}(\delta_{qq'}Z^{pp'} + \delta_{pp'}Z^{q'q})(t)\rangle_U$, for transitions $a = pq$, $b = p'q'$. In analogy with this, we introduce the notation

$$R_{a\bar{b}}(t) = i\left(\delta_{qq'}\frac{\delta}{\delta U_{pp'}(t)} + \delta_{pp'}\frac{\delta}{\delta U_{q'q}(t)}\right). \tag{3.32}$$

Using all the notation thus introduced, we can write the equation of motion for a the Green function $G_{a\bar{b}}(t,t')$ in the Hamiltonian system $\mathcal{H} = \sum_{\mathbf{k}\sigma}\varepsilon_{\mathbf{k}\sigma}n_{\mathbf{k}\sigma} + \sum_p E_p h^p + \sum_{\mathbf{k}\sigma a}[v_{\mathbf{k}\sigma}(d_\sigma)^a c^\dagger_{\mathbf{k}\sigma}X^a + H.c.]$ as

$$\begin{aligned}&[i\partial_t - \Delta_{\bar{a}}]G_{a\bar{b}}(t,t') - \sum_{\xi c} U_\xi(t)\varepsilon_c^{a\xi}G_{c\bar{b}}(t,t') \\ &\quad = \delta(t-t')P_{a\bar{b}}(t) + \sum_{cd}[P_{a\bar{c}}(t^+) + R_{a\bar{c}}(t^+)]\int_C V_{\bar{c}d}(t,t'')G_{d\bar{b}}(t'',t')dt'',\end{aligned} \tag{3.33}$$

where $\varepsilon_c^{a\xi}$ is a tensor defined such that $[X^a, Z^\xi] = \sum_c \varepsilon_c^{a\xi} X^c$, while $V_{\bar{c}d}(t,t') = \sum_\sigma V_\sigma(t,t')(d_\sigma^\dagger)^{\bar{c}}(d_\sigma)^d$.

The structure of the equation of motion suggests that we can write the matrix equation

$$[i\partial_t - \Delta - \mathbf{U}(t)]\mathbf{G}(t,t') = \delta(t-t')\mathbf{P}(t) + [\mathbf{P}(t^+) + \mathbf{R}(t^+)] \int_C \mathbf{V}(t,t'')\mathbf{G}(t'',t')dt''. \tag{3.34}$$

Then, we should in principle, obtain the equation

$$\mathbf{G}(t,t') = \mathbf{d}(t,t')\mathbf{P}(t') + \int_C \mathbf{d}(t,t_1)[\mathbf{P}(t_1^+) + \mathbf{R}(t_1^+)]\mathbf{V}(t_1,t_2)\mathbf{G}(t_2,t')dt_2dt_1, \tag{3.35}$$

where $\mathbf{d}(t,t')$ satisfying the equation $[i\partial_t - \Delta - \mathbf{U}(t)]\mathbf{d}(t,t') = \delta(t-t')$ is called the bare *locator*. The Green function clearly has the structure of a product of the locator function $\mathbf{d}$, or more generally the dressed locator $\mathbf{D}$, and the end-factor $\mathbf{P}$, according to $\mathbf{G} = \mathbf{DP}$. Hence, multiplying (3.35) from the right by $\mathbf{P}^{-1}(t')$ we obtain the equation for the locator $\mathbf{D}$, that is

$$\mathbf{D}(t,t') = \mathbf{d}(t,t') + \int_C \mathbf{d}(t,t_1)[\mathbf{P}(t_1^+) + \mathbf{R}(t_1^+)]\mathbf{V}(t_1,t_2)\mathbf{D}(t_2,t')dt_2dt_1. \tag{3.36}$$

Further, is we multiply this equation from the left by $d^{-1}(\tau,t)$ and from the right by $\mathbf{D}^{-1}(t',\tau')$, and integrate the equation over t and t', we obtain

$$\mathbf{D}^{-1}(\tau,\tau') = \mathbf{d}^{-1}(\tau,\tau') - \int_C \{[\mathbf{P}(\tau^+) + \mathbf{R}(\tau^+)]\mathbf{V}(\tau,t_2)\mathbf{D}(t_2,t')\}\mathbf{D}^{-1}(t',\tau')dt_2dt'. \tag{3.37}$$

The parenthesis has been introduced in the integral as a reminder that the functional differentiation operator $\mathbf{R}$ acts on the locator $\mathbf{D}(t_2,t')$, however, not on the inverse locator $\mathbf{D}^{-1}(t',\tau')$. Hence, the product $\{\mathbf{R}(\tau^+)\mathbf{V}(t_1,t_2)\mathbf{D}(t_2,t')\}\mathbf{D}^{-1}(t',\tau')$ cannot be further simplified since $\mathbf{D}^{-1}$ is the inverse of $\mathbf{D}$, and not of $\mathbf{RVD}$. We can, however, write the expression in (3.37) as

$$\begin{aligned}\mathbf{D}^{-1}(\tau,\tau') &= \mathbf{d}^{-1}(\tau,\tau') - \mathbf{P}(\tau^+)\mathbf{V}(\tau,\tau') \\ &\quad - \int_C \{\mathbf{R}(\tau^+)\mathbf{V}(\tau,t_2)\mathbf{D}(t_2,t')\}\mathbf{D}^{-1}(t',\tau')dt_2dt'.\end{aligned} \tag{3.38}$$

Either this expression or the one in (3.37) suggests that one can introduce a self-energy, or rather, a self-energy operator $\mathcal{S}$, by putting

$$\mathcal{S}(\tau,\tau') = \mathbf{P}(\tau^+)\mathbf{V}(\tau,\tau') + \int_C \{\mathbf{R}(\tau^+)\mathbf{V}(\tau,t_2)\mathbf{D}(t_2,t')\}\mathbf{D}^{-1}(t',\tau')dt_2dt', \tag{3.39}$$

which reduces the equation for the locator to

$$\mathbf{D}^{-1}(t,t') = d^{-1}(t,t') - \mathcal{S}(t,t'). \tag{3.40}$$

Finally, we can also write the equation for the Green function in terms of the self-energy operator as

$$\mathbf{G}(t,t') = \mathbf{g}(t,t') + \int_C \mathbf{d}(t,\tau)\mathcal{S}(\tau,\tau')\mathbf{G}(\tau',t')d\tau d\tau', \tag{3.41}$$

where $\mathbf{g} = \mathbf{dP}$ is the bare Green function.

We will explore the framework provided by the self-energy operator in Chap. 5 for a diagrammatic expansion of the Green function. Here, we notice, however, that the simplest approximation of the Green function is obtained by omitting the functional differentiation, i.e. simply letting $\mathcal{S} \approx \mathbf{PV}$. Then, by letting the external source fields $U_\xi(t) \to 0$, in which limit we can extract the physical content of the derived equations. We obtain the Dyson-like equation

$$\mathbf{G}(t,t') = \mathbf{g}(t,t') + \int_C \mathbf{g}(t,t_1)\mathbf{V}(t_1,t_2)\mathbf{G}(t_2,t')dt_2 dt_1. \tag{3.42}$$

3.4 Non-Equilibrium Green Functions

So far, we have been working with all propagators defined on the complex time contour e.g. $[t_0, t_0 - i\beta]$, which provides a convenient environment to generate the appropriate equation s of motion, and perturbational or diagrammatic expansions of the propagators. The complex contour environment is, nevertheless, inappropriate when physical information is to be acquired from the work we have done. We need to convert our propagators into the real time domain. The fundament for this conversion was laid out by Kadanoff and Baym [1], and Keldysh [3], while Langreth [4] provided simple rules that allows simple mechanical manipulations of the real time Green functions. Here, we will make a somewhat thorough discussion of the procedures to convert the contour ordered Green functions into the corresponding real time propagators. The author also refers to [5] for an analogous discussion.

There are two basic propagators in the framework of non-equilibrium Green functions, namely the lesser and greater Green functions, here defined by

$$G^{>}_{a\bar{b}}(t,t') = -i\langle X^a(t)X^{\bar{b}}(t')\rangle, \quad t > t', \tag{3.43a}$$

$$G^{<}_{a\bar{b}}(t,t') = i\langle X^{\bar{b}}(t')X^a(t)\rangle, \quad t < t'. \tag{3.43b}$$

We have taken the case when the transitions are Fermi-like. We also have reasons to define lesser and greater Green functions with Bose-like transitions e.g.

$$K^{>}_{\xi\bar{\xi}}(t,t') = -i\langle Z^{\xi}(t)Z^{\bar{\xi}}(t')\rangle, \quad t > t', \tag{3.44a}$$

$$K^{<}_{\xi\bar{\xi}}(t,t') = -i\langle Z^{\bar{\xi}}(t')Z^{\xi}(t)\rangle, \quad t < t'. \tag{3.44b}$$

The negative sign appearing on the lesser Green function $K^{<}_{\xi\bar{\xi}}(t,t')$ is related to the use of commutator when dealing with Bose-like transitions, instead of the anti-commutator as is the case in case of Fermi-like transitions, cf. $G^{<}_{a\bar{b}}(t,t')$.

The lesser (greater) superscript is used as a reminder that the time t is less (larger) than t' on the time contour. The time contour $[t_0, t_0 - i\beta]$ for instance, begins at the time t_0, goes, slightly above the real time axis, to ∞ (whatever it means) where it circumvents the real time axis, continues slightly below the real time axis and ends at $t_0 - i\beta$, see Fig. 3.1. Hence, we must understand the meaning of $t < t'$ on the contour. In fact, we can regard t as being less than t' as long as t' lies closer to the end point of the contour, e.g. $t_0 - i\beta$, than t. Sometimes, however, it is meaningful to also regard the time contour C as constituted of two disjoint parts, the upper and lower contours, C_u and C_l, respectively. The upper contour begins at the beginning of C and ends at ∞, where it is joint with the beginning of the lower contour C_l which, in turn, ends at the end of C. Finally, in order to obtain the real time Green functions, we must let the start and end points of the contour approach $-\infty$ (in the particular case of $[t_0, t_0 - i\beta]$ we have to require that $t_0 \to -\infty$).

Some care has to be taken when our propagator is a convolution of two, or more, Green functions on the time contour. Assuming that the propagator $A(t,t') = \int_C B(t,t'')D(t'',t')dt''$, where B and D are some propagators. The lesser propaga-

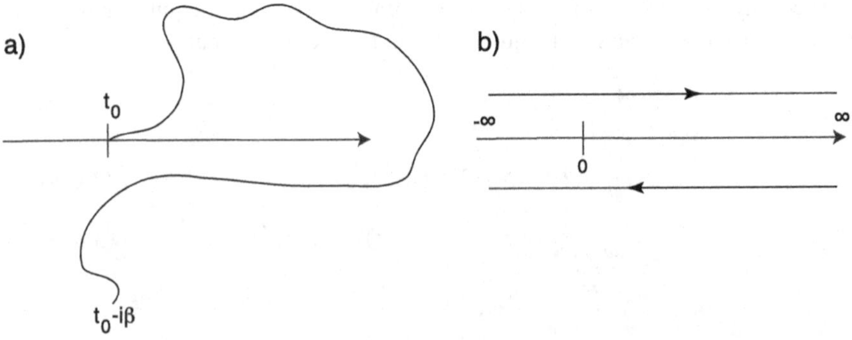

Fig. 3.1 Two examples of the complex time contour. In panel (**a**) the contour $[t_0, t_0 - i\beta]$ as suggested by Kadanoff and Baym is depicted, whereas panel (**b**) illustrates the Keldysh contour

tor $A^<$ is then extracted by the following consideration:

$$A^<(t,t') = \int_{-\infty}^{t} B^>(t,t'')D^<(t'',t')dt'' + \int_{t}^{t'} B^<(t,t'')D^<(t'',t')dt'' + \int_{t'}^{-\infty} B^<(t,t'')D^>(t'',t')dt''. \quad (3.45)$$

In the first integral, the time $t'' < t < t'$, while in the second we have $t < t'' < t'$, whereas the times $t < t' < t''$ in the last. Here, we have to think *contour-wise* and not in the usual sense. On the upper contour this introduces no problem, while in the lower contour this leads to something that might look counter-intuitive. Keeping in mind the rule given above, times closer to the end-point of the contour are *later/greater* than times closer to the starting point, such problems should not arise.

Now, the expression given above for $A^<(t,t')$ is not satisfactory and we need to shape it up. We can do the following: let the second integral be given as two, i.e.

$$\int_{t}^{t'} B^<(t,t'')D^<(t'',t')dt'' = \int_{t}^{-\infty} B^<(t,t'')D^<(t'',t')dt'' + \int_{-\infty}^{t'} B^<(t,t'')D^<(t'',t')dt''. \quad (3.46)$$

We can, then, write the lesser propagator according to

$$A^<(t,t') = \int \theta(t-t'')[B^>(t,t'') - B^<(t,t'')]D^<(t'',t')dt'' + \int B^<(t,t'')\theta(-t+t')[D^<(t'',t') - D^>(t'',t')]dt'', \quad (3.47)$$

where the integration is taken on the whole real time axis. The combinations of the correlation functions between the brackets are commonly known as the retarded and advanced Green functions, generally

$$A^r(t,t') = \theta(t-t')[A^>(t,t') - A^<(t,t')], \quad (3.48a)$$

$$A^a(t,t') = -\theta(-t+t')[A^>(t,t') - A^<(t,t')]. \quad (3.48b)$$

Using these definitions, we find for instance that

$$G^{r/a}_{a\bar{b}}(t,t') = (\mp i)\theta(\pm t \mp t')\langle\{X^{\bar{b}}(t'), X^a(t)\}\rangle, \quad (3.49a)$$

$$K^{r/a}_{\xi\bar{\zeta}}(t,t') = (\mp i)\theta(\pm t \mp t')\langle[Z^{\bar{\zeta}}(t'), Z^{\xi}(t)]\rangle, \quad (3.49b)$$

in agreement with the usual definitions for equilibrium Green functions.

Substituting the definitions given in (3.48a), (3.48b) into to (3.47), we finally find that this expression can be rewritten as

$$A^{<}(t,t') = \int [B^r(t,t'')D^{<}(t'',t') + B^{<}(t,t'')D^a(t'',t')]dt'', \tag{3.50}$$

which is most commonly found in literature.

Another interesting identity to consider is $A(t,t') = B(t,t')D(t,t')$, which gives

$$A^{</>}(t,t') = B^{</>}(t,t)D^{</>}(t,t'), \tag{3.51}$$

since $t < t'$, or $t > t'$, in both propagators B and D. Thus, when finding the retarded propagator $A^r(t,t')$, where $A(t,t') = \int_C B(t,t'')D(t'',t')dt''$, we use the definitions in (3.48a), (3.48b), and find that

$$A^r(t,t') = \theta(t-t')[B^{>}(t,t)D^{>}(t,t') - B^{<}(t,t)D^{<}(t,t')]. \tag{3.52}$$

Adding zero (!), i.e. $0 = B^{>}(t,t)D^{<}(t,t') - B^{>}(t,t)D^{<}(t,t')$, gives

$$\begin{aligned} A^r(t,t') &= \theta(t-t')[B^{>}(t,t)\{D^{>}(t,t') - D^{<}(t,t')\} \\ &\quad + \{B^{>}(t,t') - B^{<}(t,t)\}D^{<}(t,t')] \\ &= B^{>}(t,t')D^r(t,t') + B^r(t,t')D^{<}(t,t'). \end{aligned} \tag{3.53}$$

This conversion, as well as the expression in (3.50), clearly illustrates that continuation to the real time axis is not always trivial. However, relying on the basic correlation functions, the lesser and greater, always admits finding a more or less convenient final expression.

In summary, the following identifications will be useful later in the book, namely,

$$A(t,t') = \int_C B(t,t'')D(t'',t')dt'', \tag{3.54a}$$

$$\begin{cases} A^{</>}(t,t') = \int[B^r(t,t'')D^{</>}(t'',t') \\ \qquad\qquad + B^{</>}(t,t'')D^a(t'',t')]dt'', \\ A^{r/a}(t,t') = \int B^{r/a}(t,t')D^{r/a}(t,t')dt'', \end{cases}$$

$$A(t,t') = B(t,t')D(t,t'), \tag{3.54b}$$

$$\begin{cases} A^{</>}(t,t') = B^{</>}(t,t')D^{</>}(t,t'), \\ A^{r/a}(t,t') = B^{>}(t,t')D^{r/a}(t,t') + B^{r/a}(t,t')D^{<}(t,t'). \end{cases}$$

3.5 Basic Building Blocks for Non-Equilibrium Studies

Above, we have investigated the conversion of the contour ordered Green functions to real times, defining the lesser and greater Green functions (or correlation functions), and studied the structure of the propagators when they are products or convolution of two other propagators. We also defined the retarded and advanced Green functions in terms of the lesser and greater ones. This is all fine, however, when doing actual calculations we need to have explicit expressions for the basic Green functions, such as the Green function for a non-interacting electron. After all, no matter how intricate the final Green function may be in terms of expansions and diagrams, at some point we will have to insert the Green function for non-interacting electrons in order to carry out the calculations. Therefore, we spend some effort in deriving the lesser and greater Green functions for a free electron in the system $\mathcal{H}_0 = \sum_{\mathbf{k}\sigma} \varepsilon_{\mathbf{k}\sigma} c^\dagger_{\mathbf{k}\sigma} c_{\mathbf{k}\sigma}$. We also consider the bare Green functions $g_{a\bar{b}}(t,t')$ and $k_{\xi\bar{\xi}}(t,t')$ in the isolated system $\mathcal{H}_{\text{loc}} = \sum_{Nn} E_{Nn} h^n_N$.

We begin with the free electrons described by the Hamiltonian $\mathcal{H}_0$, and we define the Green function $g_{\mathbf{k}\sigma}(t,t') = (-i)\langle \mathrm{T} c_{\mathbf{k}\sigma}(t) c^\dagger_{\mathbf{k}\sigma}(t')\rangle$. This Green function satisfies the align

$$(i\partial_t - \varepsilon_{\mathbf{k}\sigma}) g_{\mathbf{k}\sigma}(t,t') = \delta(t-t'), \tag{3.55}$$

which leads to that

$$g_{\mathbf{k}\sigma}(t,t') = (-i)\mathrm{T}e^{-i\varepsilon_{\mathbf{k}\sigma}(t-t')}, \tag{3.56}$$

where T is the time-ordering operator on the contour C, as usual. How shall we find the lesser and greater counterparts of this Green function? Taking for instance $g^>_{\mathbf{k}\sigma}(t,t') = (-i)\langle c_{\mathbf{k}\sigma}(t) c^\dagger_{\mathbf{k}\sigma}(t')\rangle$, we know that $t > t'$ on the contour C, while $t < t'$ for $g^<_{\mathbf{k}\sigma}(t,t')$. To be specific, let $C = [t_0, t_0 - i\beta]$. Using the boundary conditions of the Green functions on the contour C (discussed in the first paragraph in Sect. 3.3), we find that

$$g^<_{\mathbf{k}\sigma}(t_0,t') = -e^{\mu\beta} g^>_{\mathbf{k}\sigma}(t_0 - i\beta, t'). \tag{3.57}$$

Due to the time-independence of the Hamiltonian $\mathcal{H}_0$, we have $g_{\mathbf{k}\sigma}(t,t') = g_{\mathbf{k}\sigma}(t-t')$, and Fourier transforming the above equation we find

$$g^<_{\mathbf{k}\sigma}(\omega) = -e^{-\beta(\omega-\mu)} g^>_{\mathbf{k}\sigma}(\omega). \tag{3.58}$$

Taking into account that $c_{\mathbf{k}\sigma}$ removes an electron from the system with momentum $\mathbf{k}$ and spin σ, it must remove the energy $\varepsilon_{\mathbf{k}\sigma}$ from the system, hence

$$\begin{aligned} g^<_{\mathbf{k}\sigma}(\omega) &= \int i\langle c^\dagger_{\mathbf{k}\sigma}(0) c_{\mathbf{k}\sigma}(t)\rangle e^{i\omega t} dt = i\int \langle n_{\mathbf{k}\sigma}\rangle e^{i(\omega-\varepsilon_{\mathbf{k}\sigma})t} dt \\ &= i2\pi\langle n_{\mathbf{k}\sigma}\rangle \delta(\omega - \varepsilon_{\mathbf{k}\sigma}). \end{aligned} \tag{3.59}$$

In the same way we find that $g^>_{\mathbf{k}\sigma}(\omega) = -i2\pi(1 - \langle n_{\mathbf{k}\sigma}\rangle)\delta(\omega - \varepsilon_{\mathbf{k}\sigma})$. Using these expressions along with (3.58), we arrive at the equation

$$\langle n_{\mathbf{k}\sigma}\rangle \delta(\omega - \varepsilon_{\mathbf{k}\sigma}) = e^{-\beta(\omega-\mu)}(1 - \langle n_{\mathbf{k}\sigma}\rangle)\delta(\omega - \varepsilon_{\mathbf{k}\sigma}). \tag{3.60}$$

We thus deduce that $\langle n_{\mathbf{k}\sigma}\rangle = f(\varepsilon_{\mathbf{k}\sigma}) = [e^{\beta(\varepsilon_{\mathbf{k}\sigma}-\mu)}+1]^{-1}$, the Fermi function, which is expected from e.g. equilibrium theory. Going back to time space, we find that the lesser and greater Green functions are given by

$$g^{<}_{\mathbf{k}\sigma}(t,t') = if(\varepsilon_{\mathbf{k}\sigma})e^{-i\varepsilon_{\mathbf{k}\sigma}(t-t')}, \tag{3.61a}$$

$$g^{>}_{\mathbf{k}\sigma}(t,t') = (-i)[1-f(\varepsilon_{\mathbf{k}\sigma})]e^{-i\varepsilon_{\mathbf{k}\sigma}(t-t')}, \tag{3.61b}$$

$$g^{r/a}_{\mathbf{k}\sigma}(t,t') = (\mp i)\theta(\pm t \mp t)e^{-i\varepsilon_{\mathbf{k}\sigma}(t-t')}, \tag{3.61c}$$

where we have used the definitions in (3.48a), (3.48b) to find the retarded and advanced Green functions.

We now turn our attention to the isolated system $\mathcal{H}_{\text{loc}}$ and consider the propagators $g_{a\bar{b}}(t,t')$ and $k_{\xi\bar{\zeta}}(t,t')$. The equations of motion for these propagators are given by

$$(i\partial_t - \Delta_{\bar{a}})g_{a\bar{b}}(t,t') = \delta(t-t')P_{a\bar{b}}(t), \tag{3.62a}$$

$$(i\partial_t - \Delta_{\bar{\xi}})k_{\xi\bar{\zeta}}(t,t') = \delta(t-t')Q_{\xi\bar{\zeta}}(t), \tag{3.62b}$$

where $Q_{\xi\bar{\zeta}}(t) = \langle [Z^{\xi}, Z^{\bar{\zeta}}](t)\rangle$. Following the previous discussion about the boundary conditions for Hubbard operator Green functions, it is clear that we have the boundary conditions $g^{<}_{a\bar{b}}(t_0,t) = -g^{>}_{a\bar{b}}(t_0 - i\beta, t)$ and $k^{<}_{\xi\bar{\zeta}}(t_0,t) = k^{>}_{\xi\bar{\zeta}}(t_0 - i\beta, t)$.

In order to find the lesser and greater counterparts of these Green functions we study the Fourier transforms $g^{</>}_{a\bar{b}}(\omega)$ and $k^{</>}_{\xi\bar{\zeta}}(\omega)$, under the assumption that the propagators depend only on the time difference $t-t'$. We have, for example,

$$g^{<}_{a\bar{b}}(\omega) = \int i\langle X^{\bar{b}}(0)X^{a}(t)\rangle e^{i\omega t}dt = \int i\langle X^{\bar{b}}X^{a}\rangle e^{i(\omega-\Delta_{\bar{a}})t}dt$$
$$= i2\pi\langle X^{\bar{b}}X^{a}\rangle\delta(\omega-\Delta_{\bar{a}}), \tag{3.63a}$$

$$k^{<}_{\xi\bar{\zeta}}(\omega) = \int (-i)\langle Z^{\bar{\zeta}}(0)Z^{\xi}(t)\rangle e^{i\omega t}dt = \int (-i)\langle Z^{\bar{\zeta}}Z^{\xi}\rangle e^{i(\omega-\Delta_{\bar{\xi}})t}dt$$
$$= (-i)2\pi\langle Z^{\bar{\zeta}}Z^{\xi}\rangle\delta(\omega-\Delta_{\bar{\xi}}). \tag{3.63b}$$

Here, we have used that the transition X^{a} (Z^{ξ}) changes the energy in the system by $\Delta_{\bar{a}}$ ($\Delta_{\bar{\xi}}$). In the same way, we find that $g^{>}_{a\bar{b}}(\omega) = (-i)2\pi\langle X^{a}X^{\bar{b}}\rangle\delta(\omega-\Delta_{\bar{a}})$ and $k^{>}_{\xi\bar{\zeta}}(\omega) = (-i)2\pi\langle Z^{\xi}Z^{\bar{\zeta}}\rangle\delta(\omega-\Delta_{\bar{\xi}})$. Then, turning back to time space, we can write the lesser, greater, retarded, and advanced Green functions as

$$g^{<}_{a\bar{b}}(t,t') = i2\pi \langle X^{\bar{b}} X^{a} \rangle e^{-i\Delta_{\bar{a}}(t-t')}, \tag{3.64a}$$

$$g^{>}_{a\bar{b}}(t,t') = (-i)2\pi \langle X^{a} X^{\bar{b}} \rangle e^{-i\Delta_{\bar{a}}(t-t')}, \tag{3.64b}$$

$$g^{r/a}(t,t') = (\mp i)\theta(\pm t \mp t') P_{a\bar{b}} e^{-i\Delta_{\bar{a}}(t-t')}, \tag{3.64c}$$

and

$$k^{<}_{\xi\bar{\zeta}}(\omega) = (-i)2\pi \langle Z^{\bar{\zeta}} Z^{\xi} \rangle e^{-i\Delta_{\bar{\xi}}(t-t')}, \tag{3.65a}$$

$$k^{>}_{\xi\bar{\zeta}}(\omega) = (-i)2\pi \langle Z^{\xi} Z^{\bar{\zeta}} \rangle e^{-i\Delta_{\bar{\xi}}(t-t')}, \tag{3.65b}$$

$$k^{r/a}(t,t') = (\mp i)\theta(\pm t \mp t') Q_{\xi\bar{\zeta}} e^{-i\Delta_{\bar{\xi}}(t-t')}. \tag{3.65c}$$

Thus far, we have only considered open systems in equilibrium for which there is no real reason for introducing the non-equilibrium Green functions. In the remainder of the book, however, we shall discuss open systems under non-equilibrium conditions. We will often consider the influence of a source-drain voltage, or bias voltage, which generates a charge current through the system. There are, on the other hand, systems which may be considered as being in equilibrium from certain aspects, while being in non-equilibrium from other aspects. Under such circumstances, there are also reasons to make the analysis of the physical processes within a non-equilibrium approach. Hence, from hereon the description will be based upon the conception of non-equilibrium. Therefore, let us move ahead to the next chapter.

References

1. Kadanoff, L.P., Baym, G.: Quantum Statistical Mechanics, 7th edn. Perseus Books, Reading (1998)
2. Ovchinnikov, S.G., Val'kov, V.V.: Hubbard Operators in the Theory of Strongly Correlated Electrons. Imperial College Press, London (2004)
3. Keldysh, L.V.: Sov. Phys. JETP **20**, 1018 (1965)
4. Langreth, D.C.: In: Devreese, J.T., van Doren, V.E. (eds.) Linear and Nonlinear Electron Transport in Solids. NATO Advances Study Institute, Series B: Physics, vol. 17. Plenum, New York (1976)
5. Haug, H., Jauho, A.-P.: Quantum Kinetics in Transport and Optics of Semiconductors. Springer, Berlin (1998)

Chapter 4
Non-Equilibrium Formalism

Abstract We go deeper into the general description of non-equilibrium physics, both within the occupation number approach and by means of non-equilibrium Green function. We derive expressions for the tunneling current through complexes of interacting systems coupled to thermal baths, or leads. We discuss concepts of time-dependent and stationary processes.

4.1 Occupation Number Approach

We have already seen, in Sect. 2.2, how the occupation numbers of a system can be related to each other through the conduction channels, or leads. Actually, without mentioning it, we derived the equations for the occupation numbers in a form which is prepared for non-equilibrium conditions. We obtain equilibrium by requiring that the chemical potentials $\mu_\chi = \mu_{\text{eq}}$ in all leads χ, such that there is no potential energy imbalance anywhere in the system. In this sense, equilibrium is just a special case of non-equilibrium.

When thinking of non-equilibrium, one often considers transport of some quantity, or quantities, due to the imbalance in potential energy in the system. The transported quantity may be charge, spin, or just energy in e.g. heat transport. We shall very often consider charge and spin transport in this book, although much of the content discussed here will be valid from other aspects.

In order consider the charge transport through a specific system, we take the Hamiltonian $\mathcal{H} = \sum_{r=1}^{X} \mathcal{H}_{\chi_r} + \sum_{Nn} E_{Nn} h_N^n + \sum_\nu \mathcal{H}_T^\nu$, with $\mathcal{H}_{\chi_r}$ as given in (2.1) and $\mathcal{H}_T^\nu$ as given in (2.3). We are interested in the charge that flows from lead, say χ_0, to the rest of the system. We are aware that the charge may actually flow from the rest of the system into this lead, but within the given approach such a charge flow will manifest itself as a negative flow. Therefore, there is no loss of generality in our approach. Charge flow, or charge current, is the charge rate of change and is, hence, described by the temporal derivative of the charge N_{χ_0}, i.e. we put the charge current $I_{\chi_0}(t) = -e\partial_t N_{\chi_0} = -e\partial_t \sum_{\mathbf{k}\sigma\in\chi_0} \langle n_{\mathbf{k}\sigma} \rangle$. The current may be time-dependent as a

J. Fransson, *Non-Equilibrium Nano-Physics*,
Lecture Notes in Physics 809,
DOI 10.1007/978-90-481-9210-6_4,

result of possible time-dependent variations in the potential energy imbalance in the system. Applying the methods of Chap. 2, we find that we can write the current as

$$I_{\chi_0}(t) = ie \sum_{Nnm} \sum_{\mathbf{k}\sigma} \left[v_{\mathbf{k}\sigma Nnm} \langle c^\dagger_{\mathbf{k}\sigma} X^{nm}_{NN+1} \rangle - v^*_{\mathbf{k}\sigma Nnm} \langle X^{mn}_{N+1N} c_{\mathbf{k}\sigma} \rangle \right] \quad (4.1a)$$

$$= -2e \,\mathrm{Im} \sum_{Nnm} \sum_{\mathbf{k}\sigma} v_{\mathbf{k}\sigma Nnm} \langle c^\dagger_{\mathbf{k}\sigma} X^{nm}_{NN+1} \rangle, \quad (4.1b)$$

where $v_{\mathbf{k}\sigma Nnm} = \sum_{i\nu} v_{\mathbf{k}i\sigma} (d_{\nu i\sigma})^{nm}_{NN+1}$.

Before we proceed, we notice the two different ways to write the current in (4.1a), (4.1b). The former description, (4.1a), is written as the average of the current density operator $j(t) = i \sum_{Nnm} \sum_{\mathbf{k}\sigma} v_{\mathbf{k}\sigma Nnm} c^\dagger_{\mathbf{k}\sigma} X^{nm}_{NN+1} + H.c.$, such that $I_{\chi_0}(t) = e\langle j(t) \rangle$. This form presents a very intuitive formulation of the current as the net of the transfer of electrons from the quantum dot to the lead and the transfer of electrons from the lead to the quantum dot. While the latter formulation, (4.1b), is less intuitive, it has the benefits of being mathematically convenient.

The averages $\langle c^\dagger_{\mathbf{k}\sigma} X^{nm}_{NN+1} \rangle$ and $\langle X^{mn}_{N+1N} c_{\mathbf{k}\sigma} \rangle = \langle c^\dagger_{\mathbf{k}\sigma} X^{nm}_{NN+1} \rangle^*$ can be approximated by

$$\langle c^\dagger_{\mathbf{k}\sigma} X^{nm}_{NN+1} \rangle(t) = v^*_{\mathbf{k}\sigma Nnm} \frac{f(\varepsilon_{\mathbf{k}\sigma}) N_{Nn} - f(-\varepsilon_{\mathbf{k}\sigma}) N_{N+1m}}{\varepsilon_{\mathbf{k}\sigma} - \Delta_{N+1m,Nn} + i\delta}, \quad (4.2)$$

since $f(-x) = 1 - f(x)$. Substituting back into (4.1a), (4.1b), we find that the charge current flowing from lead χ_0 is expressed by

$$I_{\chi_0} = 2\pi e \sum_{Nnm} \sum_{\mathbf{k}\sigma} |v_{\mathbf{k}\sigma Nnm}|^2 \left[f(\varepsilon_{\mathbf{k}\sigma}) N_{Nn} - f(-\varepsilon_{\mathbf{k}\sigma}) N_{N+1m} \right] \delta(\varepsilon_{\mathbf{k}\sigma} - \Delta_{N+1m,Nn}). \quad (4.3)$$

This expression is very general since we do not make any assumptions of the band structure in the leads, we have not replaced the **k** summation with energy integration over the density of states. If we do this, i.e. letting $\sum_{\mathbf{k}} \to \int \rho_{\chi_0}(\omega) d\omega$, where ρ_{χ_0} is the local density of states in the lead χ_0, we can rewrite the current as

$$I_{\chi_0} = e \sum_{Nnm} \sum_{\sigma} \Gamma^{\chi_0}_{\sigma Nnm}(\Delta_{N+1m,Nn}) \left[f_{\chi_0}(\Delta_{N+1m,Nn}) N_{Nn} - f_{\chi_0}(-\Delta_{N+1m,Nn}) N_{N+1m} \right], \quad (4.4)$$

where $\Gamma^{\chi_0}_{\sigma Nnm}(\varepsilon) = 2\pi |v_{\chi_0 \sigma Nnm}(\varepsilon)|^2 \rho_{\chi_0}$. Here, $v_{\chi_0\sigma Nnm}(\varepsilon)$ is a function of energy ε which equals $v_{\mathbf{k}\sigma Nnm}$ whenever the ε equals $\varepsilon_{\mathbf{k}\sigma}$.

Whatever approach we use in actually calculating the current, i.e. using either of (4.3) and (4.4), it is clear that the current flowing from (to) the lead χ_0 is determined by the occupations N_{Nn} and N_{N+1m} in the states $|N, n\rangle$ and $|N+1, m\rangle$, respectively, and the transition energy $\Delta_{N+1m,Nn}$ between those states. It is also clear that current depends on the density of occupied and unoccupied electron states in the lead χ_0, here signified by $f_{\chi_0}(\omega)$ and $f_{\chi_0}(-\omega)$, respectively, particularly at the transition energy $\Delta_{N+1m,Nn}$.

4.2 Green Function Approach

Now, we want to take the discussion of the charge current further, and we do this by deriving an expression for the current in terms of Green functions. Such an expression has the advantage that it opens a possibility to make a deeper analysis of the interacting region, e.g. quantum dots. Moreover, although the occupation numbers corresponding to the localized levels, in general can be calculated through a density matrix approach, as discussed in Chap. 2, these are often obtained through calculations of the Green functions for the localized levels. Therefore, this is reason enough to establish the current in terms of the Green functions directly. Some of the derivation may be thought of as repetitive, but the reason to go over the steps once more is to thoroughly describe the method using Green functions. This will also give an opportunity to make a qualitative assessment of the two different methods.

We take the same Hamiltonian as in Sect. 4.1, and calculate the charge current flowing from the lead χ_0 out to the remainder of the system, i.e.

$$I_{\chi_0}(t) = -e\partial_t \sum_{\mathbf{k}\sigma} \langle n_{\mathbf{k}\sigma}\rangle = ie \sum_{\mathbf{k}\sigma b} [v_{\mathbf{k}\sigma b}\langle c^{\dagger}_{\mathbf{k}\sigma} X^{b}\rangle - v^{*}_{\mathbf{k}\sigma b}\langle X^{\bar{b}} c_{\mathbf{k}\sigma}\rangle]. \tag{4.5}$$

In order to proceed in terms of Green functions, we identify the correlation function $\langle X^{\bar{b}}(t) c_{\mathbf{k}\sigma}(t)\rangle$ with $F^{<}_{\mathbf{k}\sigma\bar{b}}(t,t) = i\langle X^{\bar{b}}(t) c_{\mathbf{k}\sigma}(t)\rangle$. This is the lesser Green function for the process that describes the transfer of electrons between the localized levels and the lead. Here, also $v_{\mathbf{k}\sigma b} = \sum_i v_{\mathbf{k}\sigma}(d_{i\sigma})^{b}$. It is then sensible to rewrite the current as

$$I_{\chi_0}(t) = -2e\,\mathrm{Re} \sum_{\mathbf{k}\sigma b} v^{*}_{\mathbf{k}\sigma b} F^{<}_{\mathbf{k}\sigma\bar{b}}(t,t), \tag{4.6}$$

where we have made use of that the two terms are complex conjugates of one another. The expression given in (4.5) explicitly displays that the current consists of two types of transfers of electrons. One transfer goes from the lead to the localized levels, whereas the other goes in the opposite direction. The net current then, is the sum of those. The advantage with the expression in (4.6) is that it explicitly shows that we really only have to deal with one Green function at the moment, making the treatment somewhat shorter.

We define the transfer Green function

$$F_{\mathbf{k}\sigma\bar{b}}(t,t') = (-i)\langle \mathrm{T} c_{\mathbf{k}\sigma}(t) X^{\bar{b}}(t')\rangle_{U}. \tag{4.7}$$

The propagator is found to satisfy the equation

$$(i\partial_t - \varepsilon_{\mathbf{k}\sigma}) F_{\mathbf{k}\sigma\bar{b}}(t,t') = \sum_{a} v_{\mathbf{k}\sigma a} G_{a\bar{b}}(t,t'). \tag{4.8}$$

Here, the lead Green function $g_{\mathbf{k}\sigma}(t,t')$ is the one introduced in (3.55), with which we rewrite (4.8) as

$$F_{\mathbf{k}\sigma\bar{b}}(t,t') = \sum_a \int_C g_{\mathbf{k}\sigma}(t,t'')v_{\mathbf{k}\sigma a}(t'')G_{a\bar{b}}(t'',t')dt''. \tag{4.9}$$

Here, we let the hybridization coefficient $v_{\mathbf{k}\sigma a}$ remain under the integration sign since, for all we know, it may be time-dependent. Using the rules in (3.54), we thus have

$$\begin{aligned} F^{<}_{\mathbf{k}\sigma\bar{b}}(t,t) &= \sum_a \int v_{\mathbf{k}\sigma a}(t'')[g^r_{\mathbf{k}\sigma}(t,t'')G^{<}_{a\bar{b}}(t'',t) + g^{<}_{\mathbf{k}\sigma}(t,t'')G^{a}_{a\bar{b}}(t'',t)]dt'' \\ &= (-i)\sum_a \int_{-\infty}^{t} v_{\mathbf{k}\sigma a}(t'')[G^{<}_{a\bar{b}}(t'',t) - f(\varepsilon_{\mathbf{k}\sigma})G^{a}_{a\bar{b}}(t'',t)]e^{i\varepsilon_{\mathbf{k}\sigma}(t''-t)}dt'' \\ &= i\sum_a \int_{-\infty}^{t} v_{\mathbf{k}\sigma a}(t'')[f(\varepsilon_{\mathbf{k}\sigma})G^{>}_{a\bar{b}}(t'',t) \\ &\quad + f(-\varepsilon_{\mathbf{k}\sigma})G^{<}_{a\bar{b}}(t'',t)]e^{i\varepsilon_{\mathbf{k}\sigma}(t''-t)}dt''. \end{aligned} \tag{4.10}$$

The last equality is obtained by using (3.48). The final line is also very appealing since it displays the transfer of electrons between the lead and the localized region, as an electron flow between occupied states in the lead (localized region), signified by $f(\omega)$ ($G^{<}_{a\bar{b}}$), and unoccupied states in the localized region (lead), signified by $G^{>}_{a\bar{b}}$ ($f(-\omega)$). Hence, the first term can be interpreted as, if there are electrons occupying the states in the leads and simultaneously there are unoccupied localized states, there is a possibility that electrons will flow from the lead to these localized states. The second term can be interpreted as a flow of electrons in the opposite direction.

Substituting (4.10) into (4.6), we find the current in the form

$$\begin{aligned} I_{\chi_0}(t) &= -2e\,\mathrm{Im}\sum_{\mathbf{k}\sigma ab} v^*_{\mathbf{k}\sigma b}(t)\int_{-\infty}^{t} v_{\mathbf{k}\sigma a}(t') \\ &\quad \times \left[f(\varepsilon_{\mathbf{k}\sigma})G^{>}_{a\bar{b}}(t',t) + f(-\varepsilon_{\mathbf{k}\sigma})G^{<}_{a\bar{b}}(t',t)\right]e^{i\varepsilon_{\mathbf{k}\sigma}(t'-t)}dt'. \end{aligned} \tag{4.11}$$

Analogous to our interpretation of the transfer Green function in (4.10), we interpret the current as a flow between occupied and unoccupied states in the lead (localized region) and localized region (lead). Especially, if the localized region constitutes a geometrically confined part of the system, we can think in terms of electron flow in to and out from this confined space. The net current is, as always, the sum of the inflow and outflow of electrons.

We obtain a parameter expression for the tunneling current by introducing the coupling parameter

$$\Gamma^{\chi_0}_{\sigma\bar{b}a}(t,t';\omega) = 2\pi v^*_{\chi_0\sigma b}(t)v_{\chi_0\sigma a}(t')\rho_{\chi_0}(\omega), \tag{4.12}$$

see Sect. 4.1 for definitions, integrating over the density of states $\rho_{\chi 0}$, and noticing the obvious matrix product structure in the summation, that is

$$I_{\chi 0}(t) = i2e \sum_{\sigma} \mathrm{tr} \iint_{-\infty}^{t} \boldsymbol{\Gamma}_{\sigma}^{\chi 0}(t,t';\omega)$$

$$\times \big[f_{\chi 0}(\omega)\mathbf{G}^{>}(t',t) + f_{\chi 0}(-\omega)\mathbf{G}^{<}(t',t)\big]e^{i\omega(t'-t)}dt'\frac{d\omega}{2\pi}. \tag{4.13}$$

Here, we have introduced the matrices $\mathbf{G} = \{G_{a\bar{b}}\}_{ab}$ and $\boldsymbol{\Gamma}_{\sigma}^{\chi 0} = \{\Gamma_{\sigma \bar{b} a}^{\chi 0}\}_{ab}$. The trace runs over the transitions a.

Equivalent expressions are obtained for the current flowing from any other lead into the remainder of the system. Depending on the situation and circumstances that is subject to the investigation, we can make further streamlining in the formulas for the current which is convenient in the particular case. For now, however, we note that if all external forces/fields applied on the system are time-independent, the local Green functions $\mathbf{G}^{</>}(t',t) = \mathbf{G}^{</>}(t'-t)$ and $\Gamma_{\sigma}^{\chi 0}(t,t';\omega) = \Gamma_{\sigma}^{\chi 0}(\omega)$, which leads to a simplified formula for the current reading

$$I_{\chi 0} = \frac{ie}{2\pi} \sum_{\sigma} \mathrm{tr} \int \Gamma_{\sigma}^{\chi 0}(\omega)\big[f_{\chi 0}(\omega)\mathbf{G}^{>}(\omega) + f_{\chi 0}(-\omega)\mathbf{G}^{<}(\omega)\big]d\omega. \tag{4.14}$$

The current given in this expression is very similar to the one provided in (4.4), however, there is a significant difference. Here, the matrix Green functions $\mathbf{G}$ contains much more than only the diagonal transitions which are accounted for in (4.4). All off-diagonal transitions included in $\mathbf{G}$ are physical and, hence, possible, and as such they will contribute to the current. The questions is only with how much? These are questions that will be partly discussed in later chapters.

The theory introduced in this section is further discussed in [1–4].

4.3 Single-Level Quantum Dot

Let us go back to the system with a single localized level interacting with conduction channels, described in term of the Anderson model as discussed in Sect. 3.2. There, we derived the equations for the equilibrium situation, and now can do the same thing under non-equilibrium conditions. We have to introduce, at least, a second lead in order to enable a charge current to flow through the single localized level. We could of course imagine that the localized level was interacting with a single conduction channel, see Sect. 7.2.2, but for now we will think in terms of the geometrically appealing system with, at least, two leads and localized level in a quantum dot of some kind.

It is worth to mention that we are using this example in order to develop a general structure of the formalism. In the diagonal formalism, it matters less, as we have seen, whether the quantum dot contain one or more single electron levels. We have to be aware, nevertheless, that there are some simplifications appearing when using the single-level quantum dot, especially concerning the notation since we are making use of the fact that all transition matrix elements are ± 1. When we write the equations in matrix form, however, this fact does not make any difference for the structure of the equations.

Our system can be described by the Hamiltonian $\mathcal{H} = \mathcal{H}_L + \mathcal{H}_R + \mathcal{H}_{QD} + \mathcal{H}_T$, where $\mathcal{H}_{L(R)} = \sum_{\mathbf{p}(\mathbf{q})\sigma} \varepsilon_{\mathbf{p}(\mathbf{q})\sigma} c^{\dagger}_{\mathbf{p}(\mathbf{q})\sigma} c_{\mathbf{p}(\mathbf{q})\sigma}$, and $\mathcal{H}_T = \sum_{\mathbf{k}\sigma\in L\cup R} v_{\mathbf{k}\sigma} c^{\dagger}_{\mathbf{k}\sigma} [X^{0\sigma} + \sigma X^{\bar{\sigma}2}] + H.c.$ Here and in the following, will associate the momentum $\mathbf{p}$ ($\mathbf{q}$) with the left (right) lead. Formally, we can follow the procedure in Sect. 3.2, and write the equations for the Green functions $G_{0\sigma\overline{a}}$ and $G_{\bar{\sigma}2\overline{a}}$, to the same approximation, according to

$$G_{0\sigma\overline{a}}(t,t') = g_{0\sigma\overline{a}}(t,t') + \sum_b \int_C g_{0\sigma}(t,t_1) V_{\overline{0\sigma}b}(t_1,t_2) G_{b\overline{a}}(t_2,t') dt_2 dt_1, \quad (4.15a)$$

$$G_{\bar{\sigma}2\overline{a}}(t,t') = g_{\bar{\sigma}2\overline{a}}(t,t') + \sum_b \int_C g_{\bar{\sigma}2}(t,t_1) V_{\overline{0\sigma}b}(t_1,t_2) G_{b\overline{a}}(t_2,t') dt_2 dt_1. \quad (4.15b)$$

We rewrite this equation in matrix form

$$\mathbf{G}(t,t') = \mathbf{g}(t,t') + \int_C \mathbf{g}(t,t_1)\mathbf{V}(t_1,t_2)\mathbf{G}(t_2,t') dt_2 dt_1. \quad (4.16)$$

Thus, according to the rules in (3.54) we can write the lesser and greater Green function as

$$\begin{aligned} \mathbf{G}^{</>}(t,t') = \mathbf{g}^{</>}(t,t') + \int [&\mathbf{g}^r(t,t_1)\mathbf{V}^r(t_1,t_2)\mathbf{G}^{</>}(t_2,t') \\ &+ \mathbf{g}^r(t,t_1)\mathbf{V}^{</>}(t_1,t_2)\mathbf{G}^a(t_2,t') \\ &+ \mathbf{g}^{</>}(t,t_1)\mathbf{V}^a(t_1,t_2)\mathbf{G}^a(t_2,t')] dt_1 dt_2. \end{aligned} \quad (4.17)$$

We collect the terms with $\mathbf{G}^{</>}$ to the left hand side, and make use of the fact that $\int \mathbf{g}^{r,-1}(t,t_1)\mathbf{g}^{</>}(t_1,t')dt_1 = 0$. We, thus, obtain

$$\int [\mathbf{g}^{r,-1}(t,t_1) - \mathbf{V}(t,t_1)]\mathbf{G}^{</>}(t_1,t') dt_1 = \int \mathbf{V}^{</>}(t,t_1)\mathbf{G}^a(t_1,t') dt_1. \quad (4.18)$$

Making the identification $\mathbf{g}^{r,-1} - \mathbf{V} = \mathbf{G}^{r,-1}$, finally results in

$$\mathbf{G}^{</>}(t,t') = \int \mathbf{G}^r(t,t_1)\mathbf{V}^{</>}(t_1,t_2)\mathbf{G}^a(t_2,t') dt_2 dt_1. \quad (4.19)$$

Using (3.54) we also find the retarded and advanced Green functions

$$\mathbf{G}^{r/a}(t,t') = \mathbf{g}^{r/a}(t,t') + \int \mathbf{g}^{r/a}(t,t_1)\mathbf{V}^{r/a}(t_1,t_2)\mathbf{G}^{r/a}(t_2,t') dt_2 dt_1. \quad (4.20)$$

We note that we have not made any use of the particular Hamiltonian system in which the Green functions were derived and, indeed, these formulas are valid whenever the Green function is obtained in a Dyson-like equation form as in (4.16), where $\mathbf{V}$ is associated with the self-energy.

In the present particular case, the interaction potential, or self-energy, $\mathbf{V}$ reduces to a 4×4 matrix

$$\mathbf{V}(t,t') = \begin{pmatrix} \mathbf{V}'(t,t') & \sigma^z\mathbf{V}'(t,t') \\ \sigma^z\mathbf{V}'(t,t') & \mathbf{V}'(t,t') \end{pmatrix}, \tag{4.21}$$

where $\mathbf{V}'(t,t')$ is a diagonal 2×2 matrix with entries $V_\sigma(t,t') = \sum_{\mathbf{k}\sigma} v^*_{\mathbf{k}\sigma}(t) v_{\mathbf{k}\sigma}(t') \times g_{\mathbf{k}\sigma}(t,t')$. Here, the summation runs over all states in the left and right leads. Using our basic building blocks in (3.61) we find that e.g.

$$V_\sigma^{</>}(t,t') = (\pm i) \sum_{\mathbf{k}\sigma} v^*_{\mathbf{k}\sigma}(t) v_{\mathbf{k}\sigma}(t') f_\chi(\pm\varepsilon_{\mathbf{k}\sigma}) e^{-i\varepsilon_{\mathbf{k}\sigma}(t-t')}, \tag{4.22}$$

$$V_\sigma^{r/a}(t,t') = (\mp i)\theta(\pm t \mp t') \sum_{\mathbf{k}\sigma} v^*_{\mathbf{k}\sigma}(t) v_{\mathbf{k}\sigma}(t') e^{-i\varepsilon_{\mathbf{k}\sigma}(t-t')}. \tag{4.23}$$

In order to proceed with the analytical treatment, we convert the $\mathbf{k}$-summation to energy integration over the density of states $\rho_\chi(\omega)$, which enables writing the lesser Green matrix function according to

$$\mathbf{G}^<(t,t') = i \sum_\chi \int f_\chi(\omega) \mathbf{G}^r(t,t_1) \boldsymbol{\Gamma}^\chi(t_1,t_2;\omega) \mathbf{G}^a(t_2,t') e^{-i\omega(t_1-t_2)} \frac{d\omega}{2\pi} dt_1 dt_2. \tag{4.24}$$

The structure of the lesser Green function in this form suggests an interpretation of the quantum dot as a sum of contributions, each of which being in local equilibrium. This interpretation is sound, since in equilibrium the lesser Green function provides exactly the local density of occupied states, which may be displayed through the total density of states multiplied by a filling factor. Because of the non-equilibrium conditions in the system as whole, the best we can expect about the quantum dot is that it is piecewise in local equilibrium with the leads. The greater Green function $\mathbf{G}^>$ is obtained by the replacement $if_\chi(\omega) \to -if_\chi(-\omega) = -i[1 - f_\chi(\omega)]$.

Using the explicit forms of the lesser and greater Green functions we can write the current I_L flowing from the left lead as

$$I_L(t) = 2e\,\mathrm{tr} \int_{-\infty}^{t} \int \Gamma^L(t,t';\omega) \mathbf{G}^r(t',t_1) \Gamma^R(t_1,t_2;\omega') \mathbf{G}^a(t_2,t)$$
$$\times \left[f_L(\omega) - f_R(\omega')\right] e^{-i\omega(t-t')-i\omega'(t_1-t_2)} \frac{d\omega}{2\pi}\frac{d\omega'}{2\pi} dt_1 dt_2 dt'. \tag{4.25}$$

Expressing the current like this is very appealing since it suggests an interpretation of two leads being in non-equilibrium with one another. The two leads, further, has a barrier in between through which the transmission is governed by the transmission coefficient

$$\mathcal{T}(t;\omega,\omega') = \int_{-\infty}^{t} \int \Gamma^L(t,t';\omega)\mathbf{G}^r(t',t_1)\Gamma^R(t_1,t_2;\omega')\mathbf{G}^a(t_2,t) \times e^{-i\omega(t-t')-i\omega'(t_1-t_2)} dt_1 dt_2 dt', \tag{4.26}$$

giving the current simply as

$$I_L(t) = 2e\,\mathrm{tr} \int \mathcal{T}(t;\omega,\omega')\big[f_L(\omega) - f_R(\omega')\big]\frac{d\omega}{2\pi}\frac{d\omega'}{2\pi}. \tag{4.27}$$

Thus, no matter how complicated the structure is between the leads, it is only the probability for an electron in one lead to be transmitted into the other that matters. This is the general way we may look at our problems. On the other hand, since this probability is the quantity to be found, by one or another means, this simple formula is of no help if we cannot find the probability. Regardless of this, we have gained general knowledge about non-equilibrium and transport by bringing the microscopic and macroscopic pictures together.

4.3.1 Calculating the End-Factors

The above discussion has been rather general and we have made no or little reference to the Green functions being constructed of Hubbard operators. Taking this aspect into account we remember from Sect. 3.2, that the occupation numbers N_0, N_σ, and N_2 were calculated using the spectral theorem. In non-equilibrium we cannot use this approach since quantum dot Green function is not given in a single equilibrium form, cf. (4.19) and (4.24). We therefore have to come up with new formulas for the occupation numbers that is commensurate with the non-equilibrium conditions.

By straight forward calculations we find that e.g.

$$N_0(t) = \langle h^0(t)\rangle = \sum_\sigma \langle X^{0\sigma}(t)X^{\sigma 0}(t)\rangle = i\sum_\sigma G^>_{0\sigma}(t,t). \tag{4.28}$$

The summation over spin is, as we discussed in Sect. 3.2, present because the empty state couples to the singly occupied states through the transitions $|0\rangle\langle\sigma|$ and we cannot know *a priori* to which state it couples. Therefore we have to sum over all possibilities. This way of thinking becomes even more adequate in non-equilibrium since available transitions will be used although they may not couple to the ground state.

Using the above equation we find that the occupation number N_0 can be calculated through

$$N_0 = \frac{i}{2\pi}\int \sum_\sigma G^>_{0\sigma}(\omega)d\omega, \tag{4.29}$$

while the occupation numbers N_σ and N_2 are analogously found and can be written

$$N_\sigma = -\frac{i}{2\pi}\int \{G^<_{0\sigma}(\omega) - G^>_{\sigma 2}(\omega)\}d\omega, \tag{4.30}$$

$$N_2 = -\frac{i}{2\pi}\int \sum_\sigma G^<_{\sigma 2}(\omega)d\omega. \tag{4.31}$$

These formulas are easy to extend to the general situation, i.e. in other systems than the single level case.

It is interesting to compare those non-equilibrium formulas to the equilibrium ones given in Sect. 3.2. Taking for instance the lesser and greater Green functions as formulated in (4.24), and assuming that spin-flip transitions do occur anywhere in the system. The occupation number N_0 would then take the form

$$\begin{aligned} N_0 = \sum_\sigma \int & \big[\{1 - f_L(\omega)\}\Gamma^L_\sigma(t_1,t_2;\omega) + \{1 - f_R(\omega)\}\Gamma^R_\sigma(t_1,t_2;\omega)\big]e^{-i\omega(t_1-t_2)} \\ & \times \big[G^r_{0\sigma}(t,t_1) + \sigma G^r_{0\sigma 2\bar\sigma}(t,t_1)\big] \\ & \times \big[G^a_{0\sigma}(t_2,t) + \sigma G^a_{0\sigma 2\bar\sigma}(t_2,t)\big]\frac{d\omega}{2\pi}dt_1dt_2. \end{aligned} \tag{4.32}$$

The non-equilibrium conditions are explicit by the presence of the left and right Fermi functions at their corresponding chemical potentials μ_L and μ_R. Now, assuming equilibrium conditions implies that $\mu_L = \mu_R = \mu_{\text{eq}}$, which leads to that $f_L(\omega) = f_R(\omega) = f(\omega)$. Hence, the first factor in the above equation can be factorized into $\{1 - f(\omega)\}\{\Gamma^L_\sigma(t_1,t_2;\omega) + \Gamma^R_\sigma(t_1,t_2;\omega)\} = \{1 - f(\omega)\}\Gamma_\sigma(t_1,t_2;\omega)$, where $\Gamma_\sigma = \Gamma^L_\sigma + \Gamma^R_\sigma$. Moreover, in equilibrium we cannot have any time-dependence in the system, expect for possible random redistributions of the electrons which cancel on average. Hence, the coupling $\Gamma_\sigma(t_1,t_2;\omega) = \Gamma_\sigma(\omega)$ and $\mathbf{G}^{r/a}(t,t') = \mathbf{G}^{r/a}(t-t')$, so that we can write

$$\begin{aligned} N_0 &= \sum_\sigma \int \{1 - f(\omega)\}\Gamma_\sigma(\omega)\left|\int \big[G^r_{0\sigma}(t-t') + \sigma G^r_{0\sigma 2\bar\sigma}(t-t')\big]e^{i\omega(t-t')}dt'\right|^2\frac{d\omega}{2\pi} \\ &= \sum_\sigma \int \{1 - f(\omega)\}\Gamma_\sigma(\omega)\big|G^r_{0\sigma}(\omega) + \sigma G^r_{0\sigma 2\bar\sigma}(\omega)\big|^2\frac{d\omega}{2\pi}. \end{aligned} \tag{4.33}$$

The expressions for the retarded Green functions in the present approximation are given in Sect. 3.2, and using these it is a straight forward calculation showing that $|G^r_{0\sigma} + \sigma G^r_{0\sigma 2\bar\sigma}|^2(\Gamma_\sigma/2) = -\operatorname{Im} G^r_{0\sigma}$, hence, we find that

$$N_0 = -\frac{1}{\pi}\int \{1 - f(\omega)\}\operatorname{Im}\sum_\sigma G^r_{0\sigma}(\omega)d\omega. \tag{4.34}$$

We thus recover the formula obtained from the spectral theorem, and working through steps for the numbers N_σ and N_2, we recover the other two expressions as well.

We know from the definition, that e.g. $P_{0\sigma} = N_0 + N_\sigma$. Using this information we can establish an important relation concerning the total occupation. We would like to require that the total occupation naturally satisfies the conservation law $P_{0\sigma} + P_{\bar{\sigma}2} = 1$, since $P_{\bar{\sigma}2} = \langle n_{\bar{\sigma}} \rangle$ and $P_{0\sigma} = 1 - \langle n_{\bar{\sigma}} \rangle = \langle n_\sigma \rangle$. It would, on the other hand, be nice if we can prove this relation rather than postulating it. Since we now know how to related the occupation numbers to the lesser and greater Green functions, we can make a direct calculation as follows

$$\begin{aligned} P_{0\sigma} + P_{\bar{\sigma}2} &= N_0 + N_\sigma + N_{\bar{\sigma}} + N_2 \\ &= \frac{i}{2\pi} \sum_\sigma \int [G^>_{0\sigma} - G^<_{0\sigma} + G^>_{\bar{\sigma}2} - G^<_{\bar{\sigma}2}] d\omega \\ &= \frac{i}{2\pi} \sum_\sigma \int [G^r_{0\sigma} - G^a_{0\sigma} + G^r_{\bar{\sigma}2} - G^r_{\bar{\sigma}2}] d\omega \\ &= -\frac{1}{\pi} \operatorname{Im} \sum_\sigma \int [G^r_{0\sigma} + G^r_{\bar{\sigma}2}] d\omega \\ &= -\frac{1}{\pi} \operatorname{tr} \operatorname{Im} \int \mathbf{G}^r d\omega = 1. \end{aligned} \tag{4.35}$$

In the last equality we have used the fact that the trace of the integrated density of states is one. The imaginary part of the retarded Green function plays the role of a probability density, and using this interpretation we should know that the last equality follows.

4.3.2 Approximations for the Retarded Green Function

The above equations are somewhat general and we need to make further assumptions about our system in order to proceed the analytical treatment. The transmission coefficient $\mathcal{T}$ is now given in terms of the retarded and advanced Green functions and the coupling matrices. In (4.20) we see that the retarded and advanced Green functions are being expressed in term of themselves, self-consistently, hence, there is one instance in which we can make approximations.

4.3.2.1 *T*-matrix

For instance, replacing $\mathbf{G}^{r/a}$ by $\mathbf{g}^{r/a}$ in (4.20) gives a description of the quantum dot as being weakly disturbed by the presence of the leads. The levels would still have infinite life-time, as in the atomic limit, however, the disturbance from the leads allow one scattering inside the quantum dot. In this case, the retarded Green function becomes

$$\mathbf{G}^r(t,t') = \mathbf{g}^r(t,t') + \sum_\chi \int \mathbf{g}(t,t_1)\Gamma^\chi(t_1,t_2;\omega)\mathbf{g}(t_2,t')e^{-i\omega(t_1-t_2)}\frac{d\omega}{2\pi}dt_2dt_1, \tag{4.36}$$

which illustrates the idea that the population in the quantum dot (first term) is disturbed by the interaction with the leads (second term). The multiplication of two bare Green functions $\mathbf{g}$ in the second term indicates scattering between different states, where the scattering is mediated by the lead χ through the coupling Γ^χ.

We can, however, do better than only include one scattering inside the quantum dot. We shall for simplicity do this in the Markovian approximation, which means that the propagators have no memory and only depend on the time difference. This is typically applicable in the stationary, or time-independent, regime, but may also be thought of under other circumstances, e.g. Brownian motion. Under those conditions, we can Fourier transform all quantities to energy space. Using the original form of the retarded Green function in (4.20) and replacing the dressed Green function $\mathbf{G}$ on the right hand side by the whole expression to the right of the equality sign, we arrive at

$$\mathbf{G}^r(\omega) = \mathbf{g}^r(\omega) + \mathbf{g}^r(\omega)\mathbf{V}^r(\omega)\mathbf{g}^r(\omega) + \mathbf{g}^r(\omega)\mathbf{V}^r(\omega)\mathbf{g}^r(\omega)\mathbf{V}^r(\omega)\mathbf{G}^r(\omega). \tag{4.37}$$

We do not have to terminate the expansion at this step, but rather continue to arbitrary order. By continuing the procedure to infinite order we find that the dressed Green function can be algebraically written as

$$\begin{aligned}\mathbf{G}^r &= \mathbf{g}^r + \mathbf{g}^r\mathbf{V}^r\mathbf{g}^r + \mathbf{g}^r\mathbf{V}^r\mathbf{g}^r\mathbf{V}^r\mathbf{g}^r + \mathbf{g}^r\mathbf{V}^r\mathbf{g}^r\mathbf{V}^r\mathbf{g}^r\mathbf{V}^r\mathbf{g}^r + \cdots \\ &= \mathbf{g}^r + \mathbf{g}^r\mathbf{T}^r\mathbf{g}^r,\end{aligned} \tag{4.38}$$

where we have introduced the T-matrix

$$\mathbf{T}^r = \mathbf{V}^r + \mathbf{V}^r\mathbf{g}^r\mathbf{V}^r + \mathbf{V}^r\mathbf{g}^r\mathbf{V}^r\mathbf{g}^r\mathbf{V}^r + \cdots = \mathbf{V}^r[1 - \mathbf{g}^r\mathbf{V}^r]^{-1}. \tag{4.39}$$

Here, we require that $\|\mathbf{g}^r\mathbf{V}^r\| < 1$ in order to obtain a convergent series. The above expansion becomes quite involved when writing it in terms of the time-integrals. We can simplify the resulting expressions by assuming that the tunneling matrix elements $v_{\mathbf{k}\sigma}$ are time-independent, which then leads to

$$\mathbf{V}(\omega) = \sum_{\mathbf{k}\sigma} \frac{\mathbf{V_k}}{\omega - \varepsilon_{\mathbf{k}\sigma} + i\delta}, \quad \delta > 0, \tag{4.40}$$

where

$$\mathbf{V_k} = \begin{pmatrix} \mathbf{v_k} & \sigma^z\mathbf{v_k} \\ \sigma^z\mathbf{v_k} & \mathbf{v_k} \end{pmatrix}, \qquad \mathbf{v_k} = \begin{pmatrix} \delta_{\sigma\uparrow} & 0 \\ 0 & \delta_{\sigma\downarrow} \end{pmatrix} |v_{\mathbf{k}\sigma}|^2. \tag{4.41}$$

Our T-matrix thus becomes

$$\mathbf{T}^r(\omega) = \sum_{\mathbf{k}\sigma} \frac{\mathbf{V_k}}{\omega - \varepsilon_{\mathbf{k}\sigma} + i\delta}\left[\omega - \Delta - \mathbf{P}\sum_{\mathbf{k}'\sigma'} \frac{\mathbf{V_{k'}}}{\omega - \varepsilon_{\mathbf{k}'\sigma'} + i\delta}\right]^{-1}[\omega - \Delta]. \tag{4.42}$$

Parameterizing the interaction matrix defining $\Lambda = \sum_{\mathbf{k}\sigma} \mathbf{V}_{\mathbf{k}}/(\omega - \varepsilon_{\mathbf{k}\sigma})$ and $\Gamma^{\chi} = 2\pi \sum_{\mathbf{k}\sigma\in\chi} \mathbf{V}_{\mathbf{k}}\delta(\omega - \varepsilon_{\mathbf{k}\sigma})$, such that $\Gamma = \sum_{\chi} \Gamma^{\chi}$ and $\mathbf{V}^r = \Lambda - i\Gamma/2$, we write the T-matrix according to

$$\mathbf{T}^r(\omega) = \left[\Lambda - \frac{i}{2}\Gamma\right]\left[\omega - \Delta - \mathbf{P}\Lambda + \frac{i}{2}\mathbf{P}\Gamma\right]^{-1}[\omega - \Delta]. \tag{4.43}$$

The middle factor tells at which energies resonances are to be expected and we see, for instance, that the level in the quantum dot is responsible for a resonance at Δ. This resonance is, however, shifted by the amount of $\mathbf{P}\Lambda$, which is small, or even negligible, in case of metallic leads. The multiple scattering does not generate new resonances in this system. The multiple scattering do, nonetheless, cause a broadening to the existing resonances, which is seen by the presence of the imaginary part $\Gamma/2$ in the middle factor. Such broadening did not appear in our first approximation with a single scattering event inside the quantum dot. The broadening is physical and reasonable from the point of view that the quantum dot is no longer isolated from the environment, but its electrons are interacting with the electrons in the leads due to the tunneling between the leads and the quantum dot. The presence of the end-factor factoring the level broadening signifies the fact that the quantum dot level are subject to internal many-body interactions (here, Coulomb repulsion), which tend to narrow the level broadening.

The resulting Green function in the T-matrix approach becomes in this case

$$\begin{aligned}\mathbf{G}^r(\omega) &= \mathbf{g}^r(\omega)\left\{1 + \left[\Lambda - \frac{i}{2}\Gamma\right]\left[\omega - \Delta - \mathbf{P}\Lambda + \frac{i}{2}\mathbf{P}\Gamma\right]^{-1}\mathbf{P}\right\}\\ &= \left[\omega - \Delta - \mathbf{P}\Lambda + \frac{i}{2}\mathbf{P}\Gamma\right]^{-1}\mathbf{P}.\end{aligned} \tag{4.44}$$

4.3.2.2 Dyson Equation

Staying in the Markovian limit, we can also make an approximation directly from the equation for the retarded Green function in (4.20), written as

$$\mathbf{G}^r(\omega) = \mathbf{g}^r(\omega) + \mathbf{g}^r(\omega)\mathbf{V}^r(\omega)\mathbf{G}^r(\omega). \tag{4.45}$$

Gathering the terms proportional to $\mathbf{G}^r$ on the left hand side of the equation we find the result

$$\mathbf{G}^r(\omega) = \left[1 - \mathbf{g}^r(\omega)\mathbf{V}^r(\omega)\right]^{-1}\mathbf{g}^r(\omega) = \left[\omega - \Delta - \mathbf{P}\Lambda + \frac{i}{2}\mathbf{P}\Gamma\right]^{-1}\mathbf{P}. \tag{4.46}$$

Clearly, this result equals the one from the T-matrix approach, cf. (4.44). The reason if, of course, that we started from the same equation and made the same assumptions in both cases, thus, the result should be equal. Sometimes it is, however, more appealing from a physical point of view to use the T-matrix approach, while in other situations it is better employ the Dyson equation.

Both the result from the T-matrix approach as well as the one resulting from the Dyson equation, are here expressed in the *mean field* approximation. This is limit to which we will return many times, and is therefore important to give some physical interpretation about. First, and most importantly, is that the mean field theory neglects all kinds of fluctuations that may occur in the system. This implies that anything unusual that may happen in the quantum dot, is averaged out from the description. Secondly, however, since this approximation is always available in one or another form, it is very often used in literature and in research in general. It is often easy to understand some of the basic properties of a system in terms of mean field theory. On the other hand, due to the lack of precision, one immediately has to go beyond mean field theory in order to say anything about something more interesting properties. Therefore, in the next chapter we will develop a technique for many-body operator Green functions with which we can make very systematic higher order approximations beyond mean field theory.

4.4 How to Include the Bias Voltage

Up to now we have tacitly included the bias voltage via the Fermi functions for the left and right leads through $f_\chi(\omega) = f(\omega - \mu_\chi)$. Doing like this is reasonable if we imagine that the energies $\varepsilon_{\mathbf{k}\sigma}$ in the leads are given relative to their respective chemical potentials, and assuming that the leads are in (quasi-) equilibrium. Hence, the Hamiltonian for the lead χ may be written

$$\mathcal{K}_\chi = \mathcal{H}_\chi - \mu_\chi N_\chi = \sum_{\mathbf{k}\sigma} \varepsilon_{\mathbf{k}\sigma} c^\dagger_{\mathbf{k}\sigma} c_{\mathbf{k}\sigma} - \mu_\chi N_\chi = \sum_{\mathbf{k}\sigma} [\varepsilon_{\mathbf{k}\sigma} - \mu_\chi] c^\dagger_{\mathbf{k}\sigma} c_{\mathbf{k}\sigma}, \quad (4.47)$$

where $N_\chi = \sum_{\mathbf{k}\sigma} c^\dagger_{\mathbf{k}\sigma} c_{\mathbf{k}\sigma}$ is the number of electrons. Here, we use the notation $\mathcal{K}$ for the Hamiltonian which includes the chemical potential. Then, the Green function $g_{\mathbf{k}\sigma}(t, t') = (-i)\langle \mathrm{T} c_{\mathbf{k}\sigma}(t) c^\dagger_{\mathbf{k}\sigma}(t')\rangle$ satisfies the equation of motion

$$\big(i\partial_t - [\varepsilon_{\mathbf{k}\sigma} - \mu_\chi]\big) g_{\mathbf{k}\sigma}(t, t') = \delta(t - t'), \quad (4.48)$$

which leads to the retarded/advanced and lesser/greater forms

$$g^<_{\mathbf{k}\sigma}(t, t') = i f(\varepsilon_{\mathbf{k}\sigma} - \mu_\chi) e^{-i(\varepsilon_{\mathbf{k}\sigma} - \mu_\chi)(t-t')}, \quad (3.61\mathrm{a}')$$

$$g^>_{\mathbf{k}\sigma}(t, t') = -i[1 - f(\varepsilon_{\mathbf{k}\sigma} - \mu_\chi)] e^{-i(\varepsilon_{\mathbf{k}\sigma} - \mu_\chi)(t-t')}, \quad (3.61\mathrm{b}')$$

$$g^{r/a}_{\mathbf{k}\sigma}(t, t') = (\mp i)\theta(\pm t \mp t) e^{-i(\varepsilon_{\mathbf{k}\sigma} - \mu_\chi)(t-t')}. \quad (3.61\mathrm{c}')$$

Alternatively, we can transform the Hamiltonian such that the chemical potentials μ_χ goes directly into the tunneling part of the model. Using that e.g.

$$c_{\mathbf{k}\sigma}(t) = e^{i\mathcal{H}_\chi t} c_{\mathbf{k}\sigma} e^{-i\mathcal{H}_\chi t}, \tag{4.49}$$

we perform the transformation $\mathcal{H}(t) = e^{i\mathcal{H}_0 t}\mathcal{H}e^{-i\mathcal{H}_0 t}$, where $\mathcal{H}_0$ contains all the subsystems of the total structure, however, without the mutual interactions between them. The number operators commute with $\mathcal{H}_0$ which allows us to write $\mathcal{H}_0 = \mathcal{K}_0 + \sum_\chi \mu_\chi N_\chi$. If we, for instance, have $\mathcal{H} = \mathcal{H}_0 + \mathcal{H}_T$, with $\mathcal{H}_0 = \mathcal{H}_L + \mathcal{H}_R + \mathcal{H}_{QD}$, and $\mathcal{H}_T = \sum_{\mathbf{k}\sigma} v_{\mathbf{k}\sigma} c^\dagger_{\mathbf{k}\sigma} d_\sigma + H.c.$, we find that $\mathcal{H}_\chi(t) = \mathcal{H}_\chi$ and $\mathcal{H}_{QD}(t) = \mathcal{H}_{QD}$. The interaction term, on the other hand, has to be treated with some care. We obtain

$$\begin{aligned}
\mathcal{H}_T(t) &= e^{i\mathcal{H}_0 t}\mathcal{H}_T e^{-i\mathcal{H}_0 t} = e^{i(\mathcal{K}_0+\mu_L N_L+\mu_R N_R)t}\mathcal{H}_T e^{-i(\mathcal{K}_0+\mu_L N_L+\mu_R N_R)t} \\
&= e^{i\mathcal{K}_0 t} e^{i(\mu_L N_L+\mu_R N_R)t}\mathcal{H}_T e^{-i(\mu_L N_L+\mu_R N_R)t} e^{-i\mathcal{K}_0 t} \\
&= e^{i\mathcal{K}_0 t}\sum_{\mathbf{k}\sigma}\big(v_{\mathbf{k}\sigma} c^\dagger_{\mathbf{k}\sigma} d_\sigma e^{i(\mu_\chi+\mu_{QD})t} + H.c.\big)e^{-i\mathcal{K}_0 t}.
\end{aligned} \tag{4.50}$$

Using this description, we can again use the Green functions for the leads as formulated in (3.61) and the reference to the bias, or chemical potentials of the various subsystems, appears very explicitly in the formulation of the current.

References

1. Meir, Y., Wingreen, N.S.: Phys. Rev. Lett. **68**, 2512 (1992)
2. Jauho, A.-P., Wingreen, N.S., Meir, Y.: Phys. Rev. B **50**, 5528 (1994)
3. Fransson, J., Eriksson, O., Sandalov, I.S.: Phys. Rev. B **67**, 195319 (2003)
4. Fransson, J.: Phys. Rev. B **72**, 075314 (2005)

Chapter 5
Diagram Technique

Abstract A systematic diagram technique for expanding the Green function constructed of many-body operators is developed. The general approach is outlined and a few specific examples are described in detail.

We shall do some of the treatment in this chapter quite hands on and, therefore, use a model which is simple but quite general. Occasionally we will go to e.g. the single level model we have been looking at earlier in order to more explicitly exemplify some issues. We take a model with two leads since it corresponds well to experimentally realistic systems although there may be more leads attached to the central region as well. The techniques apply for more general physical systems, however, we choose here to outline different methods in terms of something more concrete. Also, we will not go to deep into the details of the foundation of the theory for all introduced concepts but we will rather take the attitude of *how to do it*. Anyone that is interested in the more formal theory connected to the different expansion schemes may consult e.g. [1–3].

5.1 Equation of Motion and Decoupling

In this first section, we shall discuss a more conventional method for performing expansions of propagators. We start with the equation of motion technique. The benefits with this method is that it permits a very hands on approach to the Green functions and expansions. It is also very easy to employ in simple systems, and when we want to study physical properties in the mean field approximation. We can, of course, go beyond mean field theory even with the equation of motion method but, as we shall see, the structure of the equations may become somewhat involved when higher order scattering processes are to be included.

We begin with a simple model in order to gain some feeling for the method and then we go to more complicated cases.

J. Fransson, *Non-Equilibrium Nano-Physics*,
Lecture Notes in Physics 809,
DOI 10.1007/978-90-481-9210-6_5,

5.1.1 Non-Interacting Resonant Level

We take a single level quantum dot without internal interactions, coupled to two leads and model this system by

$$\mathcal{H} = \sum_{\mathbf{k}\sigma} \varepsilon_{\mathbf{k}\sigma} c_{\mathbf{k}\sigma}^\dagger c_{\mathbf{k}\sigma} + \sum_{\sigma} \varepsilon_\sigma d_\sigma^\dagger d_\sigma + \sum_{\mathbf{k}\sigma} [v_{\mathbf{k}\sigma} c_{\mathbf{k}\sigma}^\dagger d_\sigma + H.c.]. \tag{5.1}$$

For simplicity, we shall assume time-independent tunneling coefficients $v_{\mathbf{k}\sigma}$, although a generalization to the time-dependent case is straight forward. We will be interested in the physics of the quantum dot, and we therefore approach the Green function $G_{\sigma\sigma'}(t,t') = (-i)\langle \mathrm{T} d_\sigma(t) d_{\sigma'}^\dagger(t')\rangle$. The equation of motion for $G_{\sigma\sigma'}$ is given by

$$(i\partial_t - \varepsilon_\sigma) G_{\sigma\sigma'}(t,t') = \delta(t-t')\delta_{\sigma\sigma'} + \sum_{\mathbf{k}} v_{\mathbf{k}\sigma}^* (-i)\langle \mathrm{T} c_{\mathbf{k}\sigma}(t) d_{\sigma'}^\dagger(t')\rangle. \tag{5.2}$$

We continue by looking at the equation of motion for the propagator $F_{\mathbf{k}\sigma\sigma'}(t,t') = (-i)\langle \mathrm{T} c_{\mathbf{k}\sigma}(t) d_{\sigma'}^\dagger(t')\rangle$, which becomes

$$(i\partial_t - \varepsilon_{\mathbf{k}\sigma}) F_{\mathbf{k}\sigma\sigma'}(t,t') = v_{\mathbf{k}\sigma}(-i)\langle \mathrm{T} d_\sigma(t) d_{\sigma'}^\dagger(t')\rangle = v_{\mathbf{k}\sigma} G_{\sigma\sigma'}(t,t'). \tag{5.3}$$

Hence, solving for $F_{\mathbf{k}\sigma\sigma'}$ and putting the result back into (5.2) we find that we can close the equation for $G_{\sigma\sigma'}$, i.e.

$$(i\partial_t - \varepsilon_\sigma) G_{\sigma\sigma'}(t,t') = \delta(t-t')\delta_{\sigma\sigma'} + \sum_{\mathbf{k}} |v_{\mathbf{k}\sigma}|^2 \int_C g_{\mathbf{k}\sigma}(t,t_1) G_{\sigma\sigma'}(t_1,t') dt_1, \tag{5.4}$$

or alternatively

$$G_{\sigma\sigma'}(t,t') = \delta_{\sigma\sigma'} g_\sigma(t,t') + \int_C g_\sigma(t,t_1) V_\sigma(t_1,t_2) G_{\sigma\sigma'}(t_2,t') dt_1. \tag{5.5}$$

Here, g_σ and $g_{\mathbf{k}\sigma}$ satisfy the equations $(i\partial_t - \varepsilon_\sigma) g_\sigma(t,t') = \delta(t-t')$ and $(i\partial_t - \varepsilon_{\mathbf{k}\sigma}) g_{\mathbf{k}\sigma}(t,t') = \delta(t-t')$, respectively, whereas $V_\sigma(t,t') = \sum_{\mathbf{k}} |v_{\mathbf{k}\sigma}|^2 g_{\mathbf{k}\sigma}(t,t')$.

Suppose that we are seeking the retarded quantum dot Green function. Then, we know from our earlier discussions that

$$G_{\sigma\sigma'}^r(t,t') = \delta_{\sigma\sigma'} g_\sigma^r(t,t') + \int g_\sigma^r(t,t_1) V_\sigma^r(t_1,t_2) G_{\sigma\sigma'}^r(t_2,t') dt_2 dt_1. \tag{5.6}$$

Using what we know about $g_{\mathbf{k}\sigma}$ from (3.61), noticing that

$$(-i)\theta(t_1-t_2)\sum_{\mathbf{k}}|v_{\mathbf{k}\sigma}|^2 e^{-i\varepsilon_{\mathbf{k}\sigma}(t_1-t_2)}$$
$$=\int\sum_{\mathbf{k}}\frac{|v_{\mathbf{k}\sigma}|^2}{\omega-\varepsilon_{\mathbf{k}\sigma}+i\delta}e^{-i\omega(t_1-t_2)}\frac{d\omega}{2\pi}$$
$$=\int\left(\Lambda_\sigma(\omega)-\frac{i}{2}\Gamma_\sigma(\omega)\right)e^{-i\omega(t_1-t_2)}\frac{d\omega}{2\pi} \tag{5.7}$$

and assuming that Λ_σ and Γ_σ are slowly varying functions of the energy so they be replaced by constants, we have $\sum_{\mathbf{k}}|v_{\mathbf{k}\sigma}|^2 e^{-i\varepsilon_{\mathbf{k}\sigma}(t_1-t_2)}\approx\delta(t_1-t_2)[\Lambda_\sigma-i\Gamma_\sigma/2]$ Going back to the differential equation, we then find

$$\left[i\partial_t-\varepsilon_\sigma-\Lambda_\sigma+\frac{i}{2}\Gamma_\sigma\right]G^r_{\sigma\sigma'}(t,t')=\delta_{\sigma\sigma'}\delta(t-t'), \tag{5.8}$$

from which we finally obtain

$$G^r_{\sigma\sigma'}(t,t')=(-i)\theta(t-t')\delta_{\sigma\sigma'}e^{-i(\varepsilon_\sigma+\Lambda_\sigma-i\Gamma_\sigma/2)(t-t')}. \tag{5.9}$$

The advanced form is found analogously.

Using the same assumptions for $V^<_\sigma$ as for V^r_σ, we can then go on to find e.g. the lesser Green function

$$G^<_{\sigma\sigma'}(t,t')=\int G^r_{\sigma\sigma''}(t,t_1)V^<_{\sigma''}(t_1,t_2)G^a_{\sigma''\sigma'}(t_2,t')dt_1dt_2$$
$$=i\delta_{\sigma\sigma'}\sum_\chi\int f_\chi(\omega)\Gamma^\chi_\sigma\int_{-\infty}^{t}e^{-i(\varepsilon_\sigma+\Lambda_\sigma-i\Gamma_\sigma/2)(t-t_1)-i\omega t_1}dt_1$$
$$\times\int_{t'}^{\infty}e^{-i(\varepsilon_\sigma+\Lambda_\sigma+i\Gamma_\sigma/2)(t_2-t')+i\omega t_2}dt_2\frac{d\omega}{2\pi}$$
$$=i\delta_{\sigma\sigma'}\sum_\chi\int\frac{f_\chi(\omega)\Gamma^\chi_\sigma}{|\omega-\varepsilon_\sigma-\Lambda_\sigma+i\Gamma_\sigma/2|^2}e^{-i\omega(t-t')}\frac{d\omega}{2\pi}. \tag{5.10}$$

We see from those last two expressions that we could have made a Fourier transform of the equation of motion of e.g. (5.8), such that

$$\left(\omega-\varepsilon_\sigma-\Lambda_\sigma+\frac{i}{2}\Gamma_\sigma\right)G^r_{\sigma\sigma'}(\omega)=\delta_{\sigma\sigma'} \tag{5.11}$$

which is easily solved and gives

$$G^r_{\sigma\sigma'}(i\omega)=\frac{\delta_{\sigma\sigma'}}{\omega-\varepsilon_\sigma-\Lambda_\sigma+i\Gamma_\sigma/2}. \tag{5.12}$$

We recognize this expression in (5.10). Indeed, by recalling that $V_\sigma^<(\omega) = i\sum_\chi f_\chi(\omega)\Gamma_\sigma^\chi$, we find that the lesser Green function can be written as

$$G_{\sigma\sigma'}^<(\omega) = i\delta_{\sigma\sigma'}\frac{f_L(\omega)\Gamma_\sigma^L + f_R(\omega)\Gamma_\sigma^R}{|\omega - \varepsilon_\sigma - \Lambda_\sigma + i\Gamma_\sigma/2|^2}$$
$$= \sum_\chi \frac{f_\chi(\omega)\Gamma_\sigma^\chi}{\Gamma_\sigma}(-2\,\mathrm{Im}\,|G_{\sigma\sigma'}^r(\omega)|^2). \tag{5.13}$$

We could, of course, gone directly to the Fourier transformed solution by noticing that the Hamiltonian in (5.1) is translational invariant in time (and space). This implies that the Green functions actually are functions of the time-difference $t - t'$, as was confirmed in our temporal Green functions. We took the opportunity to derive the Green functions in time-space since we will often use this approach later in the book.

The occupation number $\langle n_\sigma \rangle$ is given by

$$\langle n_\sigma \rangle = -\frac{i}{2\pi}\int G_{\sigma\sigma}^<(\omega)d\omega \tag{5.14}$$

and for low temperatures we have

$$\langle n_\sigma \rangle = \frac{1}{\pi}\sum_\chi \frac{\Gamma_\sigma^\chi}{\Gamma_\sigma}\left(\arctan\frac{\mu_\chi - \varepsilon_\sigma - \Lambda_\sigma}{\Gamma_\sigma/2} - \arctan\frac{D^\chi - \varepsilon_\sigma - \Lambda_\sigma}{\Gamma_\sigma/2}\right), \tag{5.15}$$

where $D^\chi > 0$ is the band width in the lead χ. From this we can see that $\langle n_\uparrow \rangle = \langle n_\downarrow \rangle = 1/2$ whenever the parameters $\Lambda_\uparrow^\chi = \Lambda_\downarrow^\chi$, $\Gamma_\uparrow^\chi = \Gamma_\downarrow^\chi$, and $\varepsilon_\uparrow = \varepsilon_\downarrow$, that is, when the system is spin-degenerate. For any type of disturbance of this symmetry there will be a spin-polarization induced on the local electrons.

Inserting (5.13), and the analogous one for the greater Green function, into the current in (4.14), using the stationary regime requirement $I_R = -I_L$ such that $I = (I_L - I_R)/2$, we can write the current through the resonant level as

$$I = \frac{e}{4\pi}\sum_\sigma \int \frac{\Gamma_\sigma^L\Gamma_\sigma^R}{|\omega - \varepsilon_\sigma - \Lambda_\sigma + i\Gamma_\sigma/2|^2}[f_L(\omega) - f_R(\omega)]d\omega. \tag{5.16}$$

This current was discussed by Larkin and Matveev [4] in connection with resonant tunneling of electrons via impurity states in short semiconducting contacts.

The solution for the Green function expressed in e.g. (5.8), as well as all other expressions displayed in this section, is the expected one for the localized level interacting through tunneling, or hybridization, with one or several conduction channels. The model given in (5.1) is called the Fano-Anderson model and can, for example, be used as an example of solvable models. Physically it describes, as mentioned, the interactions between localized and de-localized electrons through tunneling. The absence of charging interaction term for the localized electrons may be viewed as unphysical and therefore the Fano-Anderson model may be poor in the description of transport properties in nanoscale systems, where the charging energy often is larger than the couplings to the leads. It serves, however, an excellent purpose for exercises in Green function formalism. Moreover, the solutions presented here will be called for as bare solutions in other, more complicated, situations.

5.1.2 Coulomb Island

In order to make the modeling more physically sound, we introduce the first complication to our model and discuss how we should approach the physics of the localized electrons in

$$\mathcal{H}=\sum_{\mathbf{k}\sigma}\varepsilon_{\mathbf{k}\sigma}c^\dagger_{\mathbf{k}\sigma}c_{\mathbf{k}\sigma}+\sum_{\sigma}\varepsilon_\sigma d^\dagger_\sigma d_\sigma+Un_\uparrow n_\downarrow+\sum_{\mathbf{k}\sigma}[v_{\mathbf{k}\sigma}c^\dagger_{\mathbf{k}\sigma}d_\sigma+H.c.]. \tag{5.17}$$

The new term, $Un_\uparrow n_\downarrow$ accounts for the charging energy in the quantum dot, or, localized level. It is more physical than the Fano-Anderson model since it properly describes what will happen in the quantum dot when there are two electrons present. We would expect that the two electrons want to repel each other since they have the same charge and sign, which is the Coulomb repulsion, or interaction. Although we have introduced the strength of the interaction merely through a parameter U, the model clearly describes that the state with two electrons, in the atomic limit, must have the energy $\varepsilon_\uparrow+\varepsilon_\downarrow+U$.

Approaching local electrons with the equation of motion for the Green function $G_{\sigma\sigma'}$ we have

$$\begin{aligned}(i\partial_t-\varepsilon_\sigma)G_{\sigma\sigma'}(t,t')=&\,\delta_{\sigma\sigma'}\delta(t-t')+U(-i)\langle\mathrm{T}n_{\bar\sigma}(t)d_\sigma(t)d^\dagger_{\sigma'}(t')\rangle\\&+\sum_{\mathbf{k}}v^*_{\mathbf{k}\sigma}F_{\mathbf{k}\sigma\sigma'}(t,t').\end{aligned} \tag{5.18}$$

The new obstacle to our description is the two electron Green function $\mathcal{G}_{\sigma\sigma'}(t,t')=(-i)\langle\mathrm{T}n_{\bar\sigma}(t)d_\sigma(t)d^\dagger_{\sigma'}(t')\rangle$, about which we have to do something. The question is, of course, what *can* we do? First, we look at the equation of motion for this two-electron Green function. We find

$$\begin{aligned}(i\partial_t-\varepsilon_\sigma-U)\mathcal{G}_{\sigma\sigma'}(t,t')=&\,\delta_{\sigma\sigma'}\langle n_{\bar\sigma}(t)\rangle\delta(t-t')\\&+\sum_{\mathbf{k}s}v_{\mathbf{k}s}(-i)\langle\mathrm{T}[n_{\bar\sigma}d_\sigma,c^\dagger_{\mathbf{k}s}d_s](t)d^\dagger_{\sigma'}(t')\rangle,\\&+\sum_{\mathbf{k}s}v^*_{\mathbf{k}s}(-i)\langle\mathrm{T}[n_{\bar\sigma}d_\sigma,d^\dagger_s c_{\mathbf{k}s}](t)d^\dagger_{\sigma'}(t')\rangle,\end{aligned} \tag{5.19}$$

where we have used that the product $n_\sigma n_\sigma=n_\sigma$, giving the U on the left hand side of the equation. Clearly, the right hand side becomes even more complicated under this operation. The propagator in the first summand can be rewritten

$$(-i)\langle \mathrm{T}[n_{\bar{\sigma}} d_\sigma, c^\dagger_{\mathbf{k}s} d_s](t) d^\dagger_{\sigma'}(t')\rangle = -\delta_{s\bar{\sigma}}(-i)\langle \mathrm{T}(c^\dagger_{\mathbf{k}\bar{\sigma}} d_{\bar{\sigma}} d_\sigma)(t) d^\dagger_{\sigma'}(t')\rangle, \tag{5.20}$$

and, similarly, the propagator in the second summand becomes

$$(-i)\langle \mathrm{T}[n_{\bar{\sigma}} d_\sigma, d^\dagger_s c_{\mathbf{k}s}](t) d^\dagger_{\sigma'}(t')\rangle = \delta_{s\sigma}(-i)\langle \mathrm{T}(n_{\bar{\sigma}} c_{\mathbf{k}\sigma})(t) d^\dagger_{\sigma'}(t')\rangle - \delta_{s\bar{\sigma}}(-i)\langle \mathrm{T}(d^\dagger_{\bar{\sigma}} d_\sigma c_{\mathbf{k}\bar{\sigma}})(t) d^\dagger_{\sigma'}(t')\rangle. \tag{5.21}$$

Let us, for the moment, stop at this point and see what we can do. We decouple the propagators on the right hand side, such that

$$\begin{aligned}(-i)\langle \mathrm{T}(c^\dagger_{\mathbf{k}\bar{\sigma}} d_{\bar{\sigma}} d_\sigma)(t) d^\dagger_{\sigma'}(t')\rangle &= \langle c^\dagger_{\mathbf{k}\bar{\sigma}}(t) d_{\bar{\sigma}}(t)\rangle G_{\sigma\sigma'}(t,t') \\ &- \langle c^\dagger_{\mathbf{k}\bar{\sigma}}(t) d_\sigma(t)\rangle G_{\bar{\sigma}\sigma'}(t,t') \\ &+ \langle d_{\bar{\sigma}}(t) d_\sigma(t)\rangle (-i)\langle \mathrm{T} c^\dagger_{\mathbf{k}\bar{\sigma}}(t) d^\dagger_{\sigma'}(t')\rangle, \qquad (5.22a)\\ (-i)\langle \mathrm{T}(n_{\bar{\sigma}} c_{\mathbf{k}\sigma})(t) d^\dagger_{\sigma'}(t')\rangle &= \langle n_{\bar{\sigma}}(t)\rangle F_{\mathbf{k}\sigma\sigma'}(t,t') \\ &- \langle d^\dagger_{\bar{\sigma}}(t) c_{\mathbf{k}\sigma}(t)\rangle G_{\bar{\sigma}\sigma'}(t,t') \\ &+ \langle d_{\bar{\sigma}}(t) c_{\mathbf{k}\sigma}(t)\rangle (-i)\langle \mathrm{T} d^\dagger_{\bar{\sigma}}(t) d^\dagger_{\sigma'}(t')\rangle, \qquad (5.22b)\\ (-i)\langle \mathrm{T}(d^\dagger_{\bar{\sigma}} d_\sigma c_{\mathbf{k}\bar{\sigma}})(t) d^\dagger_{\sigma'}(t')\rangle &= \langle d^\dagger_{\bar{\sigma}}(t) d_\sigma(t)\rangle F_{\mathbf{k}\bar{\sigma}\sigma'}(t,t') \\ &- \langle d^\dagger_{\bar{\sigma}}(t) c_{\mathbf{k}\bar{\sigma}}(t)\rangle G_{\sigma\sigma'}(t,t') \\ &+ \langle d_\sigma(t) c_{\mathbf{k}\bar{\sigma}}(t)\rangle (-i)\langle \mathrm{T} d^\dagger_{\bar{\sigma}}(t) d^\dagger_{\sigma'}(t')\rangle. \qquad (5.22c)\end{aligned}$$

First, under the assumption that neither the leads nor the quantum dot have any superconducting properties we can safely use that any propagator, or average, with two creation or annihilation operators is negligible, if not vanishing. Second, if we wish to obtain a second order approximation in the tunneling rate, we assert that the averages of operators with equal time can be taken in the atomic limit, so that e.g. $\langle c^\dagger_{\mathbf{k}\bar{\sigma}}(t) d_\sigma(t)\rangle = 0$. We then arrive at the mean field equation for the two-electron Green function

$$(i\partial_t - \varepsilon_\sigma - U)\mathcal{G}_{\sigma\sigma'}(t,t') = \delta_{\sigma\sigma'}\langle n_{\bar{\sigma}}(t)\rangle \delta(t-t') + \sum_{\mathbf{k}} v^*_{\mathbf{k}\sigma}\langle n_{\bar{\sigma}}(t)\rangle F_{\mathbf{k}\sigma\sigma'}(t,t'). \tag{5.23}$$

The sum over **k** on the right hand side describes the transfer of electrons with spin σ between the quantum dot and the leads in presence of a spin $\bar{\sigma}$ electron in the quantum dot.

Now, in order to proceed in a lighter manner, we assume stationary conditions on the system such that we can Fourier transform our equations. We have the set of equations

$$(i\omega - \varepsilon_\sigma)G_{\sigma\sigma'}(i\omega) = \delta_{\sigma\sigma'} + U\mathcal{G}_{\sigma\sigma'}(i\omega) + \sum_{\mathbf{k}} v^*_{\mathbf{k}\sigma} F_{\mathbf{k}\sigma\sigma'}(i\omega), \quad (5.24a)$$

$$(i\omega - \varepsilon_\sigma - U)\mathcal{G}_{\sigma\sigma'}(i\omega) = \delta_{\sigma\sigma'}\langle n_{\bar{\sigma}}\rangle + \sum_{\mathbf{k}} v^*_{\mathbf{k}\sigma}\langle n_{\bar{\sigma}}\rangle F_{\mathbf{k}\sigma\sigma'}(i\omega), \quad (5.24b)$$

$$(i\omega - \varepsilon_{\mathbf{k}\sigma})F_{\mathbf{k}\sigma\sigma'}(i\omega) = v_{\mathbf{k}\sigma} G_{\sigma\sigma'}(i\omega). \quad (5.24c)$$

Solving for the single-electron Green function $G_{\sigma\sigma'}$ we arrive at

$$G_{\sigma\sigma'}(i\omega) = g_\sigma(i\omega)\big(1 + g_\sigma(i\omega - U)\langle n_{\bar{\sigma}}\rangle U\big)\big(\delta_{\sigma\sigma'} + V_\sigma(i\omega)G_{\sigma\sigma'}(i\omega)\big). \quad (5.25)$$

We now redefine our bare Green function for the localized electrons by $G_{0\sigma}(i\omega) = g_\sigma(i\omega)[1 + g_\sigma(i\omega - U)\langle n_{\bar{\sigma}}\rangle U]$, which permits us to write

$$G_{\sigma\sigma'}(i\omega) = \delta_{\sigma\sigma'}G_{0\sigma}(i\omega) + G_{0\sigma}(i\omega)V_\sigma(i\omega)G_{\sigma\sigma'}(i\omega). \quad (5.26)$$

Thus, we can obtain the retarded (advanced) and lesser (greater) Green functions using the same procedure as previously, that is

$$G^{r/a}_{\sigma\sigma'}(\omega) = \delta_{\sigma\sigma'}G^{r/a}_{0\sigma}(\omega) + G^{r/a}_{0\sigma}(\omega)V^{r/a}_\sigma(\omega)G^{r/a}_{\sigma\sigma'}(\omega), \quad (5.27a)$$

$$G^{</>}_{\sigma\sigma'}(\omega) = G^r_{\sigma\sigma''}(\omega)V^{</>}_{\sigma''}(\omega)G^a_{\sigma''\sigma'}(\omega). \quad (5.27b)$$

The retarded (advanced) bare Green function is given by

$$G^{r/a}_{0\sigma}(\omega) = \frac{\langle n_{\bar{\sigma}}\rangle}{\omega - \varepsilon_\sigma \pm i\delta} + \frac{1 - \langle n_{\bar{\sigma}}\rangle}{\omega - \varepsilon_\sigma - U \pm i\delta}, \quad (5.28)$$

and we thus obtain

$$G^{r/a}_{\sigma\sigma'}(\omega) = \frac{\delta_{\sigma\sigma'}[\omega - \varepsilon_\sigma - \langle 1 - n_{\bar{\sigma}}\rangle U]}{[\omega - \varepsilon_\sigma - \Lambda_\sigma + i\Gamma_\sigma/2][\omega - \varepsilon_\sigma - U] - \langle n_{\bar{\sigma}}\rangle U(\Lambda_\sigma - i\Gamma_\sigma/2)}. \quad (5.29)$$

The lesser Green function is, again, found to be the squared modulus of the retarded Green function, $|G^r_{\sigma\sigma}|^2$, times the lesser interaction propagator $V^<_\sigma$. However, we also have a reference to the occupation numbers $\langle n_\sigma\rangle$ which are calculated by (5.14) requiring that $\sum_\sigma\langle n_\sigma\rangle = 1$. In this way we have constructed a self-consistent solution of the Green function for the localized electrons in the quantum dot.

We can check that the obtained solution is consistent with the solution in the previous section by letting the charging energy $U \to 0$. The retarded Green function then reduces to the one given in (5.12), as we expect. In the opposite limit, i.e. $U \to \infty$, we find the solution

$$\lim_{U\to\infty} G^{r/a}_{\sigma\sigma'}(\omega) = \delta_{\sigma\sigma'}\frac{1 - \langle n_{\bar{\sigma}}\rangle}{\omega - \varepsilon_\sigma - \langle 1 - n_{\bar{\sigma}}\rangle(\Lambda_\sigma + i\Gamma_\sigma/2)}. \quad (5.30)$$

The infinite U limit corresponds to the case where there cannot be two electrons present simultaneously in the quantum dot, or island. This condition is made sure to hold by the presence of the occupation number $1 - \langle n_{\bar{\sigma}} \rangle$ in the numerator. The effective width of the level is also reduced by this same factor. In this limit and for low temperatures the occupation number $\langle n_\sigma \rangle$ can be determined from the equation

$$\frac{\langle n_\sigma \rangle}{1 - \langle n_{\bar{\sigma}} \rangle} = \frac{1}{\pi} \sum_\chi \frac{\Gamma_\sigma^\chi}{\Gamma_\sigma} \left(\arctan \frac{\mu_\chi - \tilde{\varepsilon}_\sigma}{\langle 1 - n_{\bar{\sigma}} \rangle \Gamma_\sigma / 2} + \arctan \frac{D^\chi + \tilde{\varepsilon}_\sigma}{\langle 1 - n_{\bar{\sigma}} \rangle \Gamma_\sigma / 2} \right), \tag{5.31}$$

where D^χ is the band width in the lead χ, whereas $\tilde{\varepsilon}_\sigma = \varepsilon_\sigma + \langle 1 - n_{\bar{\sigma}} \rangle \Lambda_\sigma$ is the effective position of the localized level. Again, in the spin-degenerate case we must recover that $\langle n_\sigma \rangle = 1/2$ although the island cannot host two electrons at the same time. We regard, however, the Green functions to be probability densities such that $-\mathrm{Im}\ \mathrm{tr} \int \mathbf{G}^r(\omega) d\omega / \pi = 1$. With this interpretation the occupation number $\langle n_\sigma \rangle$ give the *probability* for the level to be occupied with a spin σ electron, rather than the number of electrons.

The mean field theory for the local levels in the quantum dot provide the interaction with the electrons in the conduction channel. This interaction generates the level width Γ_σ which changes the discrete level into a Lorentzian shaped continuous distribution for the energy in the quantum dot, centered near the positions of the discrete levels from the atomic limit. The level broadening is, however, constant in energy and does not depend on any kinds of many-body interactions that do occur inside the quantum dot when coupled to leads. Therefore, we move on to a description which accounts for higher order scattering processes.

5.1.3 Beyond Mean Field Theory

Let us again look at the two-electron Green function, for which we found the equation

$$(i\partial_t - \varepsilon_\sigma - U) \mathcal{G}_{\sigma\sigma'}(t, t') = \delta_{\sigma\sigma'} \langle n_{\bar{\sigma}}(t) \rangle \delta(t - t')$$

$$- \sum_{\mathbf{k}} v_{\mathbf{k}\bar{\sigma}} (-i) \langle \mathrm{T}(d_{\bar{\sigma}} d_\sigma c_{\mathbf{k}\bar{\sigma}}^\dagger)(t) d_{\sigma'}^\dagger(t') \rangle \tag{5.32a}$$

$$+ \sum_{\mathbf{k}} v_{\mathbf{k}\sigma}^* (-i) \langle \mathrm{T}(n_{\bar{\sigma}} c_{\mathbf{k}\sigma})(t) d_{\sigma'}^\dagger(t') \rangle \tag{5.32b}$$

$$- \sum_{\mathbf{k}} v_{\mathbf{k}\bar{\sigma}}^* (-i) \langle \mathrm{T}(d_{\bar{\sigma}}^\dagger d_\sigma c_{\mathbf{k}\bar{\sigma}})(t) d_{\sigma'}^\dagger(t') \rangle. \tag{5.32c}$$

This equation holds the main difficulties in the treatment of the single level Coulomb island, because the equation of motion for each and every propagator in this equation, generates new propagators accounting for even more scattering processes. We dealt with this problem in the previous section by accepting the mean field equation.

As we discussed previously, the mean field theory excludes all fluctuations in e.g. the localized level that are caused by the interaction between the conduction channels and the local quantum dot level. Would fluctuations be expected? Of course there are reasons to believe that the couplings between the conduction channels and the local level would be able to mediate fluctuations like e.g. the local electron hopping of the quantum dot and into one of the lead, whereby another electron hops from one lead into the quantum dot and thereby replacing the first electron. The processes could occur one after the other or simultaneously. The replacing electron may have the same or opposite spin as the replaced electron. Such processes may happen even in equilibrium, but then those processes cause no net current through the system. In non-equilibrium those processes may give rise to peaks or dips in the differential conductance that cannot be ascribed to the known resonances. Hence, the higher order scattering processes may cause a build up of new resonances between the leads which may contribute to the conductance. From a fundamental point of view it is interesting to study the excitation spectrum in the quantum dot, especially if there are excitations generated through higher order scattering processes.

Here, we will proceed by making one additional step in the expansion of the two-electron Green function and also study the propagator on the right hand side in (5.32a)–(5.32c). In this way we will obtain an energy dependent self-energy to the final single electron Green function.

Considering the contributions one by one, we have for the term (5.32a) the equation

$$
\begin{aligned}
&(i\partial_t - \varepsilon_{\bar{\sigma}} - \varepsilon_\sigma + \varepsilon_{\mathbf{k}\bar{\sigma}})(-i)\langle \mathrm{T}(d_{\bar{\sigma}} d_\sigma c^\dagger_{\mathbf{k}\bar{\sigma}})(t) d^\dagger_{\sigma'}(t')\rangle \\
&\quad = \delta_{\sigma\sigma'}\delta(t-t')\langle d_{\bar{\sigma}}(t) c^\dagger_{\mathbf{k}\bar{\sigma}}(t)\rangle + U(-i)\langle \mathrm{T}(n_{\bar{\sigma}} d_{\bar{\sigma}} d_\sigma c^\dagger_{\mathbf{k}\bar{\sigma}})(t) d^\dagger_{\sigma'}(t')\rangle \\
&\quad\quad + U(-i)\langle \mathrm{T}(n_\sigma d_{\bar{\sigma}} d_\sigma c^\dagger_{\mathbf{k}\bar{\sigma}})(t) d^\dagger_{\sigma'}(t')\rangle \\
&\quad\quad + \sum_{\mathbf{k}'} \big\{ v^*_{\mathbf{k}'\sigma}(-i)\langle \mathrm{T}(c^\dagger_{\mathbf{k}\bar{\sigma}} c_{\mathbf{k}'\sigma} d_{\bar{\sigma}})(t) d^\dagger_{\sigma'}(t')\rangle \\
&\quad\quad - v^*_{\mathbf{k}'\bar{\sigma}} \big[(-i)\langle \mathrm{T}(c^\dagger_{\mathbf{k}\bar{\sigma}} c_{\mathbf{k}'\bar{\sigma}} d_\sigma)(t) d^\dagger_{\sigma'}(t')\rangle \\
&\quad\quad - (-i)\langle \mathrm{T}(d^\dagger_{\bar{\sigma}} d_{\bar{\sigma}} d_\sigma)(t) d^\dagger_{\sigma'}(t')\rangle \big] \big\},
\end{aligned}
\tag{5.33}
$$

where we have assumed that the spin-flip average $\langle d_\sigma(t) c^\dagger_{\mathbf{k}\bar{\sigma}}(t)\rangle = 0$. Here, we decouple the propagators under the same assumptions as we made in the mean field approach. We, further, take the approach to convert all propagators into their bare counterparts, expect for the ones resulting in the single electron Green function $G_{\sigma\sigma'}$. Then, we also see that the average $\langle d_{\bar{\sigma}}(t) c^\dagger_{\mathbf{k}\bar{\sigma}}(t)\rangle = 0$. The second and third terms on the right hand side of the equation reduce to contribution $U\langle (n_\sigma + n_{\bar{\sigma}})(t)\rangle = U$ in the parentheses on the left hand side. Under those assumptions, we find that the first term in the summand vanishes, while the second and third give the contributions $-v^*_{\mathbf{k}\bar{\sigma}}\langle n_{\mathbf{k}\bar{\sigma}}(t)\rangle G_{\sigma\sigma'}(t,t')$ and $-v^*_{\mathbf{k}\bar{\sigma}}\mathcal{G}_{\sigma\sigma'}$, respectively.

We thus have reduced the equation to the simpler form

$$\begin{aligned}
&(i\partial_t - \varepsilon_{\bar{\sigma}} - \varepsilon_\sigma - U + \varepsilon_{\mathbf{k}\bar{\sigma}})(-i)\langle \mathrm{T}(d_{\bar{\sigma}} d_\sigma c^\dagger_{\mathbf{k}\bar{\sigma}})(t) d^\dagger_{\sigma'}(t')\rangle \\
&\quad = v^*_{\mathbf{k}\bar{\sigma}}[\langle n_{\mathbf{k}\bar{\sigma}}(t)\rangle G_{\sigma\sigma'}(t,t') - \mathcal{G}_{\sigma\sigma'}(t,t')]. \qquad (5.34)
\end{aligned}$$

Analogously, we find the equation for (5.32b)

$$(i\partial_t - \varepsilon_{\mathbf{k}\sigma})(-i)\langle \mathrm{T}(n_{\bar{\sigma}} c_{\mathbf{k}\sigma})(t) d^\dagger_{\sigma'}(t')\rangle = v_{\mathbf{k}\sigma}\langle n_{\bar{\sigma}}(t)\rangle G_{\sigma\sigma'}(t,t'), \qquad (5.35)$$

and for (5.32c)

$$\begin{aligned}
&(i\partial_t + \varepsilon_{\bar{\sigma}} - \varepsilon_\sigma - U\langle n_{\bar{\sigma}} - n_\sigma\rangle - \varepsilon_{\mathbf{k}\sigma})(-i)\langle \mathrm{T}(d^\dagger_{\bar{\sigma}} d_\sigma c_{\mathbf{k}\bar{\sigma}})(t) d^\dagger_{\sigma'}(t')\rangle \\
&\quad = v_{\mathbf{k}\bar{\sigma}}[\langle n_{\mathbf{k}\bar{\sigma}}(t)\rangle G_{\sigma\sigma'}(t,t') - \mathcal{G}_{\sigma\sigma'}(t,t')]. \qquad (5.36)
\end{aligned}$$

It may seem inconsistent not to expressing (5.35) in terms of the two-electron Green function. We choose to make this approximation, however, in order to capture the correct result in the atomic limit.

We go over to energy space by assuming stationary conditions, and substitute (5.34)–(5.36) into (5.32a)–(5.32c). We obtain

$$\begin{aligned}
[i\omega - \varepsilon_\sigma - U]\mathcal{G}_{\sigma\sigma'}(i\omega) &= \delta_{\sigma\sigma'}\langle n_{\bar{\sigma}}\rangle + \Sigma_{2\sigma}(i\omega)\mathcal{G}_{\sigma\sigma'}(i\omega) \\
&\quad + \big[\langle n_{\bar{\sigma}}\rangle V_\sigma(i\omega) + \Sigma_{1\sigma}(i\omega)\big]G_{\sigma\sigma'}(i\omega). \qquad (5.37)
\end{aligned}$$

Here, we have introduced the self-energy

$$\begin{aligned}
\Sigma_{\alpha\sigma}(i\omega) = \sum_{\mathbf{k}} \mathcal{A}_\alpha |v_{\mathbf{k}\bar{\sigma}}|^2 \bigg(&\frac{1}{i\omega - \varepsilon_{\bar{\sigma}} - \varepsilon_\sigma - U + \varepsilon_{\mathbf{k}\bar{\sigma}}} \\
&+ \frac{1}{i\omega + \varepsilon_{\bar{\sigma}} - \varepsilon_\sigma - U\langle n_{\bar{\sigma}} - n_\sigma\rangle - \varepsilon_{\mathbf{k}\bar{\sigma}}}\bigg), \qquad (5.38)
\end{aligned}$$

where $\mathcal{A}_1 = f(\varepsilon_{\mathbf{k}\bar{\sigma}})$ and $\mathcal{A}_2 = 1$.

We stop here and conclude that the structure of the equation for the single-electron Green function begins to become rather involved. We note, however, by using (5.37) in the equation $(i\omega - \varepsilon_\sigma)G_{\sigma\sigma'} = \delta_{\sigma\sigma'} + U\mathcal{G}_{\sigma\sigma'} + V_\sigma G_{\sigma\sigma'}$, that we find the solution

$$G^{-1}_{\sigma\sigma'}(i\omega) = \delta_{\sigma\sigma'} g^{-1}_\sigma(i\omega) - \Sigma_{\sigma\sigma'}(i\omega), \qquad (5.39)$$

where we have defined the self-energy [5, 6]

$$\Sigma_{\sigma\sigma'}(i\omega) = \delta_{\sigma\sigma'}\bigg(V_\sigma(i\omega) + U\frac{\langle n_{\bar{\sigma}}\rangle(i\omega - \varepsilon_\sigma) - \Sigma_{1\sigma}(i\omega)}{i\omega - \varepsilon_\sigma - \langle 1 - n_{\bar{\sigma}}\rangle U - \Sigma_{2\sigma}(i\omega)}\bigg). \qquad (5.40)$$

The retarded and advanced single-electron Green function can be found by a straight forward analytical continuation $i\omega \to \omega + i\delta$, $\delta > 0$. The lesser and greater forms are, however, not so simple to find to this order of approximation, and it is, in fact, still an open question to find suitable expressions for these quantities.

In equilibrium we can resort to the retarded and advanced form of $G_{\sigma\sigma'}$, and the solutions given by (5.39) and (5.40) provide an additional resonance near the Fermi level, Kondo resonance, in case of a nearly spin-degenerate quantum dot, i.e. $\varepsilon_\uparrow \approx \varepsilon_\downarrow$. The appearance of this resonance is due to screening of the local magnetic moment that arise in the quantum dot in equilibrium. If the quantum dot level is far below the equilibrium chemical potential, it will be hosted by an electron with a definite spin. This spin yields an effective magnetic moment locally in the system. The de-localized electrons in the conduction channels tend to screen this moment in order to equilibrate any spin-imbalance that arise, and this screening generates the additional resonance.

We shall not dwell on this type of approximation. We learn that the equation of motion technique has its benefits, that it is straight forward and easy to employ in almost any situation. In going to higher order approximations the drawbacks are obvious, that it takes quite some bookkeeping in order to perform each step of the derivations. There are other ways to proceed with a systematic approach to approximations, i.e. using Wick's theorem. We refer the reader to the excellent introductions to this subject given in e.g. [1–3].

5.2 Expanding the Hubbard Operator Green Functions

From now on we shall assume that we have a model of the type $\mathcal{H} = \mathcal{H}_L + \mathcal{H}_R + \mathcal{H}_C + \mathcal{H}_T$, where $\mathcal{H}_{L/R}$ are as usual, whereas $\mathcal{H}_C = \sum_{Nn} E_{Nn} h^n_N$ and $\mathcal{H}_T = \sum_{\mathbf{k}\sigma a} v_{\mathbf{k}\sigma a} c^\dagger_{\mathbf{k}\sigma} X^a + H.c.$, where a is an arbitrary single-electron transition. We will also need a source potential $\mathcal{H}'(t) = U_\xi(t) Z^\xi$, where we sum over repeated indices. The source fields which are necessary in the present context, correspond to the possible Bose-like transitions that arise in the system due to multiplications of Fermi-like transitions.

We start off as before and consider the equation of motion for the Green function $G_{a\bar{b}}(t,t') = (-i)\langle \mathrm{T} X^a(t) X^{\bar{b}}(t')\rangle_U$. The equation of motion is written

$$(i\partial_t - \Delta_{\bar{a}}) G_{a\bar{b}}(t,t') = \delta(t-t') P_{a\bar{b}}(t') + \varepsilon^{ac}_\xi \big[+ v_{\mathbf{k}\sigma c}(-i)\langle \mathrm{T} Z^\xi(t) c^\dagger_{\mathbf{k}\sigma}(t) X^{\bar{b}}(t')\rangle_U \\ + v^*_{\mathbf{k}\sigma c}(-i)\langle \mathrm{T} Z^\xi(t) c_{\mathbf{k}\sigma}(t) X^{\bar{b}}(t')\rangle_U \big]. \tag{5.41}$$

Recall that the tensor ε^{ac}_ξ is defined from the anti-commutation relation $\{X^a, X^c\} = \varepsilon^{ac}_\xi Z^\xi$.

One approach to study this equation would be using conventional perturbation theory, expanding the propagators according to

$$(-i)\langle \mathrm{T}A(t)B(t')\rangle$$
$$=\sum_{n=0}^{\infty}(-i)^{n+1}\int_C\cdots\int_C\langle \mathrm{T}A(t)B(t')\mathcal{H}_T(t_1)\cdots\mathcal{H}_T(t_n)\rangle_0 dt_n\cdots dt_1. \quad (5.42)$$

Each of those propagators inside the integrals have to be decoupled to enable calculations of the propagator on the left hand side. Here, we shall not further discuss this method and instead refer the reader to [7], in which a Wick's theorem approach is extensively discussed. A fundamental reason for not taking this discussion is that Wick's theorem for Hubbard operators is inapplicable under non-equilibrium conditions.

Instead, we shall approach the problem with a technique that is reminiscent of the method introduced by Kadanoff and Baym [8] using functional derivatives. The idea was introduced already in (3.31), for a propagator with three Hubbard operators. We can, of course, do the same thing with the mixed propagators we have in the equation of motion above. Using the same approach that was employed in deriving (3.31), we find

$$(-i)\langle \mathrm{T}Z^{\xi}(t)c^{\dagger}_{\mathbf{k}\sigma}(t)X^{\bar{b}}(t')\rangle_U=\left(\langle \mathrm{T}Z^{\xi}(t)\rangle_U+i\frac{\delta}{\delta U_{\xi}(t)}\right)F^{\dagger}_{\mathbf{k}\sigma\bar{b}}(t,t'), \quad (5.43a)$$

$$(-i)\langle \mathrm{T}Z^{\xi}(t)c_{\mathbf{k}\sigma}(t)X^{\bar{b}}(t')\rangle_U=\left(\langle \mathrm{T}Z^{\xi}(t)\rangle_U+i\frac{\delta}{\delta U_{\xi}(t)}\right)F_{\mathbf{k}\sigma\bar{b}}(t,t'), \quad (5.43b)$$

where we have defined $F^{\dagger}_{\mathbf{k}\sigma\bar{b}}(t,t')=(-i)\langle \mathrm{T}c^{\dagger}_{\mathbf{k}\sigma}(t)X^{\bar{b}}(t')\rangle_U$. Unless we are considering the lead or the central region to be in the superconducting state, this propagator is negligible. It is therefore omitted in the following, since we are focusing our efforts on metallic leads and central regions not being in the superconducting state.

Substituting the above expressions into (5.41) gives

$$(i\partial_t-\Delta_{\bar{a}})G_{a\bar{b}}(t,t')=\delta(t-t')P_{a\bar{b}}(t')+v^{*}_{\mathbf{k}\sigma c}[P_{a\bar{c}}(t^{+})+R_{a\bar{c}}(t^{+})]F_{\mathbf{k}\sigma\bar{b}}(t,t'), \quad (5.44)$$

where $P_{a\bar{c}}(t)=\varepsilon^{ac}_{\xi}\langle \mathrm{T}Z^{\xi}(t)\rangle_U=\langle \mathrm{T}\{X^{a},X^{\bar{c}}\}(t)\rangle_U$ and $R_{a\bar{c}}(t)=i\varepsilon^{ac}_{\xi}\delta/\delta U_{\xi}(t)$, as defined in Chap. 3. Using (4.8) for the propagator $F_{\mathbf{k}\sigma\bar{b}}$ we finally arrive at the equation

$$(i\partial_t-\Delta_{\bar{a}})G_{a\bar{b}}(t,t')=\delta(t-t')P_{a\bar{b}}(t')$$
$$+[P_{a\bar{c}}(t^{+})+R_{a\bar{c}}(t^{+})]\int_C V_{\bar{c}d}(t,t_1)G_{d\bar{b}}(t_1,t'), \quad (5.45)$$

where $V_{\bar{c}d}(t,t')=\sum_{\mathbf{k}\sigma}v^{*}_{\mathbf{k}\sigma c}v_{\mathbf{k}\sigma d}g_{\mathbf{k}\sigma}(t,t')$.

In Chap. 3 we saw how we can write the equation for the matrix Green function $\mathbf{G}=\{G_{a\bar{b}}\}_{ab}$ according to

$$\mathbf{G}(t,t')=\mathbf{d}(t,t')\mathbf{P}(t')+\int_C\mathbf{d}(t,t_1)\mathcal{S}(t_1,t_2)\mathbf{G}(t_2,t')dt_2dt_1, \quad (5.46)$$

where the self-energy operators $\mathcal{S}$ is defined through

$$\mathcal{S}(t,t') = \mathbf{P}(t^+)\mathbf{V}(t,t') + \int_C \{\mathbf{R}(t^+)\mathbf{V}(t,t_1)\mathbf{D}(t_1,t_2)\}\mathbf{D}^{-1}(t_2,t')dt_1dt_2, \quad (5.47)$$

and the inverse of the locator as

$$\mathbf{D}^{-1}(t,t') = \mathbf{d}^{-1}(t,t') - \mathcal{S}(t,t'). \quad (5.48)$$

It is important to note that the functional derivative in the integrand only acts on the propagators in the braces. The way in which we have constructed our framework implies that the propagator $\mathbf{V}$ is unaffected by the functional derivative, however, the matrices $\mathbf{R}$ and $\mathbf{V}$ are non-commuting. Further developments are, thus, most conveniently performed for the components of the equation, i.e.

$$\mathcal{S}_{a\bar{b}}(t,t') = P_{a\bar{c}}(t^+)V_{\bar{c}b}(t,t') + \int_C \{R_{a\bar{c}}(t^+)V_{\bar{c}d}(t,t_1)D_{d\bar{e}}(t_1,t_2)\}D^{-1}_{e\bar{b}}(t_2,t')dt_1dt_2. \quad (5.49)$$

Here, we use that $\mathbf{D}\mathbf{D}^{-1} = I = \mathbf{D}^{-1}\mathbf{D}$ giving $0 = \delta(\mathbf{D}\mathbf{D}^{-1}) = (\delta\mathbf{D})\mathbf{D}^{-1} + \mathbf{D}(\delta\mathbf{D}^{-1})$, hence, $(\delta\mathbf{D})\mathbf{D}^{-1} = -\mathbf{D}(\delta\mathbf{D}^{-1})$. The self-energy operators can thus be written as

$$\mathcal{S}_{a\bar{b}}(t,t') = P_{a\bar{c}}(t^+)V_{\bar{c}b}(t,t') - \int_C V_{\bar{c}d}(t,t_1)D_{d\bar{e}}(t_1,t_2)\{R_{a\bar{c}}(t^+)D^{-1}_{e\bar{b}}(t_2,t')\}dt_1dt_2. \quad (5.50)$$

We are now in position to begin discussing approximation schemes.

5.2.1 Hubbard-I-Approximation

The Hubbard-I-approximation is the simplest approximation in the model and is given when omitting the second term in the self-energy operator, i.e. letting $\mathcal{S} = \mathbf{P}\mathbf{V}$. The resulting equation then becomes

$$\mathbf{G}(t,t') = \mathbf{g}(t,t') + \int_C \mathbf{g}(t,t_1)\mathbf{V}(t_1,t_2)\mathbf{G}(t_2,t')dt_1dt_2, \quad (5.51)$$

as was discussed in Chap. 3. This is a mean-field approximation and, as such, it neglects any type of fluctuation in the system. Hence, it is typically a good approximation under conditions where the influence of fluctuations is suppressed, e.g. under influence of strong external fields or at high temperatures.

Graphically we write this equation according to

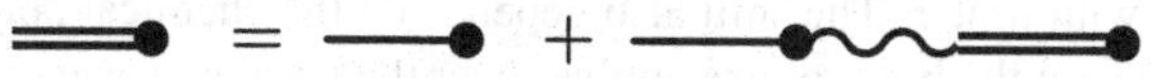

(5.51′)

where the single and double lines denote the bare and dressed locators d and D, respectively, whereas the dots mark the end-factors P, and the wiggles denote the interaction propagator V. This graphical representation explicitly illustrate the Green function being a product between the locator and the end-factor.

5.2.2 One-Loop-Approximation

The first step beyond the Hubbard-I-approximation is given by performing one functional differentiation in the self-energy operator. We, thus, have to consider the quantity $R_{a\bar{c}}(t^+)D^{-1}_{e\bar{b}}(t_2,t')$. In order to only perform one functional differentiation, we see that the inverse locator $\mathbf{D}^{-1}$ has to be approximated by $\mathbf{d}^{-1} - \mathbf{PV}$, i.e. letting $\mathcal{S} \approx \mathbf{PV} + \{\mathbf{RVD}\}[\mathbf{d}^{-1} - \mathbf{PV}]$, hence,

$$R_{a\bar{c}}(t^+)D^{-1}_{e\bar{b}}(t_2,t') \approx R_{a\bar{c}}(t^+)d^{-1}_{e\bar{b}}(t_2,t') - R_{a\bar{c}}(t^+)P_{e\bar{f}}(t_2^+)V_{\bar{f}b}(t_2,t'). \tag{5.52}$$

The inverse of the bare locator is given by

$$d^{-1}_{e\bar{b}}(t_2,t') = \delta(t_2-t')[\delta_{eb}(i\partial_{t_2} - \Delta_{\bar{e}}) - \varepsilon_b^{e\xi}U_\xi(t_2)], \tag{5.53}$$

hence, acting with the functional differentiation operator results in

$$\begin{aligned} R_{a\bar{c}}(t^+)d^{-1}_{e\bar{b}}(t_2,t') &= -i\delta(t_2-t')\varepsilon_\zeta^{ac}\varepsilon_b^{e\xi}\frac{\delta U_\xi(t_2)}{\delta U_\zeta(t^+)} \\ &= -i\delta(t_2-t')\delta(t-t_2)\delta_{\zeta\xi}\varepsilon_\zeta^{ac}\varepsilon_b^{e\xi}. \end{aligned} \tag{5.54}$$

The self-energy resulting from this contribution acquires the form

$$\Sigma^{(1a)}_{a\bar{b}}(t,t') = (-i)\delta(t-t')\int_C \varepsilon_\xi^{ac}V_{\bar{c}d}(t,t_1)D_{d\bar{e}}(t_1,t)\varepsilon_b^{e\xi}\,dt_1, \tag{5.55}$$

or

i  (5.55′)

where the bottom straight lines denote where the bubble should be connected to the incoming and outgoing propagators. This contribution arise due to electrons residing in different states in the central region and interacting with one another through the conduction channel. The appearance of this self-energy is due to proper many-body interactions between electrons in the central region which has a tendency to increase the energy separation between the singly and doubly occupied states. The real part of this contributions provides a non-negligible shift of the local levels which, in general, varies with time t. The shift also depends on the chemical potentials in the leads and, hence, on the bias voltage applied over the system, as we shall see in the next section.

Here we can, however, perform a general calculation of the self-energy by the following observation. The time variable t_1 is integrated along the contour C which we divide into the two pieces $C^< = (-\infty, t)$ and $C^> = (t, -\infty)$ on which

$t_1 < t$ and $t_1 > t$, respectively. Assuming time-independent conditions we have e.g. $V^<_{\bar{c}d}(t,t') = i\sum_{\mathbf{k}\sigma} v^*_{\mathbf{k}\sigma c} v_{\mathbf{k}\sigma d} f(\varepsilon_{\mathbf{k}\sigma}) \exp[-i\varepsilon_{\mathbf{k}\sigma}(t-t')]$, which leads to

$$\begin{aligned}
&\Sigma^{(1a)}_{a\bar{b}}(t,t') \\
&\quad = (-i)\delta(t-t')\varepsilon^{ac}_{\xi} \sum_{\mathbf{k}\sigma} v^*_{\mathbf{k}\sigma c} v_{\mathbf{k}\sigma d} \Bigg[\int \frac{f(-\varepsilon_{\mathbf{k}\sigma}) D^<_{d\bar{e}}(\omega) + f(\varepsilon_{\mathbf{k}\sigma}) D^>_{d\bar{e}}(\omega)}{\omega - \varepsilon_{\mathbf{k}\sigma}} \frac{d\omega}{2\pi} \\
&\quad - \frac{i}{2} \{ f(-\varepsilon_{\mathbf{k}\sigma}) D^<_{d\bar{e}}(\varepsilon_{\mathbf{k}\sigma}) + f(\varepsilon_{\mathbf{k}\sigma}) D^>_{d\bar{e}}(\varepsilon_{\mathbf{k}\sigma}) \} \Bigg] \varepsilon^{e\xi}_{b}.
\end{aligned} \tag{5.56}$$

Approximating the locator by its bare counterpart, e.g. $D^<_{d\bar{e}}(\omega) \approx d^<_{d\bar{e}}(\omega)$, where $d^<_{d\bar{e}}(\omega) = i2\pi\delta_{de}\delta(\omega - \Delta_{\bar{d}})$, we obtain

$$\begin{aligned}
\Sigma^{(1a)}_{a\bar{b}}(t,t') = \delta(t-t')\varepsilon^{ac}_{\xi} \sum_{\mathbf{k}\sigma} v^*_{\mathbf{k}\sigma c} v_{\mathbf{k}\sigma d} \Bigg(&\frac{1-2f(\varepsilon_{\mathbf{k}\sigma})}{\Delta_{\bar{d}} - \varepsilon_{\mathbf{k}\sigma}} \\
&- i\pi[1-2f(\Delta_{\bar{d}})]\delta(\varepsilon_{\mathbf{k}\sigma} - \Delta_{\bar{d}}) \Bigg) \varepsilon^{d\xi}_{b}.
\end{aligned} \tag{5.57}$$

The same result is obtained using frequency summation methods [9]. Note that this self-energy is purely real, which means that is only provides a shift to the transition energy $\Delta_{\bar{a}}$ in the Green function $G_{a\bar{b}}$. We thus write the dressed transition energy as $\Delta^*_{\bar{a}} = \Delta_{\bar{a}} + \delta\Delta_{\bar{a}}$, where we identify the shift $\delta\Delta_{\bar{a}} = \Sigma^{(1a)}_{a\bar{b}}$.

We can proceed further by assuming low temperatures and large band-widths W in the leads. Then, by replacing the $\mathbf{k}$-summation with an integration weighted by the density of state, we obtain

$$\delta\Delta_{\bar{a}} \approx \frac{1}{\pi} \varepsilon^{ac}_{\xi} \sum_{\chi} \Gamma^{\chi}_{cd} \ln \frac{|\mu_\chi - \Delta_{\bar{d}}|}{W} \varepsilon^{d\xi}_{b}. \tag{5.58}$$

This shift to the transition energy is small unless the transition energy $\Delta_{\bar{d}}$ lies in the proximity of the chemical potential μ_χ. Hence, for varying bias voltages the distance $|\mu_\chi - \Delta_{\bar{d}}|$ may very well become small which leads to a significant shift of the transition energy. Effectively, the conduction electrons screens the localized electrons which results in that the dressed transition energy is pushed away from being in resonance with the chemical potentials of the leads. The apparent divergence of the shift is non-physical and should be viewed as an artifact of the approximation. Recall that we replaced the locator by its bare counterpart. It is this replacement which leads to the undesired divergence. The kinematic shift $\delta\Delta_{\bar{a}}$ can, therefore, not become arbitrarily large and there is a cross-over value of the chemical potential μ_χ where the dressed transition energy goes from being below to above μ_χ, for decreasing values of μ_χ.

In the following discussion we will let the kinematic shift to the transition energies be absorbed in the transition energy itself, i.e. letting $\Delta_{\bar{a}} \to \Delta_{\bar{a}} + \delta\Delta_{\bar{a}}$. Since the kinematic shift depends on the bias voltage, it will be become time-dependent for time-varying biases. Although the shift is important under many stationary conditions, which will be discussed in detail later, it is only under time-varying external conditions we need to be slightly more explicit about the presence of the kinematic shift. Hence, we can safely let the shift be part of the definition of the transition energy.

The second term in (5.52) contains the factor $R_{a\bar{c}}(t^+)P_{e\bar{f}}(t_2^+)$ which is calculated through

$$\begin{aligned}
R_{a\bar{c}}(t^+)P_{e\bar{f}}(t_2^+) &\equiv i\varepsilon_\xi^{ac}\frac{\delta}{\delta U_\xi(t^+)}\frac{\langle \mathrm{T}S\{X^e, X^{\bar{f}}\}(t_2^+)\rangle}{\langle \mathrm{T}S\rangle} \\
&= i\varepsilon_\xi^{ac}\big((-i)\langle \mathrm{T}SZ^\xi(t^+)\{X^e, X^{\bar{f}}\}(t_2^+)\rangle_U \\
&\quad - (-i)\varepsilon_\xi^{ac}\langle \mathrm{T}SZ^\xi(t^+)\rangle_U\langle \mathrm{T}S\{X^e, X^{\bar{f}}\}(t_2^+)\rangle_U\big) \\
&= i\varepsilon_\xi^{ac}\varepsilon_\zeta^{ef}(-i)\langle \mathrm{T}Z^\xi(t^+)Z^{\bar{\zeta}}(t_2^+)\rangle_U - P_{a\bar{c}}(t^+)P_{e\bar{f}}(t_2^+) \\
&= i\varepsilon_\xi^{ac}\varepsilon_\zeta^{ef}K_{\xi\bar{\zeta}}(t^+, t_2^+) - P_{a\bar{c}}(t^+)P_{e\bar{f}}(t_2^+),
\end{aligned} \tag{5.59}$$

where we have introduced the propagator $K_{\xi\bar{\zeta}}(t,t') = (-i)\langle \mathrm{T}Z^\xi(t)Z^{\bar{\zeta}}(t')\rangle_U$ for the Bose-like transitions Z^ξ and Z^ζ, cf. definitions in (3.44). The resulting self-energy from this contribution is

$$\begin{aligned}
\Sigma_{a\bar{b}}^{(1b)}(t,t') &= \int_C V_{\bar{c}d}(t,t_1)D_{d\bar{e}}(t_1,t_2)\big[i\varepsilon_\xi^{ac}\varepsilon_\zeta^{ef}K_{\xi\bar{\zeta}}(t,t_2) \\
&\quad - P_{a\bar{c}}(t)P_{e\bar{f}}(t_2)\big]V_{\bar{f}b}(t_2,t')dt_2dt_1 \\
&= \int_C \big[V_{\bar{c}d}(t,t_1)D_{d\bar{e}}(t_1,t_2)i\varepsilon_\xi^{ac}\varepsilon_\zeta^{ef}K_{\xi\bar{\zeta}}(t,t_2) \\
&\quad - P_{a\bar{c}}(t)V_{\bar{c}d}(t,t_1)G_{d\bar{f}}(t_1,t_2)\big]V_{\bar{f}b}(t_2,t')dt_2dt_1
\end{aligned} \tag{5.60}$$

since removing the superscript "+"-signs does not change the value of the integral. Graphically this self-energy can be written

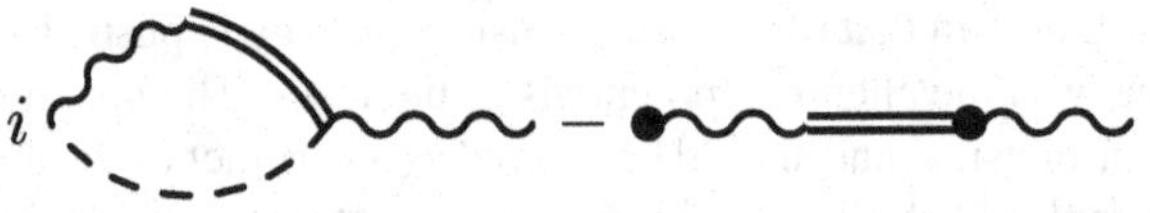

(5.60′)

where the dashed line signifies K. The labels have been omitted to the benefit of the graphical appearance of the diagrams, and the reader is encouraged to insert the labels.

The two self-energies we derived for in the one-loop-approximation are generated by effecting the functional differentiation to the locator. However, when examining the expression in (5.46), one realizes that the functional derivative actually should act on the Green function **G** rather than on only the locator **D**. The Green function $\mathbf{G} = \mathbf{DP}$ and therefore the variation $\delta\mathbf{G} = (\delta\mathbf{D})\mathbf{P} + \mathbf{D}(\delta\mathbf{P})$. Hence, in order to obtain the full equation of motion for the Green function in the one-loop-approximation, we have to insert the expression given in (5.59) into the equation. Effectively, this amounts to adding the contribution

$$\delta P_{a\bar{b}}(t,t') = \int_C V_{\bar{c}d}(t,t_1) D_{d\bar{e}}(t_1,t')\Big(i\varepsilon_{\xi}^{ac} K_{\xi\bar{\zeta}}(t_1,t')\varepsilon_{\zeta}^{eb} - P_{c\bar{d}}(t^+)P_{f\bar{b}}(t')\Big)dt_1 \tag{5.61}$$

and graphically

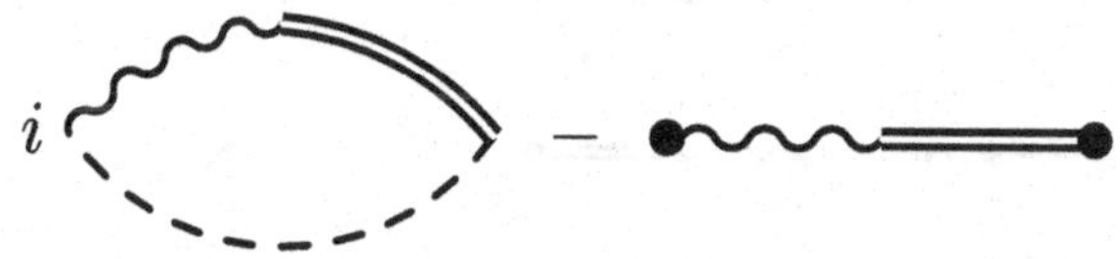

(5.61′)

Examining this result, in the equation of motion we find the term

$$-\int_C g_{a\bar{d}}(t,t_1) V_{\bar{d}e}(t_1,t_2) G_{e\bar{b}}(t_2,t') dt_2 dt_1 \tag{5.62}$$

which exactly cancels the self-energy provided by the first contribution to the diagrammatic expansion, i.e. the result given in the Hubbard-I-approximation. Hence, the mean-field contribution to the Green function is replaced by the term

$$i\int_C d_{a\bar{c}}(t,t_1)\varepsilon_{\xi}^{cd} V_{\bar{d}e}(t_1,t_2) D_{e\bar{f}}(t_2,t') K_{\xi\bar{\zeta}}(t_1,t')\varepsilon_{\zeta}^{fb} dt_1 t_2. \tag{5.63}$$

In this expression the propagator $K_{\xi\bar{\zeta}}(t_1,t')$ replaces the end-factors on the locators. The propagator $K_{\xi\bar{\zeta}}(t_1,t')$ provides a correlation between the occupation numbers calculated at different times rather than only the product between them. This generalization is important under conditions when we cannot consider the occupation numbers at the different times as being independent of one another, i.e. whenever there is a *memory* in the system one has to calculate the correlation function. On the other hand, when the memory can be neglected, e.g. stationary conditions, we can employ the Markovian approximation which results in the decoupling of the propagator $i\varepsilon_{\xi}^{cd}\varepsilon_{\zeta}^{fb}K_{\xi\bar{\zeta}}(t_1,t')$ into simply $P_{c\bar{d}}(t_1)P_{f\bar{b}}(t')$ from which we recover the mean-field contribution to the equation. This is the same as to say that the end-factors at different times are independent of one another.

The one-loop equation of motion for the Green function $G_{a\bar{b}}$ is given by

$$\begin{aligned} G_{a\bar{b}}(t,t') = g_{a\bar{b}}(t,t') &+ i\int_C d_{a\bar{c}}(t,t_1)\varepsilon_{\xi}^{cd} V_{\bar{d}e}(t_1,t_2) D_{e\bar{f}}(t_2,t') K_{\xi\bar{\zeta}}(t_1,t')\varepsilon_{\zeta}^{fb} dt_2 dt_1 \\ &+ i\int_C d_{a\bar{c}}(t,t_1)\varepsilon_{\xi}^{ce} V_{\bar{e}f}(t_1,t_2) D_{f\bar{h}}(t_2,t_3) K_{\xi\bar{\zeta}}(t_1,t_3)\varepsilon_{\zeta}^{hg} \\ &\times V_{\bar{g}d}(t_3,t_4) G_{d\bar{b}}(t_4,t') dt_4 dt_3 dt_2 dt_1 \end{aligned}$$

$$
- \int_C g_{a\bar{c}}(t,t_1) V_{\bar{c}e}(t_1,t_2) G_{e\bar{f}}(t_2,t_3) V_{\bar{f}d}(t_3,t_4) \times G_{d\bar{b}}(t_4,t') dt_4 dt_3 dt_2 dt_1, \tag{5.64}
$$

that is,

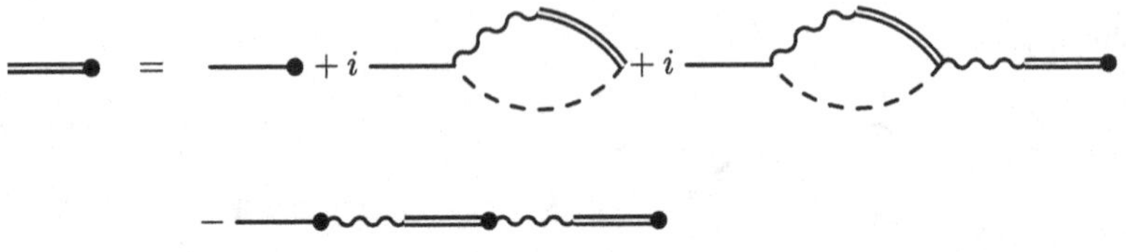

(5.64′)

Here, the first and second diagrams on the right hand side provides the bare Green function, however, with a dressed end-factor. Here we also see quite clearly that the approximation $K \to -iPP$ turns the second diagram into the mean-field diagram as shown in the Hubbard-I-approximation, cf. (5.51′). Simultaneously, in this approximation the third diagram equals the fourth which leads to that the two diagrams cancel exactly. This is expected since we cannot allow the mean-field theory to be described by different sets of diagrams. It is noticeable, however, that (5.64) already describes the physics of the quantum dot to the fourth order in the tunneling rate $v_{\mathbf{k}\sigma}$, which illustrates the power, or, efficiency of the approach using functional derivatives.

We can obtain a more general Dyson equation by making use of the diagrams that cancel each other. First we define the dressed end-factor

(5.65)

Then, we can rewrite the (5.64′) according to

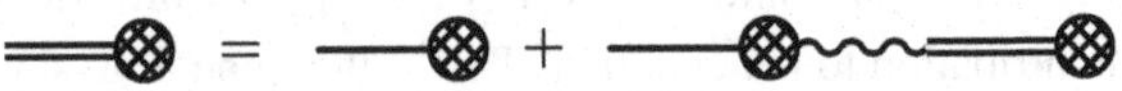

(5.66)

This compact notation suggests that we have considered the end-factor to its full extent in the sense that no higher order contributions will be added to the total end-factor $\mathbb{P} = P + \delta P$. The end-factor depends, however, on the Bose-like propagator $K_{\xi\bar{\zeta}}$ and the complexity of the processes involved into this is arbitrary. Below, we study the first order appearance of this propagator.

Up to now we have not discussed anything about the Bose-like propagator $K_{\xi\bar{\zeta}}$, and how it should be treated. However, the equation of motion for the operator Z^{ξ}

appears to be

$$(i\partial_t - \Delta_{\bar{\xi}})Z^{\xi} = \sum_{\mathbf{k}\sigma} \left(v_{\mathbf{k}\sigma a} c^{\dagger}_{\mathbf{k}\sigma} [Z^{\xi}, X^{a}] + v^{*}_{\mathbf{k}\sigma a} [Z^{\xi}, X^{\bar{a}}] c_{\mathbf{k}\sigma} \right)$$

$$= \sum_{\mathbf{k}\sigma} \varepsilon^{\xi a}_{b} \left(v_{\mathbf{k}\sigma b} c^{\dagger}_{\mathbf{k}\sigma} X^{b} + v^{*}_{\mathbf{k}\sigma b} X^{\bar{b}} c_{\mathbf{k}\sigma} \right). \tag{5.67}$$

Notice that we are using the commutator, instead of the anti-commutator, between the Z- and X-operators. For Bose-like operators one should use commutator, which is motivated by e.g. $[h^p, h^q] = \delta_{pq}$. On the other hand, since the X- and Z-operators are neither Fermi- nor Bose-operators we may employ anyone (of [,] and {, }) being most convenient. In this book we have chosen the commutator between Bose- and Fermi-like operators, e.g. $[X, Z]$ and $[c, Z]$. Equation (5.67) can be compared with the equations for the occupation numbers discussed in Chap. 2. Here we are interested in the equation for the propagator $K_{\xi\bar{\zeta}}$, however, which becomes

$$(i\partial_t - \Delta_{\bar{\xi}})K_{\xi\bar{\zeta}}(t,t') = \delta(t-t')Q_{\xi\bar{\zeta}} + \sum_{\mathbf{k}\sigma} \varepsilon^{\xi a}_{b} \left(v_{\mathbf{k}\sigma a}(-i)\langle \mathrm{T}(c^{\dagger}_{\mathbf{k}\sigma} X^{b})(t) Z^{\bar{\xi}}(t')\rangle_U \right.$$
$$\left. + v^{*}_{\mathbf{k}\sigma a}(-i)\langle \mathrm{T}(X^{\bar{b}} c_{\mathbf{k}\sigma})(t) Z^{\bar{\xi}}(t')\rangle_U \right). \tag{5.68}$$

In order to be consistent within the approach we have developed for the Green functions G, we rewrite this equation by again using the relation in (3.31). We need to consider first, however, which Green function we would like to operate on with the functional differentiation. We may, of course, rewrite the second propagator in the summand according to

$$(-i)\langle \mathrm{T}(X^{\bar{b}} c_{\mathbf{k}\sigma})(t) Z^{\bar{\xi}}(t')\rangle_U = \left(\langle \mathrm{T}X^{\bar{b}}(t)\rangle_U + i\frac{\delta}{\delta U_{\bar{b}}(t)} \right)(-i)\langle \mathrm{T}c_{\mathbf{k}\sigma}(t) Z^{\bar{\xi}}(t')\rangle_U . \tag{5.69}$$

Then, on the other hand, we encounter some problems, since the Green function $(-i)\langle \mathrm{T}c_{\mathbf{k}\sigma}(t) Z^{\bar{\xi}}(t')\rangle_U$ cannot be rewritten in a form that closes neither of the equations for K and G. Moreover, the averages $\langle \mathrm{T}X^{\bar{b}}(t)\rangle_U$ would describe the probability amplitude for an electron being in the transition $\bar{b}$, which is not well defined. Therefore, it does not make any sense to calculate this average. Thus, we have to employ some other strategy. In (5.43a), (5.43b), we have already worked out how we should treat propagators of the kind being present in the equation for K, let be that order of the operators is nor the same here. Utilizing that the c- and X-operators anti-commute, we can write

$$(-i)\langle \mathrm{T}(X^{\bar{b}} c_{\mathbf{k}\sigma})(t) Z^{\bar{\xi}}(t')\rangle_U = -\left(\langle \mathrm{T}Z^{\bar{\xi}}(t')\rangle_U + i\frac{\delta}{\delta U_{\bar{\zeta}}(t')} \right) F_{\mathbf{k}\sigma\bar{b}}(t,t^{-}). \tag{5.70}$$

We know that $F_{\mathbf{k}\sigma\bar{b}}(t,t^{-})$ can be rewritten according to $v_{\mathbf{k}\sigma a} g_{\mathbf{k}\sigma} G_{a\bar{b}}$, hence, the equation for G can be closed by taking this approach. Furthermore, we do not fall into any weird discussions on whether we can calculate the averages $\langle \mathrm{T}X^{a}\rangle_U$, we just do not have to consider them at all. As a result, we obtain the equation for K according to

$$(i\partial_t - \Delta_{\bar{\xi}})K_{\xi\bar{\zeta}}(t,t') = \delta(t-t')Q_{\xi\bar{\zeta}} - \left(\langle \mathrm{T}Z^{\bar{\zeta}}(t')\rangle_U + \frac{\delta}{\delta U_{\bar{\zeta}}(t')}\right) \times \sum_{\mathbf{k}\sigma a} \varepsilon_b^{\xi a} \int_C \left(G_{b\bar{c}}(t,t'')V_{\bar{c}a}(t'',t^-) + V_{\bar{a}c}(t,t'')G_{c\bar{b}}(t'',t^-)\right). \tag{5.71}$$

Here, we do not discard the first term in the summand since it constitutes a sensible sequence of transitions even in the case of normal metallic leads.

The simplest approximation, beyond the atomic limit, is given by the Hubbard-I-approximation, corresponding to omitting the functional differentiation, for which we obtain

$$(i\partial_t - \Delta_{\bar{\xi}})K_{\xi\bar{\zeta}}(t,t') = \delta(t-t')Q_{\xi\bar{\zeta}} - \langle \mathrm{T}Z^{\bar{\zeta}}(t')\rangle_U \sum_{\mathbf{k}\sigma a} \varepsilon_b^{\xi a} \int_C \left(G_{b\bar{c}}(t,t'')V_{\bar{c}a}(t'',t^-) + V_{\bar{a}c}(t,t'')G_{c\bar{b}}(t'',t^-)\right). \tag{5.72}$$

Graphically, this equation is written as

(5.72′)

Here, the dotted lines have been added to remind that the last dots are connected to the corresponding diagrams. Note that the two last diagrams account for different processes that couple to the Bose-like transition Z^{ξ}.

Finally, whenever the right hand side is zero, or approximately zero, K is a constant of motion. This *is* the same as to say that the operators Z^{ξ} and $Z^{\bar{\zeta}}$ are uncorrelated in time, i.e. independent of one another. This, therefore, motivates us to replace the Green function $K(t,t')$ by the product $-iP(t)P(t')$.

We have now seen how we should act with the functional derivatives appearing in the equation of motion. On the other hand, the discussion so far has been made on an abstract system and it is therefore desirable to now turn our heads to something more concrete and hands on. Hence, we will not go any further into the details of the diagrammatic expansion in general terms. Instead, we will being analyzing some of the effects from the various contributions derived thus far in a simple example, while more examples are provides in Chap. 6.

5.3 Single-Level Quantum Dot

We again look at the single-level quantum dot, since it serves as a simple example from where we can pick up some intuition, and here we will learn more about the diagrammatic technique itself. We will repeat some of the technical issues already discussed in the previous sections, in order to clarify some details that might have been hidden while making the theory as general as possible. Thus, here we take the Coulomb island model, i.e. $\mathcal{H} = \mathcal{H}_L + \mathcal{H}_R + \sum_p E_p h^p + \sum_{\mathbf{k}\sigma}(v_{\mathbf{k}\sigma} c^\dagger_{\mathbf{k}\sigma}[X^{0\sigma} + \sigma X^{\bar{\sigma}}_2] + H.c.)$, and begin our analysis in the usual manner, that is, through the equation of motion. As we now by know, this can be written as

$$\begin{aligned}&\big(i\partial_t - \Delta^0_{\sigma 0} - \Delta U_{\sigma 0}(t)\big)G_{0\sigma\bar{a}}(t,t') - U_{\sigma\bar{\sigma}}(t)G_{0\bar{\sigma}\bar{a}}(t,t')\\ &\quad = \delta(t-t')P_{0\sigma\bar{a}}(t) + \big(P_{0\sigma\bar{b}}(t^+) + R_{0\sigma\bar{b}}(t^+)\big)\int_C V_{\bar{b}c}(t,t_1)G_{c\bar{a}}(t_1,t')dt_1,\end{aligned} \tag{5.73a}$$

$$\begin{aligned}&\big(i\partial_t - \Delta^0_{2\bar{\sigma}} - \Delta U_{2\bar{\sigma}}(t)\big)G_{\bar{\sigma}2\bar{a}}(t,t') + U_{\sigma\bar{\sigma}}(t)G_{\sigma 2\bar{a}}(t,t')\\ &\quad = \delta(t-t')P_{\bar{\sigma}2\bar{a}}(t) + \big(P_{\bar{\sigma}2\bar{b}}(t^+) + R_{\bar{\sigma}2\bar{b}}(t^+)\big)\int_C V_{\bar{b}c}(t,t_1)G_{c\bar{a}}(t_1,t')dt_1.\end{aligned} \tag{5.73b}$$

The corresponding matrix equation then becomes

$$\begin{aligned}&\big(i\partial_t - \Delta^0 - \mathbf{U}(t)\big)\mathbf{G}(t,t')\\ &\quad = \delta(t-t')\mathbf{P}(t) + \big(\mathbf{P}(t^+) + \mathbf{R}(t^+)\big)\int_C \mathbf{V}(t,t_1)\mathbf{G}(t_1,t')dt_1,\end{aligned} \tag{5.74}$$

where $\Delta^0 = \text{diag}\{\Delta^0_{\uparrow 0}, \Delta^0_{\downarrow 0}, \Delta^0_{2\downarrow}, \Delta^0_{2\uparrow}\}$ contains the bare transition energies $\Delta^0_{\sigma 0} = E_\sigma - E_0$ and $\Delta^0_{2\sigma} = E_2 - E_\sigma$. The interaction matrix $\mathbf{V}$ was introduced in Chap. 4. The source fields are contained in the matrix $\mathbf{U} = \text{diag}\{\mathbf{U}_1, \mathbf{U}_2\}$, where $\mathbf{U}_i$, $i = 1, 2$, are 2×2-matrices defined by

$$\mathbf{U}_1 = \begin{pmatrix} \Delta U_{\uparrow 0} & U_{\uparrow\downarrow} \\ U_{\downarrow\uparrow} & \Delta U_{\downarrow 0}\end{pmatrix}, \qquad \mathbf{U}_2 = \begin{pmatrix} \Delta U_{2\downarrow} & -U_{\uparrow\downarrow} \\ -U_{\downarrow\uparrow} & \Delta U_{2\uparrow}\end{pmatrix}. \tag{5.75}$$

In this model the end-factor $\mathbf{P} = \text{diag}\{P_{0\uparrow}, P_{0\downarrow}, P_{\downarrow 2}, P_{\uparrow 2}\}$, where $P_{pq} \equiv P_{pqqp} = P_{pq\overline{pq}}$, since there is no term in the Hamiltonian that supports direct spin-flip transitions. There might be higher order processes, however, that involve spin-flip scattering, but those will not cause the off-diagonal end-factors like e.g. $P_{0\uparrow\downarrow 0}$ to become finite.

The functional differentiation matrix operator $\mathbf{R} = \text{diag}\{\mathbf{R}_1, \mathbf{R}_2\}$ is defined by

$$\mathbf{R}_1 = \begin{pmatrix} R_{0\uparrow} & R_{0\uparrow\downarrow 0} \\ R_{0\downarrow\uparrow 0} & R_{0\downarrow}\end{pmatrix}, \qquad \mathbf{R}_2 = \begin{pmatrix} R_{\downarrow 2} & R_{\downarrow 22\uparrow} \\ R_{\uparrow 22\downarrow} & R_{2\uparrow}\end{pmatrix}, \tag{5.76}$$

with the elements

$$R_{pqq'p'}(t) = i\left(\delta_{qq'}\frac{\delta}{\delta U_{pp'}(t)} + \delta_{pp'}\frac{\delta}{\delta U_{q'q}(t)}\right), \tag{5.77}$$

where we also use the convention $R_{pq} \equiv R_{pqqp}$.

Although the matrix representation provides a very nice appearance of the equation of motion we cannot perform the diagrammatic expansion in terms of matrix algebra, since the matrices $\mathbf{R}$ and $\mathbf{V}$ do not commute. Hence, we need to study the components one by one. The second term in the integrand of (5.73a) is then rewritten as (omitting the integrals for simplicity)

$$\begin{aligned} &R_{0\sigma\bar{b}}(t^+)V_{\bar{b}c}(t,t_1)G_{c\bar{a}}(t_1,t') \\ &= -V_{\bar{b}c}(t,t_1)D_{c\bar{d}}(t_1,t_2)\{R_{0\sigma\bar{b}}(t^+)D^{-1}_{\bar{d}\bar{e}}(t_2,t_3)\}G_{e\bar{a}}(t_3,t') \\ &+ V_{\bar{b}c}(t,t_1)D_{c\bar{d}}(t_1,t')\{R_{0\sigma\bar{b}}(t^+)P_{d\bar{a}}(t')\}. \end{aligned} \tag{5.78}$$

First, we replace the dressed inverse locator $D^{-1}_{\bar{d}\bar{e}}$ by $d^{-1}_{\bar{d}\bar{e}} - P_{\bar{d}\bar{f}}V_{\bar{f}e}$ in order or obtain the one-loop-approximation, and we use that $R_{a\bar{b}}(t^+)\mathbf{d}^{-1}(t_2,t_3) = -\delta(t_2-t_3)\delta(t^+-t_3)R_{a\bar{b}}\mathbf{U}$. It is then straight forward to see that

$$R_{0\sigma\sigma'0}\mathbf{U} = -i\begin{pmatrix} \delta_{\sigma\sigma'} - \delta_{\sigma\uparrow}\delta_{\sigma'\uparrow} & -\delta_{\sigma\downarrow}\delta_{\sigma'\uparrow} & 0 & 0 \\ -\delta_{\sigma\uparrow}\delta_{\sigma'\downarrow} & \delta_{\sigma\sigma'} - \delta_{\sigma\downarrow}\delta_{\sigma'\downarrow} & 0 & 0 \\ 0 & 0 & \delta_{\sigma\downarrow}\delta_{\sigma'\downarrow} & \delta_{\sigma\downarrow}\delta_{\sigma'\uparrow} \\ 0 & 0 & \delta_{\sigma\uparrow}\delta_{\sigma'\downarrow} & \delta_{\sigma\uparrow}\delta_{\sigma'\uparrow} \end{pmatrix}, \tag{5.79a}$$

$$R_{\sigma22\sigma'}\mathbf{U} = i\begin{pmatrix} \delta_{\sigma\uparrow}\delta_{\sigma'\uparrow} & \delta_{\sigma\uparrow}\delta_{\sigma'\downarrow} & 0 & 0 \\ \delta_{\sigma\downarrow}\delta_{\sigma'\uparrow} & \delta_{\sigma\downarrow}\delta_{\sigma'\downarrow} & 0 & 0 \\ 0 & 0 & \delta_{\sigma\sigma'} - \delta_{\sigma\downarrow}\delta_{\sigma'\downarrow} & -\delta_{\sigma\uparrow}\delta_{\sigma'\downarrow} \\ 0 & 0 & -\delta_{\sigma\downarrow}\delta_{\sigma'\uparrow} & \delta_{\sigma\sigma'} - \delta_{\sigma\uparrow}\delta_{\sigma'\uparrow} \end{pmatrix}. \tag{5.79b}$$

This is an efficient way of keeping book of our calculations. We also see that writing out the first (last) two diagonal elements in (5.79a (b)) is not superfluous, since those elements are not always zero, e.g. $R_{0\downarrow}\mathbf{U} = -i\,\mathrm{diag}\{1\ 0\ 1\ 0\}$.

The second term in the integrand of (5.73a) can now be calculated, using (5.79a), according to

$$\begin{aligned} &R_{0\sigma\bar{b}}(t^+)V_{\bar{b}c}(t,t_1)G_{c\sigma0}(t_1,t') \\ &= i\,V_{\bar{\sigma}0c}(t,t_1)D_{c\bar{\sigma}0}(t_1,t)G_{\sigma0}(t,t') \\ &\quad + V_{\sigma0c}(t,t_1)D_{c\sigma0}(t_1,t')R_{0\sigma}(t^+)P_{0\sigma}(t') \\ &\quad + i\,V_{\bar{\sigma}0c}(t,t_1)D_{c2\sigma}(t_1,t)G_{\bar{\sigma}2\sigma0}(t,t') \\ &\quad + V_{\bar{\sigma}0c}(t,t_1)D_{c\bar{\sigma}0}(t_1,t')R_{0\sigma\bar{\sigma}0}(t^+)P_{0\bar{\sigma}\sigma0}(t'). \end{aligned} \tag{5.80}$$

Here, we recall that the self-energy $\delta\Delta_{\sigma 0}(t) = iV_{\bar{\sigma}0\bar{c}}(t,t_1)D_{c\bar{\sigma}0}(t_1,t)$ is real and, thus, only provides a shift to the position of the pole of the Green function. As this shift depends on one time only (remember the integration over t_1), we can simply move this term to the left hand side of the equation. We do the same with the shift $\delta\Delta_{2\bar{\sigma}}(t) = iV_{\bar{\sigma}0\bar{c}}(t,t_1)D_{c2\sigma}(t_1,t)$. Summing up for all elements of the matrix equation, we can write our effective equation of motion in the one-loop-approximation according to (letting the external sources $U_\xi \to 0$)

$$\big(i\partial_t - \Delta(t)\big)\mathbf{G}(t,t') = \mathbb{P}(t,t') + \int_C \mathbb{P}(t,t_1)\mathbf{V}(t_1,t_2)\mathbf{G}(t_2,t')dt_1dt_2, \tag{5.81}$$

where $\Delta(t) = \Delta^0 + \delta\Delta(t)$, and

$$\delta\Delta = \begin{pmatrix} \delta\Delta_{10} & \delta\Delta_{21} \\ \delta\Delta_{10} & \delta\Delta_{21} \end{pmatrix}, \tag{5.82}$$

with $\delta\Delta_{10} = \mathrm{diag}\{\delta\Delta_{\uparrow 0}\ \delta\Delta_{\downarrow 0}\}$, and $\delta\Delta_{21} = \mathrm{diag}\{\delta\Delta_{2\downarrow}\ \delta\Delta_{2\uparrow}\}$, whereas the matrix $\mathbb{P}(t,t') = \delta(t-t')\mathbf{P}(t') + \delta\mathbf{P}$, with $\delta\mathbf{P}(t,t')$ defined through

$$\delta P_{a\bar{b}}(t,t') = \int_C V_{\bar{c}d}(t,t_1)D_{d\bar{e}}(t_1,t')R_{a\bar{c}}(t)P_{e\bar{b}}(t'). \tag{5.83}$$

5.3.1 *Excluding Fluctuations in* P

First, we omit the correction $\delta\mathbf{P}$ to the end-factor and study the resulting expression for the Green function. Since the bare end-factor is diagonal, it is possible to solve (5.81) analytically by taking advantage of its block-diagonal structure, at least in the time-independent domain. Therefore, by Fourier transforming the equation we find [9]

$$G_{0\sigma}(i\omega) = \frac{i\omega - \Delta_{2\bar{\sigma}} - P_{\bar{\sigma}2}V_\sigma(i\omega)}{H_\sigma(i\omega)}P_{0\sigma}, \tag{5.84a}$$

$$G_{\bar{\sigma}2}(i\omega) = \frac{i\omega - \Delta_{\sigma 0} - P_{0\sigma}V_\sigma(i\omega)}{H_\sigma(i\omega)}P_{\bar{\sigma}2}, \tag{5.84b}$$

$$G_{\bar{\sigma}2\sigma 0}(i\omega) = \frac{\sigma P_{\bar{\sigma}2}V_\sigma(i\omega) + \bar{\sigma}\delta\Delta_{\sigma 0}}{i\omega - \Delta_{2\bar{\sigma}} - P_{\bar{\sigma}2}V_\sigma(i\omega)}G_{0\sigma}(i\omega), \tag{5.84c}$$

$$G_{\sigma 0\bar{\sigma}2}(i\omega) = \frac{\sigma P_{0\sigma}V_\sigma(i\omega) + \bar{\sigma}\delta\Delta_{2\bar{\sigma}}}{i\omega - \Delta_{\sigma 0} - P_{0\sigma}V_\sigma(i\omega)}G_{\bar{\sigma}2}(i\omega), \tag{5.84d}$$

where $H_\sigma(i\omega) = [i\omega - \Delta^0_{\sigma 0} - V_\sigma(i\omega)][i\omega - \Delta^0_{2\bar{\sigma}} - \delta\Delta_{\sigma 0}] - U[P_{\bar{\sigma}2}V_\sigma(i\omega) - \delta\Delta_{\sigma 0}]$. Recalling that $d_\sigma = X^{0\sigma} + \sigma X^{\bar{\sigma}2}$, we find for the single-electron Green function

$G_\sigma(t,t') = (-i)\langle \mathrm{T} d_\sigma(t) d_\sigma^\dagger(t')\rangle$, the solution

$$G_\sigma(i\omega) = G_{0\sigma}(i\omega) + \sigma[G_{0\sigma 2\bar{\sigma}}(i\omega) + G_{\bar{\sigma}2\sigma 0}(i\omega)] + G_{\bar{\sigma}2}(i\omega)$$
$$= \frac{i\omega - \Delta^0_{\sigma 0} - \delta\Delta_{\sigma 0} - \delta\Delta_{2\bar{\sigma}} - UP_{0\sigma}}{[i\omega - \Delta^0_{\sigma 0} - V_\sigma(i\omega)][i\omega - \Delta^0_{2\bar{\sigma}} - \delta\Delta_{\sigma 0}] - U[P_{\bar{\sigma}2}V_\sigma(i\omega) - \delta\Delta_{\sigma 0}]}. \tag{5.85}$$

Here, we have been using that $P_{0\sigma} + P_{\bar{\sigma}2} = 1$, $\sigma^2 = 1$, and that $\sigma\bar{\sigma} = -1$.

We see the resemblance with the Hubbard-I-approximation by letting $\delta\Delta_{\bar{a}} \to 0$, and recalling that $P_{0\sigma} = 1 - \langle n_\sigma\rangle$ and $P_{\bar{\sigma}2} = \langle n_{\bar{\sigma}}\rangle$, cf. (5.29). More importantly, is that we from this approximation again capture the correct results in the non-interacting ($U \to 0$) and atomic limits ($v_{\mathbf{k}\sigma} \to 0$), of which only the former is non-trivial. However, for $U \to 0$, the denominator

$$H_\sigma(i\omega) = [i\omega - \Delta^0_{\sigma 0} - V_\sigma(i\omega)][i\omega - \Delta^0_{\sigma 0} - \delta\Delta_{\sigma 0} - \delta\Delta_{2\bar{\sigma}}] \tag{5.86}$$

from which we deduce

$$G_\sigma(i\omega) = \frac{i\omega - \Delta^0_{\sigma 0} - \delta\Delta_{\sigma 0} - \delta\Delta_{2\bar{\sigma}}}{[i\omega - \Delta^0_{\sigma 0} - V_\sigma(i\omega)][i\omega - \Delta^0_{\sigma 0} - \delta\Delta_{\sigma 0} - \delta\Delta_{2\bar{\sigma}}]}$$
$$= \frac{1}{i\omega - \Delta^0_{\sigma 0} - V_\sigma(i\omega)}, \tag{5.87}$$

as expected.

Here, it is justified to ask what happens with the corrections $\delta\Delta_{\bar{a}}$ in the non-interacting limit. We may have the attitude that is does not matter since these corrections do not enter the expressions for the Green functions, hence, we do not need to know anything about them. On the other hand, from a physical point of view it would be quite unsatisfactory if the corrections diverge in this limit, since it would suggest some violent fluctuations in the system which, nonetheless, would not make any contribution to the local properties of the quantum dot. Fortunately, the sum $\delta\Delta_{\sigma 0} + \delta\Delta_{2\bar{\sigma}} \to 0$ as $U \to 0$, as would be expected for weakly correlated electrons. Using (5.56), we find (recalling that the imaginary part vanishes) that

$$\delta\Delta_{\sigma 0} + \delta\Delta_{2\bar{\sigma}} \sim -\mathrm{Im}\,\frac{\Delta^0_{\bar{\sigma}0} - \Delta^0_{2\sigma}}{H^r_\sigma(\omega)} = \mathrm{Im}\,\frac{U}{H^r_\sigma(\omega)}, \tag{5.88}$$

which clearly approaches zero as the correlation U vanishes.

Finally, we approach the last limit, that is, $U \to \infty$. In the present case we cannot expect to reproduce the result from the Hubbard-I-approximation, since here we also have to handle the renormalization. The form of the final Green function should be similar as in the Hubbard-I-approximation, though, since the energy of the doubly occupied state tends to infinity whereas its corresponding population number $N_2 \to 0$. Indeed, dividing the Green function in (5.85) by U, the resulting numerator becomes

$$\frac{i\omega - \Delta^0_{\sigma 0} - \delta\Delta_{\sigma 0} - \delta\Delta_{2\bar{\sigma}}}{U} - P_{0\sigma}, \tag{5.89}$$

whereas the corresponding denominator equals

$$\frac{(i\omega - \Delta^0_{\sigma 0} - V_\sigma)(i\omega - \Delta^0_{2\bar{\sigma}} - \delta\Delta_{\sigma 0} - \delta\Delta_{2\bar{\sigma}})}{U} - P_{\bar{\sigma}2} V_\sigma - \delta\Delta_{\sigma 0}. \tag{5.90}$$

From (5.56) it follows that $\delta\Delta_{\bar{a}}$ is finite for all U. Moreover, since $\delta\Delta_{2\bar{\sigma}}/U \to 1$ as $U \to \infty$, the final result becomes

$$\lim_{U\to\infty} G_\sigma(i\omega) = \frac{P_{0\sigma}}{i\omega - \Delta^0_{\sigma 0} - \delta\Delta_{\sigma 0} - P_{0\sigma} V_\sigma} = \frac{P_{0\sigma}}{i\omega - \Delta_{\sigma 0} - P_{0\sigma} V_\sigma}. \tag{5.91}$$

From the last expression, we see that this result is reminiscent of the corresponding result in the Hubbard-I-approximation. Here, also the corrected transition energy is given by

$$\Delta_{\sigma 0} = \Delta^0_{\sigma 0} + \frac{1}{2\pi} \sum_{\mathbf{k}} |v_{\mathbf{k}\bar{\sigma}}|^2 \frac{2f(\varepsilon_{\mathbf{k}\bar{\sigma}}) - 1}{\varepsilon_{\mathbf{k}\bar{\sigma}} - \Delta_{\bar{\sigma}0}}, \tag{5.92}$$

where we have used the approximation of the locator introduced below (5.56). In this limit it also fairly simple to find an analytical expression for the population number $\langle n_\sigma \rangle$, since $N_2 \to 0$ leaves $\langle n_\sigma \rangle = P_{\sigma 2} \to N_\sigma$. Using $N_\sigma = \text{Im} \int (G^<_{0\sigma} - G^>_{\sigma 2}) d\omega/2\pi$ we then obtain, at zero temperature, the non-linear equation

$$\langle n_\sigma \rangle = \frac{P_{0\sigma}}{\Gamma_\sigma} \sum_{\chi=L,R} \Gamma^\chi_\sigma \left\{ \frac{1}{\pi} \arctan \frac{\mu_\chi - \Delta_{\sigma 0}}{P_{0\sigma}\Gamma_\sigma/2} + \frac{1}{2} \right\}. \tag{5.93}$$

Letting $\mu_\chi \to -\infty$, or $\Delta_{\sigma 0} \to \infty$, the right hand side of the equation vanishes giving $\langle n_\sigma \rangle = 0$, which is expected for an unoccupied quantum dot. In the opposite limit, i.e. $\mu_\chi \to \infty$, or $\Delta_{\sigma 0} \to -\infty$, the sum equals Γ_σ, hence, the resulting equation to solve is simply

$$N_\sigma = \langle n_\sigma \rangle = P_{0\sigma} = N_0 + N_\sigma, \tag{5.94}$$

which can only be satisfied by requiring that $N_0 = 0$, that is $\sum_\sigma N_\sigma = 1$. There is exactly one electron in the QD. This electron may, however, be distributed in some way among the two spin-projections.

5.3.2 *Including Fluctuations in* **P**

Now, we include the fluctuations of the population numbers, i.e. $\delta\mathbf{P}$. Considering e.g. (5.80), the two contributions $V_{\sigma 0c}(t,t_1) D_{c\sigma 0}(t_1,t') R_{0\sigma}(t) P_{0\sigma\bar{a}}(t')$ and $V_{\bar{\sigma}0c}(t,t_1) D_{c\bar{\sigma}0}(t_1,t') R_{0\sigma\bar{\sigma}0}(t) P_{0\bar{\sigma}\bar{a}}(t')$ appear in the equation for $G_{\sigma 0\bar{a}}$. Operating with R on the end-factor P results in the Bose-like Green function K. Let us remain within the simplest order of approximation for this Green function. Then, only the propagators $K_{\sigma\bar{\sigma}}$ and the end-factors $P_{0\sigma}$ are non-vanishing in general. We then have for, say, $a = 0\sigma$

$$\begin{aligned}\delta P_{0\sigma}(t,t') = {} & i K_{\bar{\sigma}\sigma}(t,t') \int_C \sum_a V_{\bar{\sigma}0a}(t,t_1) D_{a\sigma 0}(t_1,t') dt_1 \\ & + \left(\langle \mathrm{T}(h^0 + h^\sigma)(t)(h^0 + h^\sigma)(t') \rangle_U - P_{0\sigma}(t) P_{0\sigma}(t') \right) \\ & \times \int_C \sum_a V_{\sigma 0a}(t,t_1) D_{a\sigma 0}(t_1,t') dt_1, \end{aligned} \quad (5.95)$$

where $K_{\bar{\sigma}\sigma}(t,t') = (-i)\langle \mathrm{T}Z^{\bar{\sigma}\sigma}(t) Z^{\sigma\bar{\sigma}}(t')\rangle_U$. To the lowest order of approximation $K_{\bar{\sigma}\sigma}(t,t') = (-i)(N_{\bar{\sigma}} - N_\sigma)\mathrm{T}e^{-i\Delta_{\sigma\bar{\sigma}}(t-t')} = (-i)(P_{0\bar{\sigma}} - P_{0\sigma})\mathrm{T}e^{-i\Delta_{\sigma\bar{\sigma}}(t-t')}$, where the last equality is clear since $P_{0\sigma} = N_0 + N_\sigma$ by definition. Clearly, in this approximation the fluctuations in the populations of the different spin states may provide some effect to the properties of the quantum dot only in the case of a spin-polarized quantum dot such that $N_\sigma \neq N_{\bar{\sigma}}$.

The last contribution is proportional to $[\delta(t - t') - P_{0\sigma}(t)]P_{0\sigma}(t')$ under conditions for which the diagonal operators h^p are constant, while it vanishes if those operators are independent, since then $\langle \mathrm{T}(h^0 + h^\sigma)(t)(h^0 + h^\sigma)(t')\rangle_U = P_{0\sigma}(t)P_{0\sigma}(t')$. In either case, this contribution is negligible at this stage. Later we will encounter situations when the finiteness of this contributions, although small, is essential for the description of correlation effects.

It is instructive to study the graphical appearance of this dressing to the end-factor, here given by

(5.95′)

We can, thus, interpret the first term in $\delta P_{0\sigma}$ as spin $\bar{\sigma}$ electrons in the leads (wiggle) that tunnel into the quantum dot (double line), a process which is assisted by a spin-flip transition in the quantum dot (dashed line).

Also using the simplification $D_{a\bar{b}}(t,t') = (-i)\delta_{ab}\mathrm{T}e^{-i\Delta_{\bar{a}}(t-t')}$, we can find the lesser form of δP according to

$$\begin{aligned}\delta P^<_{0\sigma}(t,t') = {} & i \int \left(V^r_{\bar{\sigma}0}(t,t_1) D^<_{0\bar{\sigma}}(t_1,t') + V^<_{\bar{\sigma}0}(t,t_1) D^a_{0\bar{\sigma}}(t_1,t') \right) dt_1 K^<_{\bar{\sigma}\sigma}(t,t') \\ & - \int \left(V^r_{\sigma 0}(t,t_1) D^<_{0\sigma}(t_1,t') + V^<_{\sigma 0}(t,t_1) D^a_{0\sigma}(t_1,t') \right) dt_1 P_{0\sigma}(t) P_{0\sigma}(t') \\ = {} & N_\sigma \sum_{\mathbf{k}} |v_{\mathbf{k}\bar{\sigma}}|^2 \left(\int_{-\infty}^{t} e^{-i\varepsilon_{\mathbf{k}\bar{\sigma}}(t-t_1) - i\Delta_{\bar{\sigma}0}(t_1-t')} dt_1 \right. \\ & \left. - \int_{-\infty}^{t'} f(\varepsilon_{\mathbf{k}\bar{\sigma}}) e^{-i\varepsilon_{\mathbf{k}\bar{\sigma}}(t-t_1) - i\Delta_{\bar{\sigma}0}(t_1-t')} dt_1 \right) e^{-i\Delta_{\sigma\bar{\sigma}}(t-t')} \\ = {} & i N_\sigma \sum_{\mathbf{k}} |v_{\mathbf{k}\bar{\sigma}}|^2 \frac{f(\varepsilon_{\mathbf{k}\bar{\sigma}}) e^{-i(\varepsilon_{\mathbf{k}\bar{\sigma}} + \Delta_{\sigma\bar{\sigma}})(t-t')} - e^{-i\Delta_{\sigma 0}(t-t')}}{\varepsilon_{\mathbf{k}\bar{\sigma}} - \Delta_{\bar{\sigma}0} - i\delta}, \end{aligned} \quad (5.96)$$

where we have used that $\Delta_{\sigma\bar{\sigma}} + \Delta_{\bar{\sigma}0} = \Delta_{\sigma 0} - \Delta_{\bar{\sigma}0} + \Delta_{\bar{\sigma}0} = \Delta_{\sigma 0}$, which is true also in the one-loop-approximation. We notice that the first term accounts for spin-flips of the electrons on the quantum dot, in presence of the spin $\bar{\sigma}$ (remember that we are considering the spin σ-channel) electrons in the leads.

We obtain the greater correction $\delta P^{>}_{0\sigma}$ analogously, i.e.

$$\delta P^{>}_{0\sigma}(t,t') = -iN_{\bar{\sigma}} \sum_{\mathbf{k}} |v_{\mathbf{k}\bar{\sigma}}|^2 \frac{[1 - f(\varepsilon_{\mathbf{k}\bar{\sigma}})]e^{-i(\varepsilon_{\mathbf{k}\bar{\sigma}} + \Delta_{\sigma\bar{\sigma}})(t-t')} - e^{-i\Delta_{\sigma 0}(t-t')}}{\varepsilon_{\mathbf{k}\bar{\sigma}} - \Delta_{\bar{\sigma}0} - i\delta}. \tag{5.97}$$

With the help of these expressions we can finally find the retarded/advanced form of $\delta P_{0\sigma}$, using $A^{r/a}(t,t') = \pm\theta(\pm t \mp t')[A^{>}(t,t') - A^{<}(t,t')]$, that is,

$$\delta P^{r/a}_{0\sigma}(t,t') = (\mp i)\theta(\pm t \mp t') \sum_{\mathbf{k}} \frac{|v_{\mathbf{k}\bar{\sigma}}|^2}{\varepsilon_{\mathbf{k}\bar{\sigma}} - \Delta_{\bar{\sigma}0} - i\delta} \Big(\{N_{\bar{\sigma}} - f(\varepsilon_{\mathbf{k}\bar{\sigma}})(N_{\bar{\sigma}} - N_{\sigma})\} \times e^{-i(\varepsilon_{\mathbf{k}\bar{\sigma}} + \Delta_{\sigma\bar{\sigma}})(t-t')} - (N_{\bar{\sigma}} + N_{\sigma})e^{-i\Delta_{\sigma 0}(t-t')}\Big). \tag{5.98}$$

In the stationary regime we go over to energy space, for which we obtain the lesser correction

$$\begin{aligned} \delta P^{<}_{0\sigma}(\omega) &= i2\pi N_{\sigma} \sum_{\mathbf{k}} |v_{\mathbf{k}\bar{\sigma}}|^2 \frac{f(\varepsilon_{\mathbf{k}\bar{\sigma}})\delta(\omega - \varepsilon_{\mathbf{k}\bar{\sigma}} - \Delta_{\sigma\bar{\sigma}}) - \delta(\omega - \Delta_{\sigma 0})}{\varepsilon_{\mathbf{k}\bar{\sigma}} - \Delta_{\bar{\sigma}0} - i\delta} \\ &\approx iN_{\sigma} \sum_{\chi} \Gamma^{\chi}_{\bar{\sigma}} \left(\frac{f_{\chi}(\omega - \Delta_{\sigma\bar{\sigma}})}{\omega - \Delta_{\sigma 0}} - i\pi[1 - f_{\chi}(\Delta_{\bar{\sigma}0})]\delta(\omega - \Delta_{\sigma 0}) \right), \end{aligned} \tag{5.99}$$

whereas the greater correction becomes

$$\delta P^{>}_{0\sigma}(\omega) \approx -iN_{\bar{\sigma}} \sum_{\chi} \Gamma^{\chi}_{\bar{\sigma}} \left(\frac{1 - f_{\chi}(\omega - \Delta_{\sigma\bar{\sigma}})}{\omega - \Delta_{\sigma 0}} - i\pi f_{\chi}(\Delta_{\bar{\sigma}0})\delta(\omega - \Delta_{\sigma 0}) \right). \tag{5.100}$$

Finally, we also have the retarded/advanced correction given by

$$\begin{aligned} \delta P^{r}_{0\sigma}(\omega) = \frac{P_{0\bar{\sigma}} - P_{0\sigma}}{\omega - \Delta_{\sigma 0}} \sum_{\chi} \frac{\Gamma^{\chi}_{\bar{\sigma}}}{2\pi} \Bigg\{ \log\left| \frac{\omega - \mu_{\chi} - \Delta_{\sigma\bar{\sigma}}}{\mu_{\chi} - \Delta_{\bar{\sigma}0}} \right| \\ - i\pi\big[f_{\chi}(\Delta_{\bar{\sigma}0}) - f_{\chi}(\omega - \Delta_{\sigma\bar{\sigma}})\big] \Bigg\} - \frac{i}{2}\Gamma_{\bar{\sigma}} \frac{N_{\bar{\sigma}} + N_{\sigma}}{\omega - \Delta_{\sigma 0} + i\delta}, \end{aligned} \tag{5.101}$$

where we have used that $N_{\bar{\sigma}} - N_{\sigma} = P_{0\bar{\sigma}} - P_{0\sigma}$.

The derivations, so far, have been made for the Green function $G_{0\sigma}$. The derivations for $G_{\bar{\sigma}2}$ are very analogous and are left as an exercise for the reader.

At this stage we have everything we need in order to obtain the non-equilibrium Green functions for the single-level quantum dot. We only need to find the expression for the lesser and greater Green functions, which would have been easy if it were not for the additional contribution in the end-factor. Recall that the equation to the Green function reads

$$\big(i\partial_t - \Delta(t)\big)\mathbf{G}(t,t') = \mathbb{P}(t,t') + \int_C \mathbb{P}(t,t_1)\mathbf{V}(t_1,t_2)\mathbf{G}(t_2,t')dt_1dt_2. \quad \text{(R5.81)}$$

Algebraically, we thus have the equation

$$\mathbf{G}(t,t') = \mathbf{d}(t,t_1)\mathbb{P}(t_1,t') + \mathbf{d}(t,t_1)\mathbb{P}(t_1,t_2)\mathbf{V}(t_2,t_3)\mathbf{G}(t_3,t'). \quad (5.102)$$

Applying the rules for analytical continuation we obtain the equation for the lesser Green function according to

$$\mathbf{G}^< = \mathbf{d}^r\mathbb{P}^< + \mathbf{d}^<\mathbb{P}^a + \mathbf{d}^r\mathbb{P}^r\mathbf{V}^r\mathbf{G}^< + \mathbf{d}^r\mathbb{P}^r\mathbf{V}^<\mathbf{G}^a + \mathbf{d}^r\mathbb{P}^<\mathbf{V}^a\mathbf{G}^a + \mathbf{d}^<\mathbb{P}^a\mathbf{V}^a\mathbf{G}^a. \quad (5.103)$$

Due to the increased complexity in the end-factor, it depends on two times instead of a single, we do not get away as easy as we did previously in the derivation of $\mathbf{G}^<$. Collecting the $G^<$-terms on the left hand side and using that $d^{r,-1}d^< = 0$ identically, we have

$$\mathbf{G}^<(t,t') = \int \mathbf{G}^r(t,t_1)\mathbf{V}^<(t_1,t_2)\mathbf{G}^a(t_2,t')dt_1dt_2 + \int \mathbf{D}^r(t,t_1)\Big[\mathbb{P}^<(t_1,t') + \int \mathbb{P}^<(t_1,t_2)\mathbf{V}^a(t_2,t_3)\mathbf{G}^a(t_3,t')dt_2dt_3\Big]dt_1, \quad (5.104)$$

and analogously for the greater Green function.

The first term (formally) remains the same as in mean-field theory, although the (retarded/advanced) Green functions themselves are given beyond mean-field. The increased complexity in the end-factor, however, gives rise to a correction term which accounts for fluctuations in the occupation on the quantum dot, which would not be present in any mean-field type of theory.

5.3.3 Higher Order Approximations

The basic equations for the single level quantum dot is given by (5.73a), (5.73b), where the diagrammatic expansion of the Green function is generated in the second term of the integrand of each equation. For e.g. (5.73a), the starting point for higher order approximations can be written as in (5.78), and analogously for (5.73b). In order to obtain higher order approximations compared to the ones that we already studied, we go back to (5.78), here given again for convenience,

$$\begin{aligned} &R_{0\sigma\bar{b}}(t^+)V_{\bar{b}c}(t,t_1)G_{c\bar{a}}(t_1,t') \\ &\quad = -V_{\bar{b}c}(t,t_1)D_{c\bar{d}}(t_1,t_2)\{R_{0\sigma\bar{b}}(t^+)D^{-1}_{\bar{d}\bar{e}}(t_2,t_3)\}G_{e\bar{a}}(t_3,t') \\ &\qquad + V_{\bar{b}c}(t,t_1)D_{c\bar{d}}(t_1,t')\{R_{0\sigma\bar{b}}(t^+)P_{d\bar{a}}(t')\}. \end{aligned} \quad \text{(R5.78)}$$

Let us concentrate our efforts to the first term. Although we could dwell on the various approximations of δP, the second term is in principle dealt with to full extent, and can thus be left without further considerations. Also, in order not to make the following discussion to cumbersome in terms of notation, we assume infinite Coulomb repulsion in the quantum dot, i.e. $U \to \infty$. Our discussion is straight forwardly generalized to finite correlation strength, although the algebraic manipulations may become slightly more involved.

The inverse of the locator

$$D^{-1}_{d\bar{e}}(t_2,t_3) = d^{-1}_{d\bar{e}}(t_2,t_3) - \mathcal{S}_{d\bar{e}}(t_2,t_3), \tag{R5.48$'$}$$

with the self-energy operator

$$\begin{aligned}\mathcal{S}_{d\bar{e}}(t_2,t_3) &= P_{d\bar{c}}(t_2)V_{\bar{c}e}(t_2,t_3)\\ &\quad - \int_C V_{\bar{c}f}(t_2,t_4)D_{f\bar{g}}(t_4,t_5)\{R_{d\bar{c}}(t_2)D^{-1}_{g\bar{e}}(t_5,t_3)\}dt_4dt_5. \end{aligned}\tag{R5.50$'$}$$

We recall that the one-loop approximation was obtained by omitting the second term in $\mathcal{S}_{d\bar{e}}$. Thus, we obtain the next order of approximation by keeping this term, however, replacing the inverse locator inside the integral by its corresponding bare locator. Putting $b = 0\bar{\sigma}$ and using that V, D, and G all are diagonal in the limit $U \to \infty$, we find for the first term in (R5.78)

$$\begin{aligned}
&-\int_C V_{\bar{\sigma}}(t,t_1)D_{0\bar{\sigma}}(t_1,t_2)\{R_{0\sigma\bar{\sigma}0}(t^+)D^{-1}_{0\bar{\sigma}\sigma 0}(t_2,t_3)\}G_{0\sigma}(t_3,t')dt_1dt_2dt_3\\
&= -\int_C V_{\bar{\sigma}}(t,t_1)D_{0\bar{\sigma}}(t_1,t_2)\Bigg\{R_{0\sigma\bar{\sigma}0}(t^+)\Bigg[d^{-1}_{0\bar{\sigma}\sigma 0}(t_2,t_3) - P_{0\bar{\sigma}\sigma 0}(t_2)V_{\sigma'}(t_2,t_3)\\
&\quad + \int_C V_s(t_2,t_4)D_{0ss'0}(t_4,t_5)\{R_{0\bar{\sigma}s0}(t_2)d^{-1}_{0s'\sigma 0}(t_5,t_3)\}dt_4dt_5\Bigg]\Bigg\}\\
&\quad \times G_{0\sigma}(t_3,t')dt_1dt_2dt_3\\
&\approx -\int_C V_{\bar{\sigma}}(t,t_1)D_{0\bar{\sigma}}(t_1,t_2)\Bigg\{-i[\delta(t^+ + t_2)\delta(t_2-t_3) + K_{\bar{\sigma}\sigma}(t^+,t_2)V_\sigma(t_2,t_3)]\\
&\quad - \int_C V_s(t_2,t_4)D_{0s}(t_4,t_6)\{R_{0\sigma\bar{\sigma}0}(t^+)d^{-1}_{0ss'0}(t_6,t_7)\}D_{0s'}(t_7,t_5)\\
&\quad \times \{R_{0\sigma s0}(t_2)d^{-1}_{0s'\sigma 0}(t_5,t_3)\}dt_4dt_5dt_6dt_7\Bigg\}G_{0\sigma}(t_3,t')dt_1dt_2dt_3\\
&= \delta\Delta_{\sigma 0}(t)G_{0\sigma}(t,t') + \int_C\big[\Sigma^{(1b)}_{0\sigma}(t,t_1) + \Sigma^{(2b)}_{0\sigma}(t,t_1)\big]G_{0\sigma}(t_1,t')dt_1,
\end{aligned}\tag{5.105}$$

where the first and second terms are given according to the previous discussion, whereas the last contribution is of higher order and given by

$$\Sigma^{(2b)}_{0\sigma}(t,t') = \int_C V_{\bar{\sigma}}(t,t_1)D_{0\bar{\sigma}}(t_1,t')V_{\bar{\sigma}}(t',t_3)D_{0\bar{\sigma}}(t_3,t)D_{0\sigma}(t,t')dt_1dt_3. \tag{5.106}$$

Instead letting $b = 0\sigma$ in (R5.78) we obtain

$$-\int_C V_\sigma(t,t_1)D_{0\sigma}(t_1,t_2)\{R_{0\sigma\sigma 0}(t^+)D^{-1}_{0\sigma\sigma' 0}(t_2,t_3)\}G_{0\sigma'\sigma 0}(t_3,t')dt_1dt_2dt_3$$

$$= \delta_{\sigma\sigma'}\int_C \Sigma^{(2a)}_{0\sigma}(t,t_2)G_{0\sigma}(t_2,t'), \tag{5.107}$$

where

$$\Sigma^{(2a)}_{0\sigma}(t,t') = \int_C V_\sigma(t,t_1)D_{0\sigma}(t_1,t')V_{\bar\sigma}(t',t_3)D_{0\bar\sigma}(t_3,t)D_{0\bar\sigma}(t,t')dt_1dt_3. \tag{5.108}$$

Graphically those contributions can be drawn as

$\Sigma^{(2a)}_{0\sigma} =$ 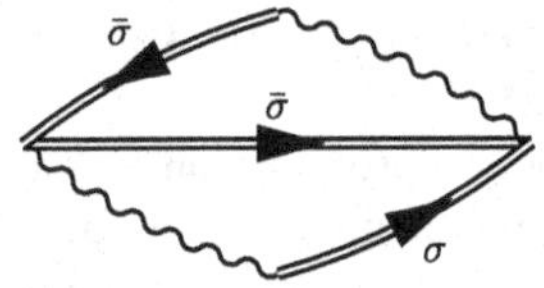

(5.108′)

$\Sigma^{(2b)}_{0\sigma} =$ 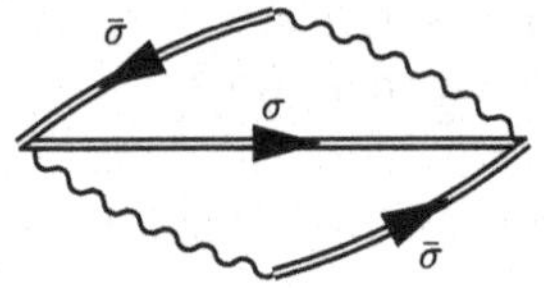

(5.106′)

which illustrate the propagation of a quantum dot electron from left to the right in the presence of electrons tunneling between the leads and quantum dot. Be reminded that there should be quantum dot electron lines with spin σ attached to the left and right hand sides of each self-energy. Then, the first self-energy, $\Sigma^{(2a)}_{0\sigma}$, represents that an incoming electron flips its spin at the left vertex, and propagates to the right vertex where it flips its spin back to σ. This propagation from the left to the right vertex takes place under influence of an spin σ electron propagating in the lead which tunnels into the quantum dot, flips its spin at the right vertex while it tunnels back into the lead and then again into the quantum dot. The second self-energy, $\Sigma^{(2b)}_{0\sigma}$, represents propagation from the left to the right vertex of a spin σ electron in the quantum dot, which scatters off spin $\bar\sigma$ electrons at the vertices, electrons which tunnel back and forth between the leads and the quantum dot.

Making use of that the expressions for the self-energies, cf. (5.108) and (5.106), are given as two separate integrals, we convert the self-energy, e.g. $\Sigma^{(2a)}$, into its lesser counterpart according to

$$\Sigma^{(2a),<}_{0\sigma}(t,t') = D^<_{0\bar\sigma}(t,t')\int\big(V^r_\sigma(t,t'')D^<_{0\sigma}(t'',t') + V^<_{\bar\sigma}(t,t'')D^a_{0\bar\sigma}(t'',t')\big)dt''$$

$$\times\int\big(V^r_{\bar\sigma}(t',t'')D^>_{0\bar\sigma}(t'',t) + V^>_{\bar\sigma}(t',t'')D^a_{0\bar\sigma}(t'',t)\big)dt''. \tag{5.109}$$

Using the previous approximations for V and D, we obtain the lesser self-energy

$$\Sigma_{0\sigma}^{(2a),<}(t,t') = ie^{-i\Delta_{\bar{\sigma}0}(t-t')} \sum_{\mathbf{k}} |v_{\mathbf{k}\sigma}|^2 \frac{f(\varepsilon_{\mathbf{k}\sigma})e^{-i\varepsilon_{\mathbf{k}\sigma}(t-t')} - e^{-i\Delta_{\sigma0}(t-t')}}{\varepsilon_{\mathbf{k}\sigma} - \Delta_{\sigma0} - i\delta}$$
$$\times \sum_{\mathbf{p}} |v_{\mathbf{p}\bar{\sigma}}|^2 \frac{[1-f(\varepsilon_{\mathbf{p}\bar{\sigma}})]e^{i\varepsilon_{\mathbf{p}\bar{\sigma}}(t-t')} - e^{i\Delta_{\bar{\sigma}0}(t-t')}}{\varepsilon_{\mathbf{p}\bar{\sigma}} - \Delta_{\bar{\sigma}0} - i\delta}. \quad (5.110)$$

Going back in the derivation of the self-energies $\Sigma^{(2a)}$ and $\Sigma^{(2b)}$, we simplified the inverse of the locators by its bare counterparts. Therefore, we proceed by reinserting the dressed locators inside the functional differentiations. The we find that we first have a product of first order derivatives, similar to what we have in the self-energies $\Sigma^{(2a)}$ and $\Sigma^{(2b)}$. In addition, we also have to account for a second order derivative appearing due to the product structure of the self-energy operator. Consider the derivative $R_{0\sigma\bar{\sigma}0}(t^+)D^{-1}_{0\bar{\sigma}\sigma0}(t_2,t_3)$, expand the inverse locator, and consider the last term in the first order expansion, i.e.

$$\begin{aligned}
&R_{0\sigma\bar{\sigma}0}(t^+)\big[\big\{R_{0\bar{\sigma}s0}(t_2)V_s(t_2,t_4)D_{0ss'0}(t_4,t_5)\big\}D^{-1}_{0s'\sigma0}(t_5,t_3)\big] \\
&= -R_{0\sigma\bar{\sigma}0}(t^+)\big[V_s(t_2,t_4)D_{0ss'0}(t_4,t_5)\big\{R_{0\bar{\sigma}s0}(t_2)D^{-1}_{0s'\sigma0}(t_5,t_3)\big\}\big] \\
&= V_s(t_2,t_4)D_{0s}(t_4,t_6)\big\{R_{0\sigma\bar{\sigma}0}(t^+)D^{-1}_{0ss'0}(t_6,t_7)\big\} \\
&\quad\times D_{0s'}(t_7,t_5)\big\{R_{0\bar{\sigma}s0}(t_2)D^{-1}_{0s'\sigma0}(t_5,t_3)\big\} \\
&\quad - V_s(t_2,t_4)D_{0s}(t_4,t_5)\big\{R_{0\sigma\bar{\sigma}0}(t^+)R_{0\bar{\sigma}s0}(t_2)D^{-1}_{0s\sigma0}(t_5,t_3)\big\}.
\end{aligned} \quad (5.111)$$

Expanding the inverse locator once more, i.e. $D^{-1} = d^{-1} - PV - [RVD]D^{-1}$, however, omitting the last term, give the self-energy contribution

$$\begin{aligned}
&-V_{\bar{\sigma}}(t,t_1)D_{0\bar{\sigma}}(t_1,t_2)R_{0\sigma\bar{\sigma}0}(t^+)\big[\big\{R_{0\bar{\sigma}s0}(t_2)V_s(t_2,t_4)D_{0ss'0}(t_4,t_5)\big\}D^{-1}_{0s'\sigma0}(t_5,t_3)\big] \\
&= \Sigma^{(2a)}_{0\sigma}(t,t_3) + \Sigma^{(2b)}_{0\sigma}(t,t_3) + V_{\bar{\sigma}}(t,t_1)D_{0\bar{\sigma}}(t_1,t_2)\big(V_s(t_2,t_4)D_{0s}(t_4,t_6) \\
&\quad\times\big[\big\{R_{0\sigma\bar{\sigma}0}(t^+)d^{-1}_{0ss'0}(t_6,t_7)\big\}\big\{R_{0\bar{\sigma}s0}(t_2)P_{0s'\sigma0}(t_5)V_\sigma(t_5,t_3)\big\} \\
&\quad+\big\{R_{0\sigma\bar{\sigma}0}(t^+)P_{0ss'0}(t_6)V_{s'}(t_6,t_7)\big\}\big\{R_{0\bar{\sigma}s0}(t_2)d^{-1}_{0s'\sigma0}(t_5,t_3)\big\} \\
&\quad-\big\{R_{0\sigma\bar{\sigma}0}(t^+)P_{0ss'0}(t_6)V_{s'}(t_6,t_7)\big\} \\
&\quad\times\big\{R_{0\bar{\sigma}s0}(t_2)P_{0s'\sigma0}(t_5)V_\sigma(t_5,t_3)\big\}\big]D_{0s'}(t_7,t_5) \\
&\quad-V_s(t_2,t_4)D_{0s}(t_4,t_5)\big\{R_{0\sigma\bar{\sigma}0}(t^+)R_{0\bar{\sigma}s0}(t_2)P_{0s'\sigma0}(t_5)V_\sigma(t_5,t_3)\big\}\big),
\end{aligned} \quad (5.112)$$

where we have used that the second derivative of the bare locator identically vanishes. Here, there are four third order contributions with respect to the interaction potential V. We study them one by one. First consider

$$\begin{aligned}
\Sigma^{(3a)}_{0\sigma}(t,t_3) &= V_{\bar{\sigma}}(t,t_1)D_{0\bar{\sigma}}(t_1,t_2)V_s(t_2,t_4)D_{0s}(t_4,t_6)\big\{R_{0\sigma\bar{\sigma}0}(t^+)d^{-1}_{0ss'0}(t_6,t_7)\big\} \\
&\quad\times D_{0s'}(t_7,t_5)\big\{R_{0\bar{\sigma}s0}(t_2)P_{0s'\sigma0}(t_5)V_\sigma(t_5,t_3)\big\}.
\end{aligned} \quad (5.113)$$

From (5.79a) we find the requirement that $s = \bar{\sigma}$ and $s' = \sigma$, which yields $R_{0\sigma\bar{\sigma}0}(t^+)d^{-1}_{0ss'0}(t_6, t_7) = i\delta_{s\bar{\sigma}}\delta_{s'\sigma}\delta(t - t_6)\delta(t_6 - t_7)$. This leads to that the second factor, $R_{0\bar{\sigma}s0}(t_2)P_{0s'\sigma 0}(t_5)V_\sigma(t_5, t_3) = -P_{0\bar{\sigma}}(t_2)P_{0\sigma}(t_5)V_\sigma(t_5, t_3)$. Hence, using that $G = DP$, we obtain

$$\Sigma^{(3a)}_{0\sigma}(t, t_3) = -V_{\bar{\sigma}}(t, t_1)G_{0\bar{\sigma}}(t_1, t_2)V_{\bar{\sigma}}(t_2, t_4)D_{0\bar{\sigma}}(t_4, t)G_{0\sigma}(t, t_5)V_\sigma(t_5, t_3) \tag{5.114}$$

or

$$\Sigma^{(3a)}_{0\sigma} = -$$

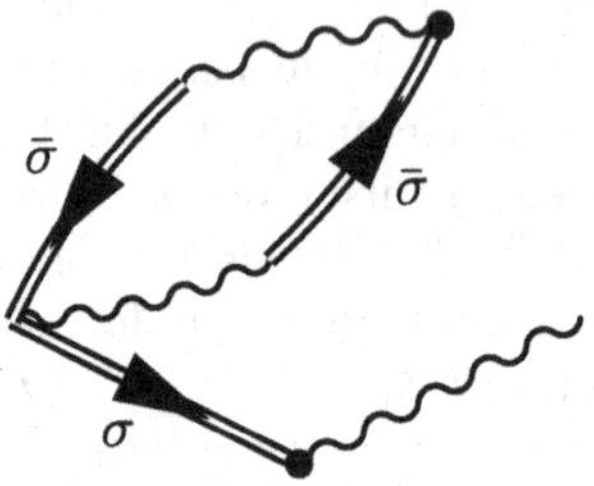

(5.114′)

which is basically a dressing of $\Sigma^{(2a)}_{0\sigma}$. The second new contribution in (5.112) also dresses a lower order diagram, since

$$\begin{aligned}
\Sigma^{(3b)}_{0\sigma}(t, t_3) &= V_{\bar{\sigma}}(t, t_1)D_{0\bar{\sigma}}(t_1, t_2)V_s(t_2, t_4)D_{0s}(t_4, t_6) \\
&\quad \times \left\{R_{0\sigma\bar{\sigma}0}(t^+)P_{0ss'0}(t_6)V_{s'}(t_6, t_7)\right\} \\
&\quad \times D_{0s'}(t_7, t_5)\left\{R_{0\bar{\sigma}s0}(t_2)d^{-1}_{0s'\sigma 0}(t_5, t_3)\right\} \\
&= V_{\bar{\sigma}}(t, t_1)D_{0\bar{\sigma}}(t_1, t_2)V_s(t_2, t_4)D_{0s}(t_4, t_6)iK_{\bar{\sigma}\sigma}(t^+, t_6)V_{s'}(t_6, t_7) \\
&\quad \times D_{0s'}(t_7, t_5)\left[-i\delta_{s\bar{\sigma}}\delta_{s'\sigma}\delta(t_2 - t_5)\delta(t_5 - t_3)\right] \\
&= V_{\bar{\sigma}}(t, t_1)D_{0\bar{\sigma}}(t_1, t_3)V_{\bar{\sigma}}(t_3, t_4)D_{0\bar{\sigma}}(t_4, t_6)K_{\bar{\sigma}\sigma}(t^+, t_6) \\
&\quad \times V_\sigma(t_6, t_7)D_{0\sigma}(t_7, t_3)
\end{aligned} \tag{5.115}$$

which has the graphical appearance

$$\Sigma^{(3b)}_{0\sigma} =$$

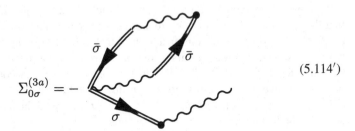

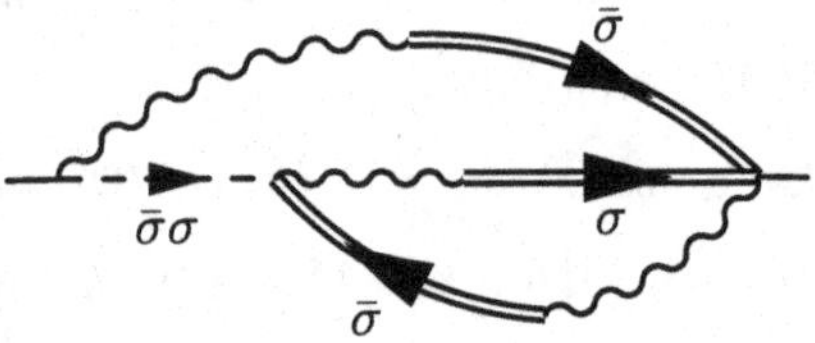

(5.115′)

The third new contribution in (5.112) gives

$$\begin{aligned}
\Sigma^{(3c)}_{0\sigma}(t, t_3) &= -V_{\bar{\sigma}}(t, t_1)D_{0\bar{\sigma}}(t_1, t_2)V_s(t_2, t_4)D_{0s}(t_4, t_6) \\
&\quad \times \left\{R_{0\sigma\bar{\sigma}0}(t^+)P_{0ss'0}(t_6)V_{s'}(t_6, t_7)\right\} \\
&\quad \times D_{0s'}(t_7, t_5)\left\{R_{0\bar{\sigma}s0}(t_2)P_{0s'\sigma 0}(t_5)V_\sigma(t_5, t_3)\right\}
\end{aligned}$$

$$= i V_{\bar{\sigma}}(t, t_1) G_{0\bar{\sigma}}(t_1, t_2) V_{\bar{\sigma}}(t_2, t_4) D_{0\bar{\sigma}}(t_4, t_6) K_{\bar{\sigma}\sigma}(t, t_6) V_{\sigma}(t_6, t_7)$$
$$\times G_{0\sigma}(t_7, t_5) V_{\sigma}(t_5, t_3) \qquad (5.116)$$

or

$$\Sigma^{(3c)}_{0\sigma} = i$$

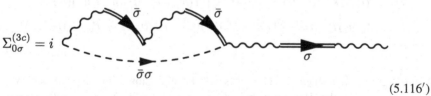

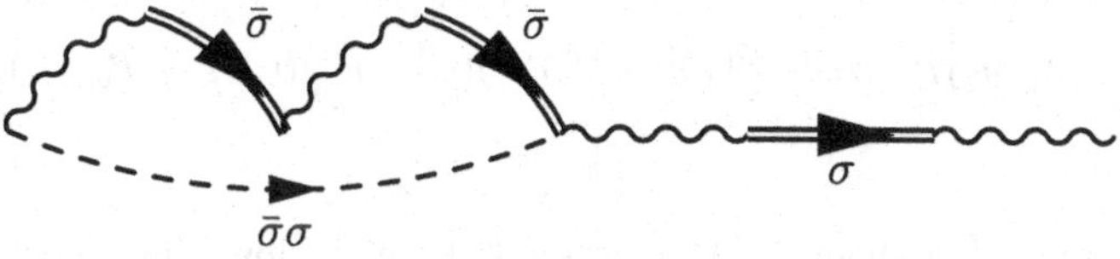

(5.116′)

which is of fourth order with respect to the interaction potential V.

All the three above contributions provide a dressing to lower order diagram and do not introduce anything significantly new to our description. The last contribution in (5.112) does, however, bring forward pieces of physics which is not contained in the lower order diagrams. The second order derivative acting on the end-factor takes a higher order moment of the fluctuations in the quantum dot occupation. We have

$$\Sigma^{(3d)}_{0\sigma}(t, t_3) = -V_{\bar{\sigma}}(t, t_1) D_{0\bar{\sigma}}(t_1, t_2) V_s(t_2, t_4) D_{0s}(t_4, t_5)$$
$$\times \left\{ R_{0\sigma\bar{\sigma}0}(t^+) R_{0\bar{\sigma}s0}(t_2) P_{0s'\sigma 0}(t_5) V_{\sigma}(t_5, t_3) \right\}. \qquad (5.117)$$

We assume that the order of the functional differentiation is of no importance (one arrives at the same result when this assumption is relaxed so this is no serious simplification). Then, we consider as the "first" derivative $R_{0\sigma\bar{\sigma}0}(t) P_{0s'\sigma 0}(t_5) = i\delta_{s'\bar{\sigma}} K_{\bar{\sigma}\sigma}(t, t_5)$. The equation of motion for $K_{\bar{\sigma}\sigma}(t, t_5)$ can to the simplest order be written

$$\left(i\partial_t - \Delta_{\sigma\bar{\sigma}} - \Delta U_{\sigma\bar{\sigma}}(t)\right) K_{\bar{\sigma}\sigma}(t, t_5) = \delta(t - t_5)\left(P_{0\bar{\sigma}} - P_{0\sigma}\right). \qquad (5.118)$$

We here note two things which are crucial in order to proceed with the self-energy $\Sigma^{(3d)}$. First, the propagator $K_{\bar{\sigma}\sigma}$ can be written as a product of a locator and end-factor analogously to G. We, thus, write $K_{\bar{\sigma}\sigma} = L_{\bar{\sigma}\sigma}[P_{0\bar{\sigma}} - P_{0\sigma}]$, where $L_{\bar{\sigma}\sigma}$ is the locator. In principle, the product LP is a matrix product and should involve more terms, terms which we do not have to consider here since those vanish. Second, the left hand side of the equation for K also contains the source fields U_ξ which is also in analogy with the equation for G.

Making use of those observations, and making the simplest possible expansion of K, we now act with the second derivative, i.e.

$$R_{0\bar{\sigma}s0}(t_2) i\delta_{s'\bar{\sigma}} K_{\bar{\sigma}\sigma}(t, t_5) = -i L_{\bar{\sigma}\sigma}(t, t_8)[R_{0\bar{\sigma}s0}(t_2) L^{-1}_{\bar{\sigma}\sigma}(t_8, t_9)] K_{\bar{\sigma}\sigma}(t_9, t_5)$$
$$+ i L_{\bar{\sigma}\sigma}(t, t_5) R_{0\bar{\sigma}s0}(t_2)[P_{0\bar{\sigma}}(t_5) - P_{0\sigma}(t_5)]. \qquad (5.119)$$

A quick look at (5.118) gives at hand that the first term in the expression for RK vanishes unless $s = \bar{\sigma}$. This requirement is also necessary for the second contribution since an off-diagonal functional derivative acting on the diagonal end-factors vanishes, in the present model. Hence, we find the following expression for the second order derivative:

$$
\begin{aligned}
&R_{0\bar{\sigma}s0}(t_2) i\delta_{s'\bar{\sigma}} K_{\bar{\sigma}\sigma}(t, t_5) \\
&\quad = L_{\bar{\sigma}\sigma}(t, t_2) K_{\bar{\sigma}\sigma}(t_2, t_5) + i L_{\bar{\sigma}\sigma}(t, t_5)[\langle \mathrm{T}(h^0 + h^{\bar{\sigma}})(t_2)(h^0 + h^{\bar{\sigma}})(t_5)\rangle_U \\
&\qquad - P_{0\bar{\sigma}}(t_2) P_{0\bar{\sigma}}(t_5) - \langle \mathrm{T}(h^0 + h^{\bar{\sigma}})(t_2)(h^0 + h^{\sigma})(t_5)\rangle_U + P_{0\bar{\sigma}}(t_2) P_{0\sigma}(t_5)].
\end{aligned}
\tag{5.120}
$$

Assuming that the diagonal operators h^p have a slow time-dependence, we can approximate the last term by $i L_{\bar{\sigma}\sigma}(t, t_5)[\delta(t_2 - t_5) N_{\bar{\sigma}}(t_2) - \{P_{0\bar{\sigma}}(t_5) - P_{0\sigma}(t_5)\} P_{0\bar{\sigma}}(t_2)]$. Finally, we then have the self-energy

$$
\begin{aligned}
\Sigma^{(3d)}_{0\sigma}(t, t_3) = &-V_{\bar{\sigma}}(t, t_1) D_{0\bar{\sigma}}(t_1, t_2) V_{\bar{\sigma}}(t_2, t_4) D_{0\bar{\sigma}}(t_4, t_5) \big[L_{\bar{\sigma}\sigma}(t, t_2) K_{\bar{\sigma}\sigma}(t_2, t_5) \\
&+ i L_{\bar{\sigma}\sigma}(t, t_5) \big\{ \delta(t_2 - t_5) N_{\bar{\sigma}}(t_2) - \{P_{0\bar{\sigma}}(t_5) - P_{0\sigma}(t_5)\} P_{0\bar{\sigma}}(t_2) \big\} \big] \\
&\times V_{\sigma}(t_5, t_3)
\end{aligned}
\tag{5.121}
$$

and graphically

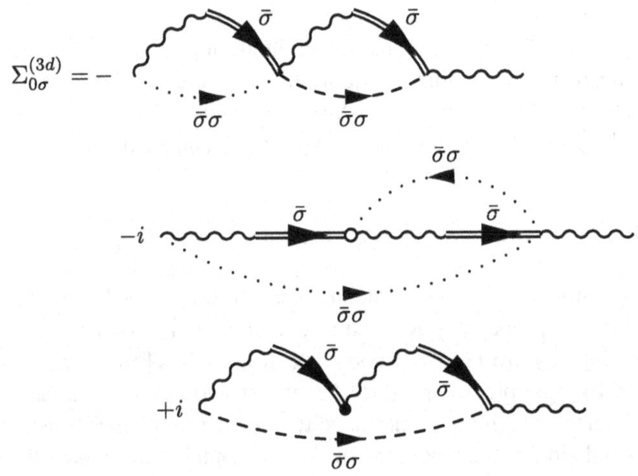

(5.121′)

where the dotted line denotes $L_{\bar{\sigma}\sigma}$, whereas the empty circle in the second diagram stands for $N_{\bar{\sigma}}$. It should also be noticed in the second diagram, that the small loop to the right is actually contracted to a single time due to $\delta(t_2 - t_5)$.

It is interesting to see that, despite all other diagram involving a functional derivative of the end-factor vanish in the spin-degenerate case, this last self-energy contributes for all conditions of spin-polarization. While the first and last diagrams are finite whenever the spin-degeneracy is broken, due to the presence of $P_{0\sigma} - P_{0\bar{\sigma}}$, the second diagram is finite also under spin-degenerate conditions, due to the absence of $P_{0\sigma} - P_{0\bar{\sigma}}$. We also note, that there is a set of diagrams also when considering the derivative $R_{0\sigma}(t^+)D_{0\sigma}^{-1}$, which are reminiscent of those discussed here and the derivation of those are left as an exercise.

The middle diagram in the self-energy $\Sigma_{0\sigma}^{(3d)}$ contributes even under spin-degenerate conditions. Specifically, we have

$$\begin{aligned}
&(-i)\int_C V_{\bar{\sigma}}(t,t_1)D_{0\bar{\sigma}}(t_1,t_2)V_{\bar{\sigma}}(t_2,t_4)D_{0\bar{\sigma}}(t_4,t_2)L_{\bar{\sigma}\sigma}(t,t_2)\\
&\quad\times N_{\bar{\sigma}}(t_2)V_\sigma(t_2,t_3)dt_1dt_2dt_4\\
&=(-i)\int_C \mathcal{D}_{\bar{\sigma}}(t,t_2)\mathcal{D}_{\bar{\sigma}}(t_2,t_2)L_{\bar{\sigma}\sigma}(t,t_2)N_{\bar{\sigma}}(t_2)V_\sigma(t_2,t_3)dt_2,
\end{aligned}\tag{5.122}$$

where $\mathcal{D}_{\bar{\sigma}}(t,t_2)=\int_C V_{\bar{\sigma}}(t,t_1)D_{0\bar{\sigma}}(t_1,t_2)dt_1$. It is now a bit simpler to convert this contribution into the real time domain. In particular, by assuming that $t<t_3$, we obtain

$$\begin{aligned}
&(-i)\int_C \mathcal{D}_{\bar{\sigma}}(t,t_2)\mathcal{D}_{\bar{\sigma}}(t_2)L_{\bar{\sigma}\sigma}(t,t_2)N_{\bar{\sigma}}(t_2)V_\sigma(t_2,t_3)dt_2\\
&=(-i)\int_{-\infty}^{t}\big(\mathcal{D}_{\bar{\sigma}}^{>}(t,t_2)L_{\bar{\sigma}\sigma}^{>}(t,t_2)-\mathcal{D}_{\bar{\sigma}}^{<}(t,t_2)L_{\bar{\sigma}\sigma}^{<}(t,t_2)\big)\\
&\quad\times\mathcal{D}_{\bar{\sigma}}(t_2,t_2)N_{\bar{\sigma}}(t_2)V_\sigma^{<}(t_2,t_3)dt_2\\
&\quad+(-i)\int_{-\infty}^{t_3}\mathcal{D}_{\bar{\sigma}}^{<}(t,t_2)L_{\bar{\sigma}\sigma}^{<}(t,t_2)\mathcal{D}_{\bar{\sigma}}(t_2,t_2)\\
&\quad\times N_{\bar{\sigma}}(t_2)\big(V_\sigma^{<}(t_2,t_3)-V_\sigma^{>}(t_2,t_3)\big)dt_2,
\end{aligned}\tag{5.123}$$

where

$$\begin{aligned}
\mathcal{D}_{\bar{\sigma}}^{</>}(t,t_2)&=\int\big(V_{\bar{\sigma}}^{r}(t,t_1)D_{0\bar{\sigma}}^{</>}(t_1,t_2)dt_1+V_{\bar{\sigma}}^{</>}(t,t_1)D_{0\bar{\sigma}}^{a}(t_1,t_2)\big)dt_1\\
&=\pm i\sum_{\mathbf{k}}|v_{\mathbf{k}\bar{\sigma}}|^2\frac{e^{-i\Delta_{\bar{\sigma}0}(t-t_2)}-f(\pm\varepsilon_{\mathbf{k}\bar{\sigma}})e^{-i\varepsilon_{\mathbf{k}\bar{\sigma}}(t-t_2)}}{\Delta_{\bar{\sigma}0}-\varepsilon_{\mathbf{k}\bar{\sigma}}+i\delta},
\end{aligned}\tag{5.124a}$$

$$\begin{aligned}
\mathcal{D}_{\bar{\sigma}}(t_2,t_2)&=\int_{-\infty}^{t_2}\big(V_{\bar{\sigma}}^{<}(t_2,t_1)D_{0\bar{\sigma}}^{>}(t_1,t_2)-V_{\bar{\sigma}}^{>}(t_2,t_1)D_{0\bar{\sigma}}^{<}(t_1,t_2)\big)dt_1\\
&=i\sum_{\mathbf{k}}|v_{\mathbf{k}\bar{\sigma}}|^2\frac{2f(\varepsilon_{\mathbf{k}\bar{\sigma}})-1}{\Delta_{\bar{\sigma}0}-\varepsilon_{\mathbf{k}\bar{\sigma}}+i\delta}.
\end{aligned}\tag{5.124b}$$

Here, we have assumed, for simplicity, that $V_\sigma^{</>}(t,t')=(\pm i)\sum_{\mathbf{k}}f(\pm\varepsilon_{\mathbf{k}\sigma})\times e^{-i\varepsilon_{\mathbf{k}\sigma}(t-t')}$, and we have replaced the dressed locators by their bare counterparts $d_{0\sigma}^{</>}(t,t')=(\pm i)e^{-i\Delta_{\sigma0}(t-t')}$. We, also, replace the locators $L_{\bar{\sigma}\sigma}$ by $l_{\bar{\sigma}\sigma}^{</>}(t,t')=(-i)e^{-i\Delta_{\sigma\bar{\sigma}}(t-t')}$ and assume that the occupation numbers $N_{\bar{\sigma}}$ are time-independent.

We then obtain e.g.

$$
\begin{aligned}
&(-i)\int_{-\infty}^{t}\mathcal{D}_{\bar{\sigma}}^{>}(t,t_2)L_{\bar{\sigma}\sigma}^{>}(t,t_2)V_{\sigma}^{<}(t_2,t_3)dt_2\\
&\approx -i\sum_{\chi\chi'}\frac{\Gamma_{\bar{\sigma}}^{\chi}}{2\pi}\frac{\Gamma_{\sigma}^{\chi'}}{2\pi}\int f_{\chi'}(\omega)\frac{2\pi\delta(\omega-\varepsilon')}{\Delta_{\bar{\sigma}0}-\varepsilon+i\delta}\bigg(\frac{1}{\omega-\Delta_{\sigma 0}+i\delta}\\
&\quad-\frac{f_{\chi}(-\varepsilon)}{\omega-\varepsilon-\Delta_{\sigma\bar{\sigma}}+i\delta}\bigg)d\varepsilon d\varepsilon' e^{-i\omega(t-t_3)}\frac{d\omega}{2\pi}.
\end{aligned}
\tag{5.125}
$$

The first term gives a contribution of roughly $-(1/4\pi)\sum_{\chi\chi'}\Gamma_{\bar{\sigma}}^{\chi}\Gamma_{\sigma}^{\chi'}f_{\chi'}(\omega)/(\omega-\Delta_{\sigma 0}+i\delta)$, whereas the second term provides the logarithmic correction

$$
\sim\frac{i}{(2\pi)^2}\sum_{\chi\chi'}\Gamma_{\bar{\sigma}}^{\chi}\Gamma_{\sigma}^{\chi'}f_{\chi'}(\omega)\ln\left|\frac{\mu_\chi-\Delta_{\bar{\sigma}0}}{\omega-\mu_\chi-\Delta_{\sigma\bar{\sigma}}}\right|
\tag{5.126}
$$

to the self-energy. This contribution gives rise to a so-called Kondo peak around the chemical potential μ_χ for small spin-flip energies $\Delta_{\sigma\bar{\sigma}}$. More contributions of this sort are obtained by carry out the calculations for the remaining part of the lesser, and greater, forms of the middle diagram of the self-energy $\Sigma_{0\sigma}^{(3d)}$.

We have now discussed most of the issues when it comes to expand the Hubbard operator Green functions in terms of diagrams. We have not, however, taken a deeper discussion about e.g. vertex corrections and vertex equation, and other highly non-trivial matters. Those discussions will be left untouched for the benefit of considering more examples from physics in the following chapters.

References

1. Abrikosov, A.A., Gorkov, L.P., Dzyaloshinski, I.E.: Methods of Quantum Field Theory in Statistical Physics. Dover, New York (1975)
2. Mahan, G.D.: Many-Particle Physics, 2nd edn. Plenum Press, New York (1990)
3. Negele, J.W., Orland, H.: Quantum Many-Particle Systems. Perseus Books, Reading (1998)
4. Larkin, A.I., Matveev, K.A.: Sov. Phys. JETP **66**, 580–584 (1987)
5. Meir, Y., Wingreen, N.S., Lee, P.A.: Phys. Rev. Lett. **66**, 3048–3051 (1991)
6. Meir, Y., Wingreen, N.S., Lee, P.A.: Phys. Rev. Lett. **70**, 2601–2604 (1993)
7. Ovchinnikov, S.G., Val'kov, V.V.: Hubbard Operators in the Theory of Strongly Correlated Electrons. Imperial College Press, London (2004)
8. Kadanoff, L.P., Baym, G.: Quantum Statistical Mechanics, 7th edn. Perseus Books, Reading (1998)
9. Fransson, J.: Phys. Rev. B **72**, 075314 (2005)

Chapter 6
Tunneling Current in a Single Level System

Abstract We study the local properties and the tunneling current in a single level quantum dot system employing the different approximation schemes developed in the previous chapters. The impact of the one-loop approximation cause variations in the broadening of the local level and an additional explicit contribution to the current. We proceed by discussing the behavior under spin-dependent conditions.

6.1 Single Level Quantum Dot

From the simplest theory we know that the single level in the quantum dot is infinitely sharp if there is no environment coupled to it. Upon coupling the quantum dot to external bath(s), the level broadens and one cannot any longer speak of a discrete level in a strict sense. It is, nonetheless, simpler to think of the system as still being separated into pieces consisting of the quantum dot level and the delocalized environment, especially when we go to higher order approximation.

In Sects. 5.3.1 and 5.3.2 we studied the diagrammatic expansion of the single level quantum dot Green function up to the first order beyond mean-field theory, in the sense that we included fluctuations in the occupation numbers. These fluctuations result in a self-energy which is energy dependent. Moreover, including these fluctuations also resulted in an increased complexity in the lesser and greater Green functions, as we saw in (5.104).

In this section we will study the single level quantum dot in some more detail, however, we will not go into the Kondo physics since this theory is not developed at the moment in terms of Hubbard operators. We will take the full one-loop approximation and analyze the physics of the single level system under spin-degenerate as well as under spin-dependent conditions, in order to acquire as complete picture as possible. For simplicity and since we are mainly seeking a qualitative picture, however, we perform most of the analysis in the large Hubbard U limit, so that the doubly occupied state can be ignored.

The local density of states in the quantum dot is given by $-\text{tr}\,\text{Im}\,\mathbf{G}^r/\pi$, in equilibrium, where we trace over the many-body states included in $\mathbf{G}$. In the present

J. Fransson, *Non-Equilibrium Nano-Physics*,
Lecture Notes in Physics 809,
DOI 10.1007/978-90-481-9210-6_6, © Springer Science+Business Media B.V. 2010

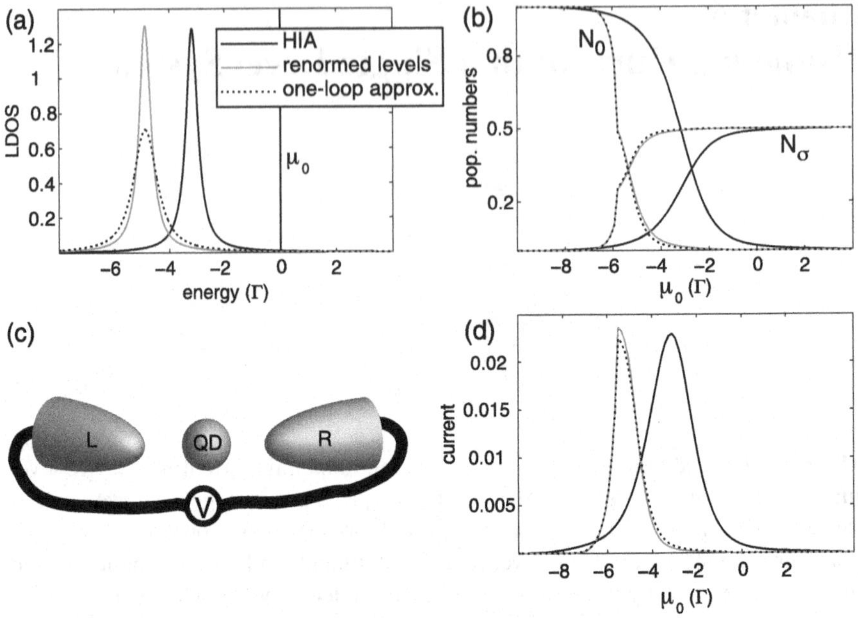

Fig. 6.1 Local equilibrium properties of the single level quantum dot system under spin-degenerate conditions. (**a**) Local density of states (LDOS) as function of the energy ω, where the line at the origin marks the location of the equilibrium chemical potential μ_0, and (**b**) population numbers for varying position of the bare local level relative to μ_0, (**c**) schematic sketch of the system, and (**d**) current at $eV = 0.01\Gamma$. Here, we have taken $k_B T = 0.5\Gamma$, and $\varepsilon_0 = -5\Gamma/\pi$

case, we can decouple the equations for spin ↑ and ↓, in the stationary regime leading to

$$-\frac{1}{\pi}\,\mathrm{Im}\,G^r_{0\sigma}(\omega) = -\frac{1}{\pi}\,\mathrm{Im}\,\frac{P_{0\sigma} + \delta P^r_{0\sigma}(\omega)}{\omega - \Delta_{\sigma 0} + i\Gamma_\sigma[P_{0\sigma} + \delta P^r_{0\sigma}(\omega)]/2}, \tag{6.1}$$

where $\delta P^r_{0\sigma}(\omega)$ is given in (5.101). In Fig. 6.1(a) we plot the total local density of states $(-\mathrm{Im}\sum_\sigma G^r_{0\sigma}/\pi)$ in three different approximations, i.e. Hubbard-I-approximation (solid), re-normalized levels (faint), and one-loop-approximation (dotted). In the Hubbard-I-approximation, we have

$$-\frac{1}{\pi}\,\mathrm{Im}\,G^{r,\mathrm{HIA}}_{0\sigma}(\omega) = \frac{\Gamma_\sigma}{2\pi}\,\frac{P^2_{0\sigma}}{(\omega - \Delta^0_{\sigma 0})^2 + (\Gamma_\sigma P_{0\sigma}/2)^2}, \tag{6.2}$$

which is a typical Lorentzian of width $\Gamma_\sigma P_{0\sigma}/2$, see Fig. 6.1(a) (solid). Adding the level renormalization does not change the functional appearance of the density of state, only the peak position is changed from $\Delta^0_{\sigma 0}$ to $\Delta_{\sigma 0}$. This is also clearly illustrated in Fig. 6.1(a) (faint).

Adding the fluctuating contribution, δP, we find

$$
\begin{aligned}
-\frac{1}{\pi}&\operatorname{Im} G^{r,\mathrm{OLA}}_{0\sigma}(\omega)\\
&=\frac{\Gamma_\sigma}{2\pi}\frac{[P_{0\sigma}+\operatorname{Re}\delta P^r_{0\sigma}]^2+[\operatorname{Im}\delta P^r_{0\sigma}]^2}{[\omega-\Delta_{\sigma 0}-\operatorname{Im}\delta P^r_{0\sigma}\Gamma_\sigma/2]^2+[(P_{0\sigma}+\operatorname{Re}\delta P^r_{0\sigma})\Gamma_\sigma/2]^2}\\
&\quad-\frac{1}{\pi}\frac{(\omega-\Delta_{\sigma 0})\operatorname{Im}\delta P^r_{0\sigma}}{[\omega-\Delta_{\sigma 0}-\operatorname{Im}\delta P^r_{0\sigma}\Gamma_\sigma/2]^2+[(P_{0\sigma}+\operatorname{Re}\delta P^r_{0\sigma})\Gamma_\sigma/2]^2}.
\end{aligned}
\tag{6.3}
$$

The first term has a functional appearance which is similar to the simpler approximations, i.e. a Lorentzian. One should bear in mind, though, that $\delta P^r_{0\sigma}(\omega)$ depends on the energy ω. The overall behavior is, thus, much more complex than a simple Lorentzian function. We note that there is an additional shift to the local energy level, $\operatorname{Im}\delta P^r_{0\sigma}\Gamma_\sigma/2$. This shift is small except for $\omega\approx\Delta_{\sigma 0}$, where it diverges, cf. (5.101). The width of the level, $(P_{0\sigma}+\operatorname{Re}\delta P^r_{0\sigma})\Gamma_\sigma/2$ is increased, which is physically reasonable since the inclusion of more electron movements back and forth should lead to a decreased lifetime of the state. This additional broadening of the local level is also seen in Fig. 6.1(a) (dotted).

The second term provides only minor modifications to the overall picture, which is seen by e.g. taking the ratio between the first and the second contribution in the above expression. We then obtain

$$
\begin{aligned}
&\frac{2}{\Gamma_\sigma}\frac{(\omega-\Delta_{\sigma 0})\operatorname{Im}\delta P^r_{0\sigma}}{[P_{0\sigma}+\operatorname{Re}\delta P^r_{0\sigma}]^2+[\operatorname{Im}\delta P^r_{0\sigma}]^2}\\
&\quad=-\frac{2}{\Gamma_\sigma}\frac{\pi(\omega-\Delta_{\sigma 0})[f(\Delta_{\bar\sigma 0})-f(\omega-\Delta_{\sigma\bar\sigma})]+[N_{\bar\sigma}+N_\sigma]\Gamma_{\bar\sigma}/2}{[P_{0\sigma}+\operatorname{Re}\delta P^r_{0\sigma}]^2+[\operatorname{Im}\delta P^r_{0\sigma}]^2},
\end{aligned}
\tag{6.4}
$$

since $\operatorname{Im}\delta P^r_{0\sigma}(\omega)=-\pi[f(\Delta_{\bar\sigma 0})-f(\omega-\Delta_{\sigma\bar\sigma})]-(\Gamma_{\bar\sigma}/2)[N_{\bar\sigma}+N_\sigma]/(\omega-\Delta_{\sigma 0})$. The divergence in $\operatorname{Im}\delta P^r_{0\sigma}$ leads to that the ratio above approaches zero, and apart from this case, the ratio is smoothly behaving.

In equilibrium the population numbers N_p are given by simply $N_0=1-N_\uparrow-N_\downarrow$ and $N_\sigma=-\operatorname{Im}\int f(\omega)G^r_{0\sigma}(\omega)d\omega/\pi$, and these are plotted in Fig. 6.1(b). Here, the quantum dot is empty ($N_0\to 1$) when $\mu_0\ll E_\sigma=\varepsilon_0$, that is, when the electron level is far above the chemical potential μ_0, while it is filled by half an electron of each spin ($N_\sigma\to 1/2$, such that $\sum_\sigma N_\sigma\to 1$) in average at the opposite end. We also see that there are kinks in the population numbers in the two approximations beyond the Hubbard-I-approximation, which are related to logarithmic divergence of the level renormalization when $\mu\to\Delta_{\sigma 0}$. This behavior is, as we previously discussed, caused by replacing the dressed locator with its bare counterpart in the level renormalization and is smoothened out when the level renormalization is calculated in a bit more sophisticated manner, see e.g. [1, 2].

In Fig. 6.1(d) we plot the near equilibrium current (we discuss in the succeeding section how the current is calculated) for varying positions of the quantum dot level relative to μ_0, using a small bias voltage. In this way we mimic the linear response measurements for the conductance of system. We find that there are no significant differences between the three approximation schemes under spin-degenerate conditions, which is not surprising since the contribution δP has very little impact in this case.

The above relations are written with explicit references to the spin degree of freedom. Let us, therefore, study what will happen with the quantum dot when there is a spin-dependence imposed on the system. We assume that the spin-polarization in the leads can be parametrized through the couplings using the relation $p_\chi = \Gamma_\uparrow^\chi - \Gamma_\downarrow^\chi$, and we set $\Gamma_\sigma^\chi = \Gamma_0^\chi(1+\sigma p_\chi)/2$.

The transition energy $\Delta_{\sigma 0}$ is coupled to $\Delta_{\bar{\sigma}0}$ through the renormalization equation, e.g.

$$\Delta_{\sigma 0} = \Delta_{\sigma 0}^0 + \sum_\chi \frac{\Gamma_{\bar{\sigma}}^\chi}{\pi} \ln \frac{|\mu_\chi - \Delta_{\bar{\sigma}0}|}{W}.$$

Hence, $\Delta_{\uparrow 0} = \Delta_{\uparrow 0}^0$ in the limit $\Gamma_\downarrow^\chi = 0$. In this limit, however, $\Gamma_\uparrow^\chi = \Gamma^\chi$ which leads to that $\Delta_{\downarrow 0} - \Delta_{\downarrow 0}^0 \approx \sum_\chi \Gamma^\chi \ln\{|\mu_\chi - \Delta_{\uparrow 0}|/W\}/\pi < 0$, since $|\mu_\chi - \Delta_{\sigma 0}|/W < 1$. Physically, this means that the degeneracy of the quantum dot level is broken, and that the spin level with the least coupling to the leads becomes the ground state. The equilibrium local density of states therefore should acquire a double peak structure, which is also seen in Fig. 6.2(a). In the full one-loop-approximation, the increased broadening tends to smear out the double peak structure.

The occupations of the spin-split levels become significantly altered from the Hubbard-I-approximation-picture, see Fig. 6.2(b), where the occupations for the $\uparrow$ and $\downarrow$ levels differ throughout the whole range in the one-loop-approximation, while there is a difference only in a limited range for the simpler approximations. Also, the linear response current displays some information about the spin-dependence in the quantum dot, see Fig. 6.2(c) and (d).

6.2 Spin-Dependent Transport in Quantum Dot Systems

We have now studied the local properties of the single level system under spin-dependent conditions in equilibrium. Here, we proceed by looking at the transport properties for these conditions and link the behavior of the differential conductance to the local density of states.

We already know what the current looks like in mean field theory. Hence, let us look at what happens with the current in the next order of approximation, when using the formula in (4.14), here rewritten to suit the present situation to read

$$I_L = \frac{ie}{2\pi}\mathrm{tr} \int \Gamma^L \Big[f_L(\omega)\mathbf{G}^>(\omega) + [1 - f_L(\omega)]\mathbf{G}^<(\omega) \Big] d\omega, \tag{4.14$'$}$$

in the stationary regime for constant coupling Γ^L. For simplicity, we also let the Coulomb repulsion $U \to \infty$, so that we only have to consider the Green func-

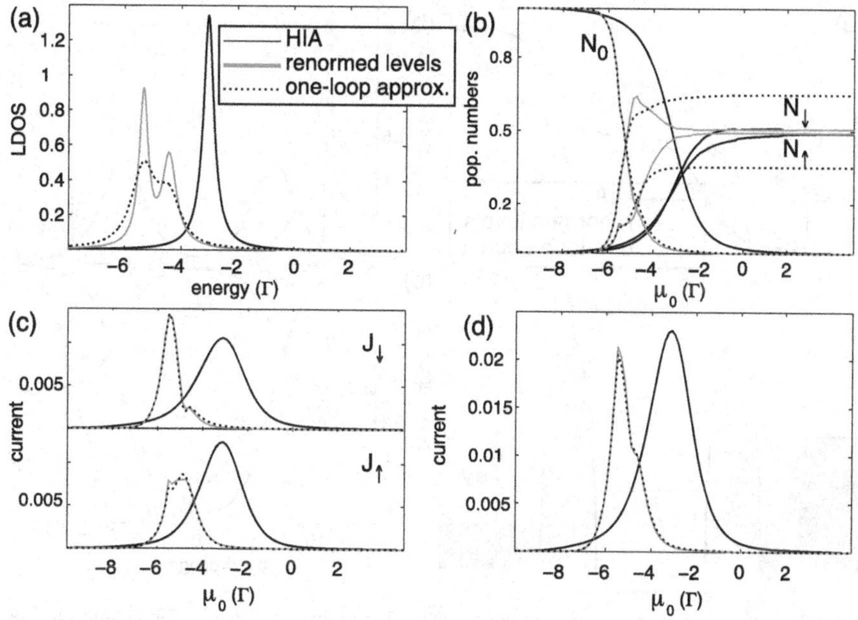

Fig. 6.2 Local equilibrium properties of the single level quantum dot system under spin-dependent conditions ($p_L = 0.5$ and $p_R = 0$). (**a**) Local density of states (LDOS) as function of the energy ω, where the line at the origin marks the location of the equilibrium chemical potential μ_0, and (**b**) population numbers, (**c**) schematic sketch of the system, and (**d**) current at $eV = 0.01\Gamma$. Here, we have taken $k_B T = 0.5\Gamma$, and $\varepsilon_0 = -5\Gamma/\pi$

tion $G_{0\sigma}$. In this case, the matrices in the formula above are diagonal. We recall that e.g. $\mathbf{G}^< = \mathbf{G}^r\mathbf{V}^<\mathbf{G}^a + \mathbf{D}^r\mathbb{P}^<(1+\mathbf{V}^a\mathbf{G}^a)$ in the one-loop-approximation, which reduces to $\mathbf{G}^< = \mathbf{G}^r\mathbf{V}^<\mathbf{G}^a$ in the simpler approximations. Let us derive the current for the one-loop-approximation, in order to be slightly more general. Using that the matrices are diagonal, we find that the second term in the expression for the lesser Green function has the entries $\mathbb{P}^<_{0\sigma}D^r_{0\sigma}(1+V^a_\sigma G^a_{0\sigma}) = \mathbb{P}^<_{0\sigma}(\omega - \Delta_{\sigma 0} - V^a_\sigma\mathbb{P}^a_{0\sigma} + V^a_\sigma\mathbb{P}^a_{0\sigma})|D^r_{0\sigma}|^2 = \mathbb{P}^<_{0\sigma}(\omega - \Delta_{\sigma 0})|D^r_{0\sigma}|^2$. The lesser, and greater, form of the end-factor is found in Sect. 5.3.2 and, moreover, using that $V^{</>}_\sigma(\omega) = \pm i f_L(\pm\omega)\Gamma^L_\sigma \pm i f_R(\pm\omega)\Gamma^R_\sigma$, we can write the current in the form

$$
\begin{aligned}
I_L = \frac{e}{2\pi}\sum_\sigma \int \Gamma^L_\sigma\Big[\Gamma^R_\sigma\big(f_L(\omega) - f_R(\omega)\big)|G^r_{0\sigma}(\omega)|^2 \\
+ \sum_\chi \Gamma^\chi_{\bar\sigma}\big\{N_{\bar\sigma} f_L(\omega)[1 - f_\chi(\omega - \Delta_{\sigma\bar\sigma})] \\
- N_\sigma[1 - f_L(\omega)] f_\chi(\omega - \Delta_{\sigma\bar\sigma})\big\}|D^r_{0\sigma}(\omega)|^2\Big] d\omega. \qquad (6.5)
\end{aligned}
$$

Formally, the first term in (6.5) appears the same as in the mean field approximation whereas the second term is a direct consequence of the fluctuations between

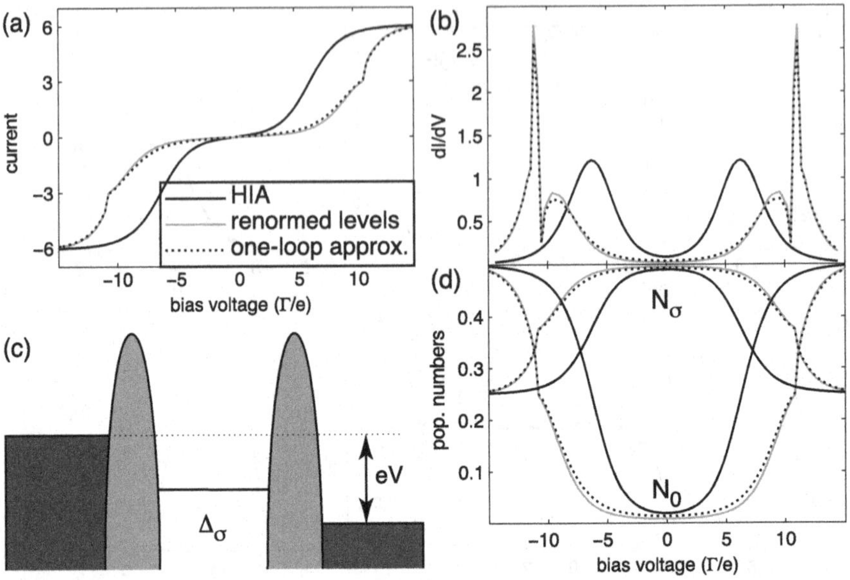

Fig. 6.3 Non-equilibrium properties of the single level quantum dot system under spin-degenerate conditions. (**a**) Total charge current through the single level quantum dot, (**b**) differential conductance, (**c**) potential diagram of the system, and (**d**) bias dependent population numbers. Here, we have taken $k_B T = 0.5\Gamma$, and $\varepsilon_0 = -5\Gamma/\pi$

the two spin states, spin-flip fluctuations. Hence, the fluctuations in the quantum dot occupation does not only enter indirectly through the Green functions, but also directly as additional terms in the current.

The second term in the current, arises due to inelastic spin-flip transitions in the quantum dot between the states $|\uparrow\rangle$ and $|\downarrow\rangle$. Here, it is reasonable to use the terminology of inelastic transitions even when $E_\uparrow = E_\downarrow$, since the state of the electron in the quantum dot is changing during the transition from one spin state to the other. In the spin-degenerate limit, e.g. $E_\uparrow = E_\downarrow$, and $N_\uparrow = N_\downarrow = N_1$, the second term goes like $(N_1/2)\sum_\sigma \int [f_L - f_R] |D^r_{0\sigma}|^2 d\omega$, which leads to that the total current is approximately proportional to $[1-(1-N_1/2)N_1/2]\sum_\sigma \int [f_L - f_R]|D^r_{0\sigma}|^2 d\omega$. This should be compared to the current in the Hubbard-I-approximation, which is approximately proportional to $(1-N_1/2)^2 \sum_\sigma \int [f_L - f_R]|D^r_{0\sigma}|^2 d\omega$. The ratio between the two expressions gives

$$\frac{1-(1-N_1/2)N_1/2}{(1-N_1/2)^2} = \frac{(1-N_1/2)^2 + N_1/2}{(1-N_1/2)^2} = 1 + \frac{N_1/2}{(1-N_1/2)^2}. \tag{6.6}$$

Thus, the current in the Hubbard-I-approximation is slightly less than the current in the one-loop-approximation, meaning that the fluctuations of the quantum dot occupation leads to a slight increase in the total current. This observation is confirmed in the plots of the current and differential conductance in Fig. 6.3(a) and (b), although the differences are small. Note that one has to compare the one-loop-approximation (dotted) with the current which includes the re-normalized level (faint), which behaves like the Hubbard-I-approximation in all respect except for the transition energy, and henceforth referred to as the re-normalized Hubbard-I-approximation. We should also note this approximate result only holds for low biases. For higher biases we need to account for the self-consistent variations in the population numbers, which is non-trivial.

The self-consistently calculated population numbers are plotted in Fig. 6.3(d), illustrating the slightly increased width of the quantum dot states through the increased population in the empty state, N_0. Hence, the smearing of the localized states due to the fluctuations results in a higher probability that the quantum dot is unoccupied in equilibrium.

Considering the spin-dependent currents in case of the left lead being ferromagnetic and the right non-magnetic changes the behavior of the system rather substantially. The spin-imbalance generates a net flow of electrons through the spin-flip channel supported by the bias voltage applied across the junction. More understanding about the current described by the expression above is given by studying its corresponding differential conductance. Some caution has to taken when doing this, for instance, the Green functions should be calculated self-consistently with respect to the occupation numbers. The occupation numbers are, in turn, expected to vary with the bias voltage which means that the derivative e.g. $(\partial/\partial V)G^r_{0\sigma} \neq 0$, in general. For an analytical qualitative analysis, albeit not entirely correct, we omit this dependence.

Then, for low temperatures, the differential conductance is approximated by

$$\begin{aligned}\frac{\partial I_L}{\partial V} = \frac{e^2}{4\pi}\sum_\sigma \int \Gamma^L_\sigma |D^r_{0\sigma}(\omega)|^2 \Big[\Gamma^R_\sigma \big(\delta(\omega - eV/2) + \delta(\omega + eV/2)\big) |\mathbb{P}^r_{0\sigma}(\omega)|^2 \\ + \frac{1}{2\pi}\Big\{ \sum_\chi \Gamma^\chi_{\bar\sigma}\{(N_{\bar\sigma}[1 - f_\chi(\omega - \Delta_{\sigma\bar\sigma})] + N_\sigma f_\chi(\omega - \Delta_{\sigma\bar\sigma})\}\delta(\omega - eV/2) \\ - \{\Gamma^L_{\bar\sigma}\delta(\omega - \Delta_{\sigma\bar\sigma} - eV/2) - \Gamma^R_{\bar\sigma}\delta(\omega - \Delta_{\sigma\bar\sigma} + eV/2)\} \\ \times \{N_{\bar\sigma} f_L(\omega) + N_\sigma[1 - f_L(\omega)]\}\Big\}\Big] d\omega. \end{aligned} \tag{6.7}$$

The interpretation of the various terms is straight forward. The first term provides the usual elastic peaks at biases corresponding to the transition energy. The second term, in braces, gives a redistribution of the current at the spin-flip energy $\Delta_{\sigma\bar\sigma}$, such that the current passing through the state N_σ ($N_{\bar\sigma}$) is increased (decreased). In the plots given in Fig. 6.4(b) and (d), which show the total and spin-resolved differential conductance, respectively, the spin-flip energies $\Delta_{\uparrow\downarrow(\downarrow\uparrow)} \approx \pm 1$. The effect on the current is small and therefore not visible in the plots. The differential conductance does, however, show several narrow peaks in the region of the transition energies. These are, as before, related to the renormalization of the transition energies, but should in a more sophisticated calculation be less pronounced.

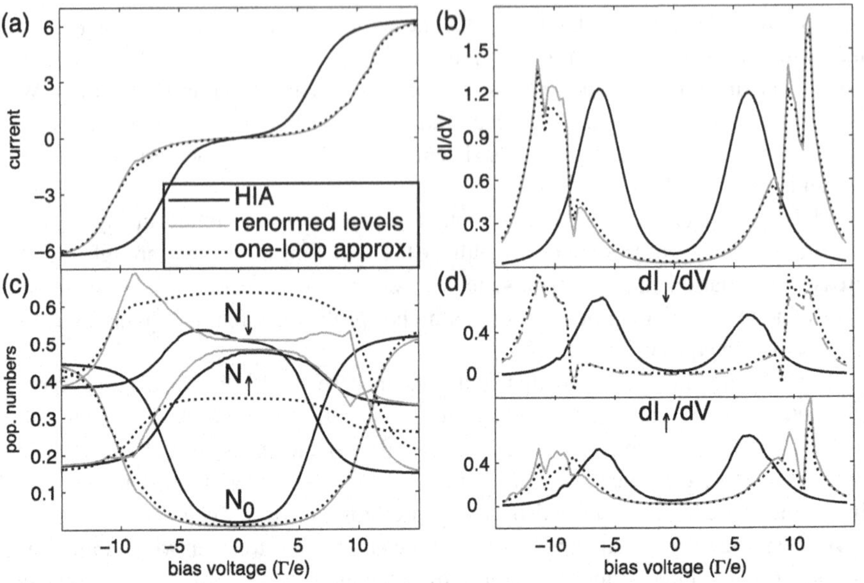

Fig. 6.4 Non-equilibrium properties of the single level quantum dot system under spin-dependent conditions ($p_L = 0.5$ and $p_R = 0$). (**a**) Total charge current through the single level quantum dot, (**b**) differential conductance, (**c**) bias dependent population numbers, and (**d**) spin-resolved differential conductance. Here, we have taken $k_B T = 0.5\Gamma$, and $\varepsilon_0 = -5\Gamma/\pi$

The question is whether we can tune the parameters Γ_σ^χ in such a way that the second contribution to the differential conductance becomes important, or alternatively put, is there a way to enhance the significance of the contribution $D^r \mathbb{P}^{</>}(1 + V^a G^a)$ in the expression for $G^{</>}$. First, we need to separate the quantum dot levels, which can be done by an external magnetic field **B** directly applied across the system, or by using ferromagnetic leads. We use the latter approach here, since we know from above that the spin-polarization in the leads induce a spin-splitting of the quantum dot levels. The spin-splitting naturally leads to a spin-polarization of the quantum dot, i.e. $N_\uparrow \neq N_\downarrow$. Moreover, we saw above that application of a single ferromagnetic lead, while the other being non-magnetic, generates an asymmetry in the populations. This asymmetry becomes apparent in the spin-resolved differential conductance, while the asymmetry in the total current is not as dramatic.

We can provide an asymmetry in the total current by attaching the leads asymmetrically in the sense that the quantum dot is stronger coupled to one of the leads, which can be expressed through the ratio Γ^L/Γ^R. Assume, for definiteness, that $\Gamma^L/\Gamma^R \gg 1$. Then, electrons coming into the quantum dot from the left lead have a hard time exiting the quantum dot to the right lead. Hence, biasing the system from the left to the right, leads to an accumulation of electron density in the quantum dot, see Fig. 6.5(b). This is also true in all three approximation schemes. In simple mean field theory, e.g. Landauer picture [3–5], however, the differential conductance, Fig. 6.5(a), is almost symmetric which can be explained by the fact that the

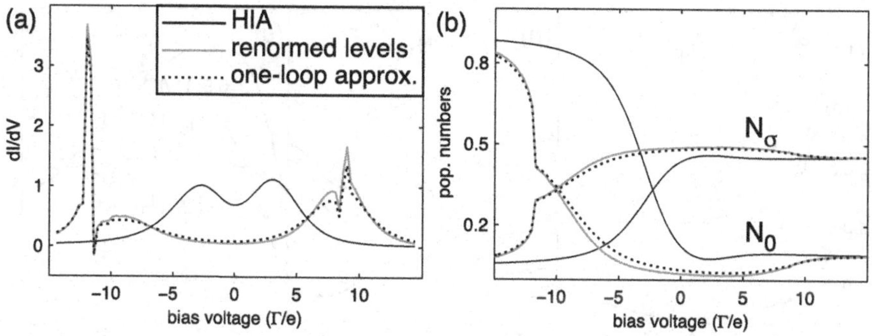

Fig. 6.5 Non-equilibrium properties of the single level quantum dot system under spin-degenerate conditions and asymmetric couplings to the leads. (**a**) Total differential conductance and (**b**) bias dependent population numbers. Here, we have taken $k_BT = 0.5\Gamma$, $\varepsilon_0 = -5\Gamma/\pi$, and $\Gamma^L/\Gamma^R = 10$

transmission in this approximation is given by $\mathcal{T} \sim \sum_\sigma \Gamma^L_\sigma \Gamma^R_\sigma |G^r_{0\sigma}|^2$. Irrespective of whether the electron is entering the quantum dot from the left or the right lead, it "sees" an available level weighted by the population numbers, which is proportional to the total coupling $\Gamma^L_\sigma \Gamma^R_\sigma$. The mean field according to the Hubbard-I-approximation is also almost symmetric, for the same reasons. The Green function, which is included in the transmission matrix, may carry some asymmetries which is brought into the current and differential conductance, see Fig. 6.5(a). It is clearly seen in Fig. 6.5(a) that the more advanced approximation schemes provide further signatures of asymmetry in the differential conductance.

Combining the asymmetric couplings with the ferromagnetic leads introduce an enhanced spin-polarization of the quantum dot in one bias direction compared to the case with symmetric couplings. In view of the previous discussion this is understandable since the accumulation of charge in the quantum dot due to the weak coupling to one lead must also be spin-polarized if couplings to the leads are spin-dependent as well. Thus, assuming e.g. $\Gamma^L > \Gamma^R$ and $\Gamma^\chi_\uparrow > \Gamma^\chi_\downarrow$, we would expect the accumulated charge to have a $\downarrow$ majority spin, which indeed is verified in the computations, see Fig. 6.6. It is also clear that the asymmetry is expected in both the re-normalized Hubbard-I-approximation and one-loop-approximation.

The main difference between the two approximation schemes is that the latter changes the functional dependence on the bias from the usual mean field appearance. This added contribution to the current tends to enhance the effect from the asymmetric couplings as well as from the spin-dependence. The result is that one spin-projections of the current becomes further suppressed in comparison to the mean field current, see Fig. 6.7. Panel (d) shows the spin-polarization of the current, i.e. $(J_\uparrow - J_\downarrow)/\sum_\sigma J_\sigma$ for the two asymmetric cases, which gives an illustration of that it is possible to almost completely suppress on spin-channel in the current simply by configuring the system in an asymmetric way in combination with spin-dependent couplings between the leads and the quantum dot.

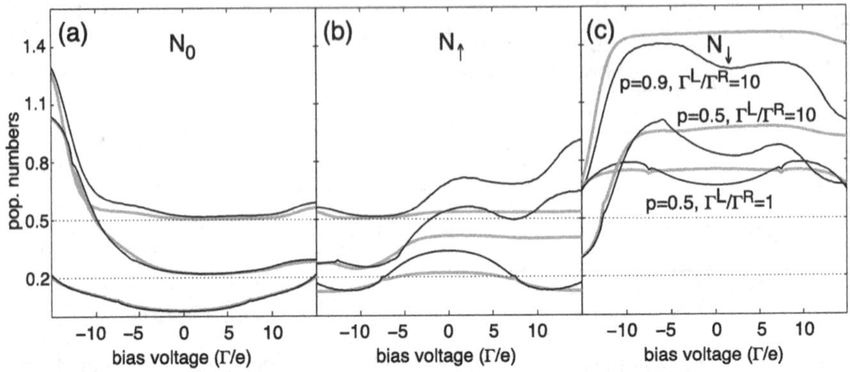

Fig. 6.6 Non-equilibrium population numbers of the single level quantum dot system under spin-dependent conditions ($p_L = p_R = p$) in the one-loop-approximation (*bold grey*) and re-normalized Hubbard-I-approximation (*solid*). (**a**) Empty state, (**b**) spin ↑, and (**c**) spin ↓. Here, we have taken $k_B T = 0.5\Gamma$, and $\varepsilon_0 = -5\Gamma/\pi$

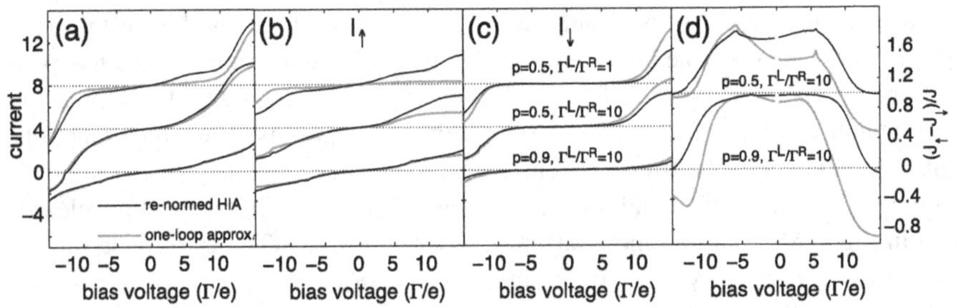

Fig. 6.7 Current under spin-dependent conditions ($p_L = p_R = p$) in the one-loop-approximation (*bold grey*) and re-normalized Hubbard-I-approximation (*solid*). (**a**) Total current, (**b**) spin ↑, and (**c**) spin ↓. Here, we have taken $k_B T = 0.5\Gamma$, and $\varepsilon_0 = -5\Gamma/\pi$. The plots are off-set for clarity

6.3 Non-Collinearly Aligned Ferromagnetic Leads

Ferromagnetic leads is a way to generate a spin-polarized current in the system, a spin-polarization that might be enhanced or reduced by a quantum dot in between. The magnetic moment of the lead may be aligned in a parallel or anti-parallel fashion, as in the previous cases. The magnetic moments may also be non-collinearly aligned.

We continue to consider a single level quantum dot. Although the magnetization directions in the two leads are noncollinear, it is useful to introduce a global reference frame, where the x-direction lies along the direction of the charge current, see Fig. 6.8. As the global z direction is arbitrary around the x axis, there is no restriction in choosing it along the magnetization direction of the left reservoir, since its magnetic moment is assumed to be fixed. The magnetization of the right reservoir is rotated by the angle ϕ in the global xz plane.

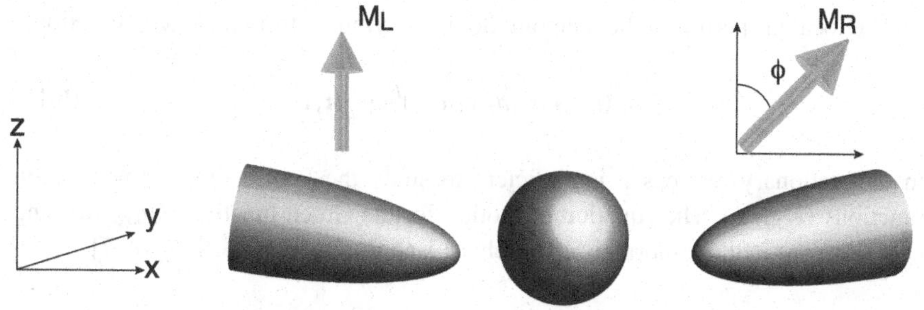

Fig. 6.8 The quantum dot coupled to noncollinearly oriented ferromagnetic leads. The global reference frame and the magnetization M_L of the *left* reservoir coincide, while the magnetization M_R of the *right* reservoir is rotated by the angle ϕ

We assume that the bare quantum dot level is spin-degenerate at the energy ε_0 and that the on-site Coulomb repulsion is U. Here, the level spacing of the quantum dot is assumed to exceed the thermal broadening. In this spirit then, the energy of the quantum dot is given by $\mathcal{H}_{QD} = \sum_\sigma \varepsilon_0 d_\sigma^\dagger d_\sigma + U n_\uparrow n_\downarrow = \sum_{p=0,\sigma,2} E_p h^p$.

The energies of the reservoirs are, for simplicity, given by $\mathcal{H}_{L/R} \sum_{ks\in L/R} \varepsilon_{\mathbf{k}s} \times c_{\mathbf{k}s}^\dagger c_{\mathbf{k}s}$. The ferromagnetism in the leads is modeled in the spirit of the Stoner theory in the sense that a strong spin asymmetry in the density of states is assumed. The density of states is, moreover, approximated to be energy independent. The tunneling interaction between the left lead and the quantum dot is given by

$$\mathcal{H}_{TL} = \sum_{\mathbf{k}\sigma\in L,a} v_{\mathbf{k}\sigma} (d_\sigma)^a c_{\mathbf{k}\sigma}^\dagger X^a + H.c., \tag{6.8}$$

since the local reference frame coincides with the global. Here, $\sum_a (d_\sigma)^a X^a = \langle 0|d_\sigma|\sigma\rangle X^{0\sigma} + \langle\bar{\sigma}|d_\sigma|2\rangle X^{\bar{\sigma}2}$. The corresponding the energy for the interaction between the quantum dot and the right reservoir if given by

$$\begin{aligned}\mathcal{H}_{TR} = \sum_{\mathbf{k}\in R,a} \Bigg(& \left[v_{\mathbf{k}+} c_{\mathbf{k}+}^\dagger \cos\frac{\phi}{2} - v_{\mathbf{k}-} c_{\mathbf{k}0-}^\dagger \sin\frac{\phi}{2} \right] (d_\uparrow)^a \\ & + \left[v_{\mathbf{k}+} c_{\mathbf{k}+}^\dagger \sin\frac{\phi}{2} + v_{\mathbf{k}-} c_{\mathbf{k}0-}^\dagger \cos\frac{\phi}{2} \right] (d_\downarrow)^a \Bigg) X^a + H.c.,\end{aligned} \tag{6.9}$$

where the rotation of the magnetization direction is included. Here, the spin indices $s = \pm$ have been used in order to distinguish between the global and local reference frames. The expression given in (6.9) can be conveniently rewritten by introducing the rotated quantum dot operators

$$\begin{pmatrix} d_+ \\ d_- \end{pmatrix} = \mathcal{R} \begin{pmatrix} d_\uparrow \\ d_\downarrow \end{pmatrix}, \quad \mathcal{R} = \begin{pmatrix} \cos(\phi/2) & \sin(\phi/2) \\ -\sin(\phi/2) & \cos(\phi/2) \end{pmatrix}. \tag{6.10}$$

Hence, (6.9) becomes simply $\mathcal{H}_{TR} = \sum_{\mathbf{k}s\in R,a} v_{\mathbf{k}s} (d_s)^a c_{\mathbf{k}s}^\dagger X^a + H.c.$

The local properties in the quantum dot is determined from the Green function

$$\mathcal{G}_{\sigma\sigma'}(t,t') = (d_\sigma)^a (d^\dagger_{\sigma'})^b G_{a\bar{b}}(t,t'). \tag{6.11}$$

In the stationary regimes it is sufficient to study the Fourier transformed Green functions $G_{a\bar{b}}(i\omega)$. The equation of motion for the Green functions $G_{0\sigma\bar{a}}(i\omega)$ and $G_{\bar{\sigma}2\bar{a}}(i\omega)$ are in the re-normalized Hubbard-I-approximation given by [6, 7]

$$\begin{aligned}\big(i\omega - \Delta_{\sigma 0} - P_{0\sigma\bar{b}}V_{\bar{b}\sigma 0}\big)G_{0\sigma\bar{a}}(i\omega) &= P_{0\sigma\bar{a}} + \sigma\big(P_{0\sigma\bar{b}}V_{\bar{b}\sigma 0} + \delta\Delta_{2\bar{\sigma}}\big)G_{\bar{\sigma}2\bar{a}}(i\omega)\\ &\quad + \Sigma_{0\sigma\bar{\sigma}0}G_{0\bar{\sigma}\bar{a}}(i\omega) + \Sigma_{\bar{\sigma}22\sigma}G_{\sigma 2\bar{a}}(i\omega),\end{aligned} \tag{6.12a}$$

$$\begin{aligned}\big(i\omega - \Delta_{2\bar{\sigma}} - P_{\bar{\sigma}2\bar{b}}V_{\bar{b}2\bar{\sigma}}\big)G_{\bar{\sigma}2\bar{a}}(i\omega) &= P_{\bar{\sigma}2\bar{a}} + \sigma\big(P_{\bar{\sigma}2\bar{b}}V_{\bar{b}2\bar{\sigma}} + \bar{\sigma}\delta\Delta_{\sigma 0}\big)G_{0\sigma\bar{a}}(i\omega)\\ &\quad + \Sigma_{0\sigma\bar{\sigma}0}G_{0\bar{\sigma}\bar{a}}(i\omega) + \Sigma_{\bar{\sigma}22\sigma}G_{\sigma 2\bar{a}}(i\omega).\end{aligned} \tag{6.12b}$$

The structure of these equations suggests that we can write the equation of motion on terms as a 4×4 matrix equation

$$[i\omega - \Delta^0 - \Sigma(i\omega)]\mathbf{G}(i\omega) = \mathbf{P}, \tag{6.13}$$

where $\Delta^0 = \mathrm{diag}\{\Delta^0_{\uparrow 0}, \Delta^0_{\downarrow 0}, \Delta^0_{2\downarrow}, \Delta^0_{2\uparrow}\}$ contains the bare transition energies, whereas the self-energy matrix $\Sigma(i\omega) = \mathbf{PV}(i\omega) + \delta\Delta$. In this form the interaction matrix is given by

$$\mathbf{V} = \mathbf{V}^L + \mathbf{V}^R = \mathbf{V}^L + \begin{pmatrix} \mathbf{v}^r & \sigma^z\mathbf{v}^r \\ \sigma^z\mathbf{v}^r & \mathbf{v}^r \end{pmatrix}, \tag{6.14a}$$

$$\mathbf{V}^L = \sum_{\mathbf{k}\sigma\in L} \frac{|v_{\mathbf{k}\sigma}|^2}{i\omega - \varepsilon_{\mathbf{k}\sigma}} \begin{pmatrix} \delta_{\sigma\uparrow} & 0 & \delta_{\sigma\uparrow} & 0 \\ 0 & \delta_{\sigma\downarrow} & 0 & -\delta_{\sigma\downarrow} \\ \delta_{\sigma\uparrow} & 0 & \delta_{\sigma\uparrow} & 0 \\ 0 & -\delta_{\sigma\downarrow} & 0 & \delta_{\sigma\downarrow} \end{pmatrix}, \tag{6.14b}$$

$$\mathbf{v}^R = \sum_{\mathbf{k}s\in R} \frac{|v_{\mathbf{k}s}|^2}{i\omega - \varepsilon_{\mathbf{k}s}} \mathcal{R}^T(\phi) \begin{pmatrix} \delta_{s+} & 0 \\ 0 & \delta_{s-} \end{pmatrix} \mathcal{R}(\phi), \tag{6.14c}$$

where σ^z is the z component of the Pauli spin vector. We define the coupling matrix $\boldsymbol{\Gamma}^{L/R} = -2\,\mathrm{Im}\,\mathbf{V}^{L/R}(\omega + i0^+)$, where the coupling constants $\Gamma^{L/R}_\sigma$ is parametrized in terms of $p_{L/R} = (\Gamma^{L/R}_\uparrow - \Gamma^{L/R}_\downarrow)/\Gamma_0$, where $\Gamma_0 = \Gamma^{L/R}_\uparrow + \Gamma^{L/R}_\downarrow$, such that we can write $\Gamma^{L/R}_\sigma = \Gamma_0(1 + \sigma p_{L/R})/2$. Here, $\Gamma^{L/R}_\sigma$ defines the coupling constant between the spin σ channel in the left/right lead and the quantum dot. By this procedure no essential physics is lost. In terms of the spin-dependent parameters p_α the coupling matrices to the left and right leads become

$$\Gamma^L = \frac{\Gamma_0}{2}\begin{pmatrix} 1+p_L & 0 & 1+p_L & 0 \\ 0 & 1-p_L & 0 & -1+p_L \\ 1+p_L & 0 & 1+p_L & 0 \\ 0 & -1+p_L & 0 & 1-p_L \end{pmatrix}, \tag{6.15a}$$

$$\Gamma^R = \begin{pmatrix} \gamma^R & \sigma^s\gamma^R \\ \sigma^z\gamma^R & \gamma^R \end{pmatrix}, \quad \gamma^R = \frac{\Gamma_0}{2}\begin{pmatrix} 1+p_R\cos\phi & p_R\sin\phi \\ p_R\sin\phi & 1-p_R\cos\phi \end{pmatrix}. \tag{6.15b}$$

Finally, it is important to note that the end-factor $\mathbf{P}$ is block diagonal $\mathbf{P} = \text{diag}\{\mathbf{P}_1, \mathbf{P}_2\}$, where

$$\mathbf{P}_1 = \begin{pmatrix} N_0+N_\uparrow & N_{\downarrow\uparrow} \\ N_{\uparrow\downarrow} & N_0+N_\downarrow \end{pmatrix}, \qquad \mathbf{P}_2 = \begin{pmatrix} N_\downarrow+N_2 & N_{\downarrow\uparrow} \\ N_{\uparrow\downarrow} & N_\uparrow+N_2 \end{pmatrix}, \tag{6.16a}$$

explicitly in terms of the population numbers of the involved transitions.

Consider the renormalization energy $\delta\Delta_{\sigma 0} = \delta\Delta^L_{\sigma 0} + \delta\Delta^R_{\sigma 0}$ for the transition $X^{0\sigma}$, where [2]

$$\delta\Delta^L_{\sigma 0} = -\frac{1}{\pi}\sum_{\mathbf{k}\in L}|v_{\mathbf{k}\bar\sigma}|^2\int\frac{f(\varepsilon_{\mathbf{k}\bar\sigma})-f(\omega)}{\varepsilon_{\mathbf{k}\bar\sigma}-\omega}\,\text{Im}\big[D^r_{0\bar\sigma}(\omega)+\bar\sigma D^r_{\sigma 2\bar\sigma 0}(\omega)\big]d\omega, \tag{6.17a}$$

whereas $\delta\Delta^R_{\sigma 0} = \delta_{\sigma\uparrow}\delta\Delta^R_{\sigma 0:\uparrow} + \delta_{\sigma\downarrow}\delta\Delta^R_{\sigma 0:\downarrow}$ with [6, 7]

$$\begin{aligned}\delta\Delta^R_{\sigma 0:\uparrow} = -\frac{1}{\pi}\sum_{\mathbf{k}\in R}\int\Bigg\{&\Bigg([f(\varepsilon_{\mathbf{k}+})-f(\omega)]\frac{|v_{\mathbf{k}+}|^2}{\varepsilon_{\mathbf{k}+}-\omega}\sin^2(\phi/2) \\ &+[f(\varepsilon_{\mathbf{k}-})-f(\omega)]\frac{|v_{\mathbf{k}-}|^2}{\varepsilon_{\mathbf{k}+}-\omega}\cos^2(\phi/2)\Bigg)\text{Im}\big[D^r_{0\downarrow}(\omega)-D^r_{\uparrow 2\downarrow 0}(\omega)\big] \\ &+\frac{\sin\phi}{2}\Bigg([f(\varepsilon_{\mathbf{k}+})-f(\omega)]\frac{|v_{\mathbf{k}+}|^2}{\varepsilon_{\mathbf{k}+}-\omega}-[f(\varepsilon_{\mathbf{k}-})-f(\omega)]\frac{|v_{\mathbf{k}-}|^2}{\varepsilon_{\mathbf{k}+}-\omega}\Bigg) \\ &\times\text{Im}\big[D^r_{0\uparrow\downarrow 0}(\omega)+D^r_{\downarrow 2\downarrow 0}(\omega)\big]\Bigg\}d\omega\end{aligned} \tag{6.17b}$$

and

$$\begin{aligned}\delta\Delta^R_{\sigma 0:\downarrow} = -\frac{1}{\pi}\sum_{\mathbf{k}\in R}\int\Bigg\{&\Bigg([f(\varepsilon_{\mathbf{k}+})-f(\omega)]\frac{|v_{\mathbf{k}+}|^2}{\varepsilon_{\mathbf{k}+}-\omega}\cos^2(\phi/2) \\ &+[f(\varepsilon_{\mathbf{k}-})-f(\omega)]\frac{|v_{\mathbf{k}-}|^2}{\varepsilon_{\mathbf{k}+}-\omega}\sin^2(\phi/2)\Bigg)\text{Im}\big[D^r_{0\uparrow}(\omega)+D^r_{\downarrow 2\uparrow}(\omega)\big] \\ &+\frac{\sin\phi}{2}\Bigg([f(\varepsilon_{\mathbf{k}+})-f(\omega)]\frac{|v_{\mathbf{k}+}|^2}{\varepsilon_{\mathbf{k}+}-\omega}-[f(\varepsilon_{\mathbf{k}-})-f(\omega)]\frac{|v_{\mathbf{k}-}|^2}{\varepsilon_{\mathbf{k}+}-\omega}\Bigg) \\ &\times\text{Im}\big[D^r_{0\downarrow\uparrow 0}(\omega)-D^r_{\uparrow 2\uparrow 0}(\omega)\big]\Bigg\}d\omega.\end{aligned} \tag{6.17c}$$

Note that those expressions were derived using frequency summation over the loop diagram in the expansion of the Green functions, resulting in slightly different expressions compared to using the non-equilibrium techniques discussed previously in this book. Except for some minor details, the above formulas are equivalent to the ones derived using non-equilibrium techniques.

Analogous expressions are, of course, present for the other shifts. In fact, the re-normalizaton term $\delta\Delta$ in the self-energy is a full 4×4 matrix, in general, and reduces to the case discussed previously only for collinear configurations of the magnetic leads, including non-magnetic leads. Hence, the spin-flip transition energies are also subject to the renormalization due to the magnetism in the leads.

More important in the present context is the strong dependence of the shifts on the magnetic properties in the reservoirs. This is clearly displayed in (6.17a)–(6.17c), which shows an explicit dependence on the couplings between the leads and the quantum dot. The induced shift of the quantum dot transition energies are (exactly) equal only when the couplings to the leads are equal. In other words are the induced shift distinct whenever the couplings differ. This is simplest illustrated in e.g. the left shift, $\delta\Delta^L_{\sigma 0}$. Assume that the magnetization directions in the two leads are collinear, so that all off-diagonal locators vanish, and assume that $U \to \infty$, for simplicity. The latter assumption leads to that all propagators involving transitions between one- and two-particle states can be neglected. Putting $D^r_{0\sigma}(\omega) = (\omega - \Delta^0_{\sigma 0} + i0^+)$, where $\Delta^0_{\uparrow 0} = \Delta^0_{\downarrow 0}$ $(= \Delta^0_{10})$ by construction, (6.17a) reduces to

$$\delta\Delta^L_{\sigma 0} = \frac{\Gamma^L_{\bar{\sigma}}}{\pi} \ln \frac{|\mu_L - \Delta^0_{10}|}{D}, \tag{6.18}$$

where $2D$ is the width of the conduction band in the lead. For sufficiently large $D \sim 1$ eV, which is reasonable for normal metals, the ratio in the logarithm lies between 0 and 1, hence, $\delta\Delta^L_{\sigma 0} < 0$. Therefore, the difference

$$\delta\Delta^L_{\uparrow 0} - \delta\Delta^L_{\downarrow 0} = \frac{1}{\pi} \ln\left|\frac{\mu_L - \Delta^0_{10}}{D}\right|^{\Gamma^L_{\downarrow} - \Gamma^L_{\uparrow}} \tag{6.19}$$

vanishes only if $\Gamma^L_{\uparrow} = \Gamma^L_{\downarrow}$, whereas the difference is negative (positive) whenever $\Gamma^L_{\uparrow} < \Gamma^L_{\downarrow}$ $(\Gamma^L_{\uparrow} > \Gamma^L_{\downarrow})$, which means that $\Delta_{\uparrow 0} < \Delta_{\downarrow 0}$ $(\Delta_{\uparrow 0} > \Delta_{\downarrow 0})$. Consequently, a spin-splitting is induced in the quantum dot when it is contacted by a magnetic lead to the left. The same argument holds for the right contribution and is valid for arbitrary U, and non-equilibrium conditions.

The spin-polarizing shifts of the transition energies induced by the left and right leads provide a combined effect, such that it becomes maximal when the reservoirs are magnetically parallel, whereas the spin-split becomes minimal for anti-parallel alignment of magnetizations in the leads, see Fig. 6.9(a). A continuous rotation of the magnetic direction in e.g. the right lead $(0 \leq \phi \leq \pi)$, yields a continuous variation of the induced spin split from its maximum to its minimum. In particular, the minimum split in equilibrium (black) is zero for $\Gamma^L_{\sigma} = \Gamma^R_{\sigma}$, whereas the finite bias

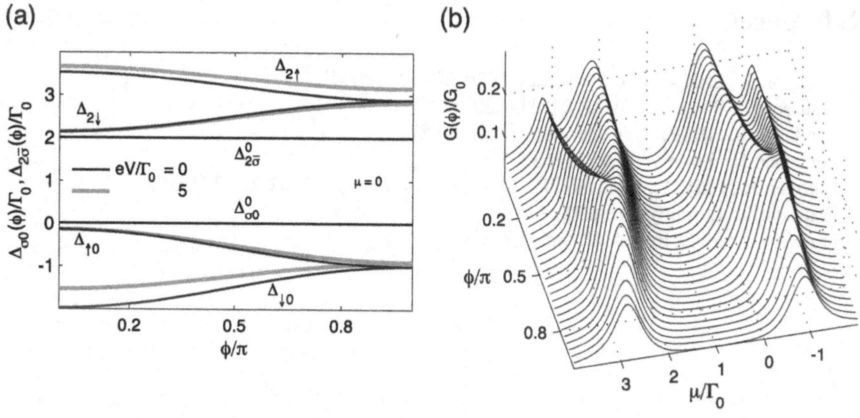

Fig. 6.9 (**a**) Dressed transition energies in equilibrium (*black*) and non-equilibrium (*grey*) at $eV/\Gamma_0 = 5$ as function of the rotation angle ϕ and (**b**) conductance $G(\phi)/G_0$ ($G_0 = e^2/h$) as function ϕ and equilibrium chemical potential μ. In (**a**), the bare transition energies $\Delta^0_{\sigma 0}$ and $\Delta^0_{2\sigma}$ are included for reference. Here $\varepsilon_0 = 0$, $U = 2$, and $\mu = 0$ in units of Γ_0, whereas the spin-asymmetry $p_{L/R} = 0.85$ at $k_B T/\Gamma_0 = 0.08$

in non-equilibrium (grey) yields an unequal shift from the left and right leads which causes a difference between the state energies even in the anti-parallel configuration which is illustrates in Fig. 6.9(a).

The equilibrium conductance $G(\phi)/G_0$ ($G_0 = e^2/h$) as function of the rotation angle ϕ and chemical potential μ for a typical set-up is displayed in Fig. 6.9(b). For $\phi = \pi$, which corresponds to parallel magnetic configuration, it is readily seen that the system is resonant at four different energies due to the spin-splitting of the quantum dot level. For increasing angle, the difference between the transition energies $\Delta_{\uparrow 0}$ ($\Delta_{2\downarrow}$) and $\Delta_{\downarrow 0}$ ($\Delta_{2\uparrow}$) decreases which is in agreement with the behavior illustrated in Fig. 6.9(a).

A striking feature of the conductance is that it is not a monotonic function of ϕ whenever the μ is in the vicinity of any of the transition energies. From any linear response mean-field theory it is expected that the conductance varies with the local current density $j(\omega, \phi) \sim \text{tr Im}\, \Gamma^L \mathbf{G}^r(\omega, \phi)\Gamma^R \mathbf{G}^a(\omega, \phi)$, at the chemical potential. This picture is not altered here. However, the varying positions of the transition energies as function of ϕ provide and additional feature, namely, that the conductance is not necessarily maximal for parallel or anti-parallel alignment of the magnetic leads. It is also clear that the conductance is a strict monotonic function of ϕ whenever μ lies either below of above the transitions energies $\Delta_{\sigma 0}$ and $\Delta_{2\sigma}$, or in the gap between them. The non-monotonic characteristics of the conductance is predicted to be a feature of the strongly coupled regime, i.e. $|\Delta_{\sigma 0} - \mu|/\Gamma_0 \ll 1$ or $|\Delta_{2\sigma} - \mu| \ll 1$. In the weakly coupled regime, i.e. Coulomb blockade, the system returns to a normal spin-valve behavior.

References

1. Fransson, J., Eriksson, O.: Phys. Rev. B **70**, 085301 (2004)
2. Fransson, J.: Phys. Rev. B **72**, 075314 (2005)
3. Landauer, R.: IBM J. Res. Develop. **1**, 223 (1957)
4. Landauer, R.: Philos. Mag. **21**, 863 (1970)
5. Büttiker, R., Imry, Y., Landauer, R., Pinhaus, S.: Phys. Rev. B **31**, 6207 (1985)
6. Fransson, J.: Europhys. Lett. **70**, 796 (2005)
7. Fransson, J.: Phys. Rev. B **72**, 045415 (2005)

Chapter 7
Coupling to Vibrational Mode

Abstract A higher degree of complexity is added by including coupling between electrons and local vibrational modes. Here, we begin by considering the theory for electronic transport through molecular quantum dots in which bosonic modes, e.g. vibrations, are present. We finally show a procedure which allows to also include the bosonic degrees of freedom into the many-body operators, thereby putting the electronic and bosonic degrees of freedom on the same level.

7.1 Introduction

In the 1960s, measurements were made on tunnel junctions that were exposed to e.g. propionic acid [$CH_3(CH_2)COOH$] and acetic acid [CH_3COOH] [1]. When compared to the tunnel junctions that were not exposed to the acid treatment, it became obvious that the molecules added in the tunnel junctions, generated additional signatures in the conductance spectrum. The additional features in the spectrum could not be explained by the usual elastic theory for the conductance, i.e. the *normal* electron levels that are due to level quantization. It was necessary to add electron levels that are caused by interactions between electronic and bosonic degrees of freedom in the tunnel junction, that is, inelastic modes.

Since the first measurements on nanoscale structures in the 1960s, e.g. tunnel junctions, there are innumerably many reports of phenomenon where couplings between electronic and bosonic degrees of freedom are pertinent. Particularly on molecular systems, where a molecular structure, e.g. C_{60} [2], H_2 [3], SC_8H_16S [4], or molecule comprising a single transition metal ion (Co^{2+}) [5], have been placed between electrodes, additional conductance peaks have been measured, conductance peaks that cannot be explained solely by the electronic structure of the molecular systems. Moreover, scanning tunneling microscopy (STM) measurements, performed on e.g. single O_2 molecule chemisorbed on Ag(110) [6] and directly on Au(111) and Cu(111) surfaces [7], reveal features in the conductance that are generated by the coupling between electrons and vibrations (vibrons).

J. Fransson, *Non-Equilibrium Nano-Physics*,
Lecture Notes in Physics 809,
DOI 10.1007/978-90-481-9210-6_7,

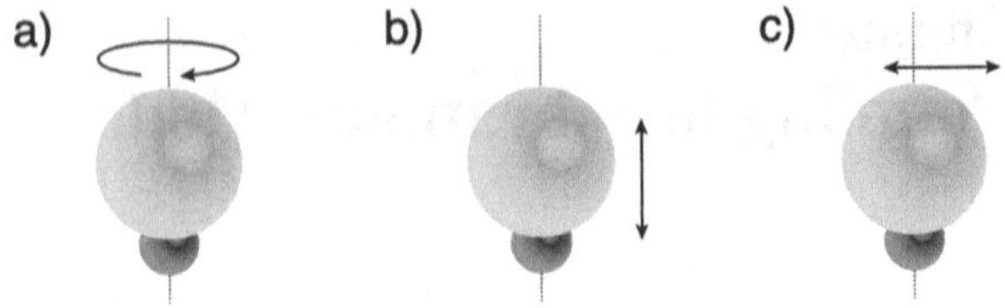

Fig. 7.1 Alternative mechanical vibrational motions in a simple diatomic molecule. (**a**) Rotational oscillation, (**b**) and (**c**) translational oscillations

Theoretically, many of these experimental evidences have been modeled within one or another framework. Persson and Baratoff [8] assumed the vibrational (vibronic) mode to be inherent in the resonant level of the adsorbate molecule in the STM set-up. One can think of the vibronic mode as a mechanical motion of the molecule, e.g. as a rotational oscillation, see Fig. 7.1(a), or translational oscillation, Fig. 7.1(b) and (c). Other types of mechanical oscillations are of course possible, e.g. breathing mode. Irrespective of the nature of the generation of the vibronic mode, by suggesting that $\varepsilon(Q)$ is the resonant level for a fixed displacement Q, and expanding to first order in Q one obtains the resonant energy $\varepsilon(Q)\approx\varepsilon_0+\varepsilon_0'Q=\varepsilon_0+\varepsilon_0'(a^\dagger+a)\hbar/\sqrt{2m^*\Omega}$, where m^* and $\Omega/\hbar$ is the reduced mass and frequency of the vibrational mode, respectively, whereas $a^\dagger$ (a) creates (annihilates) the vibrational mode. This treatment leads to the effective model for the resonant level as $\varepsilon(Q)d^\dagger d\approx\varepsilon_0 d^\dagger d+\lambda d^\dagger d(a^\dagger+a)$, where $\lambda=\varepsilon_0'\hbar/\sqrt{2m^*\Omega}$ defines the coupling energy between the electronic and vibronic modes.

This type of model for the vibrational mode included in the structure defines the starting point for many studies of the vibronic contributions to the current, e.g. in connection with STM [9, 10], and in molecular quantum dots studies of the temperature dependence of the transport [11, 12], current and shot noise [12, 13], level broadening [14], and hysteresis, switching, and negative differential conductance [15], and electron spin resonance (ESR) [16].

7.2 Local Electron Coupled to Vibrational Mode

Here we simply adopt the simple concept for the coupling between the electronic and vibronic degrees of freedom, and we model the single level molecular quantum dot coupled to leads with the Hamiltonian

$$\begin{aligned}\mathcal{H}=&\sum_{\mathbf{k}\sigma\in L,R}\varepsilon_{\mathbf{k}\sigma}c^\dagger_{\mathbf{k}\sigma}c_{\mathbf{k}\sigma}+\sum_\sigma\varepsilon_\sigma d^\dagger_\sigma d_\sigma+Un_\uparrow n_\downarrow+\omega_0 a^\dagger a\\&+\lambda\sum_\sigma d^\dagger_\sigma d_\sigma(a^\dagger+a)+\sum_{\mathbf{k}\sigma}(v_{\mathbf{k}\sigma}c^\dagger_{\mathbf{k}\sigma}d_\sigma+H.c.),\end{aligned}\tag{7.1}$$

where ω_0 is the frequency of the vibration. In principle, there is nothing that hinders us from including more vibrational modes, however, for simplicity we make our following analysis for a single one.

7.2.1 Weak Electron-Vibron Coupling

The first approach to this new type of problem is to consider a canonical transformation of the molecular subsystem

$$\mathcal{H}_{\rm mol} = \sum_\sigma \varepsilon_\sigma d_\sigma^\dagger d_\sigma + U n_\uparrow n_\downarrow + \omega_0 a^\dagger a + \lambda \sum_\sigma d_\sigma^\dagger d_\sigma (a^\dagger + a). \tag{7.2}$$

In our first attempt, we also omit the charging energy U and the spin degree of freedom, hence our (over) simplified molecule can be written

$$\mathcal{H}_{\rm mol} = \varepsilon_0 d^\dagger d + \omega_0 a^\dagger a + \lambda d^\dagger d (a^\dagger + a). \tag{7.3}$$

This model can be diagonalized through the following procedure. We wish to find a transformation $\mathcal{U}$ such that

$$e^{\mathcal{U}^\dagger} \mathcal{H}_{\rm mol} e^{\mathcal{U}} = \tilde{\varepsilon} d^\dagger d + \tilde{\omega}_0 a^\dagger a, \tag{7.4}$$

that is, the molecular model becomes diagonal, possibly at the cost of redefining the energies for the electronic and vibronic modes. The operator transformation

$$e^S A e^{-S} = A + [S, A] + \frac{1}{2!}[S,[S,A]] + \frac{1}{3!}[S,[S,[S,A]]] + \cdots, \tag{7.5}$$

using

$$S = \frac{\lambda}{\omega_0} d^\dagger d (a^\dagger - a), \tag{7.6}$$

giving $S^\dagger = -S$, such that $e^S = (e^{-S})^\dagger$, does the job. We, then, obtain the commutators

$$[S, \mathcal{H}_{\rm mol}] = -\frac{\lambda}{\omega_0}\big(\omega_0 [a^\dagger + a] + 2\lambda\big) d^\dagger d,$$

$$[S,[S,\mathcal{H}_{\rm mol}]] = 2\frac{\lambda^2}{\omega_0} d^\dagger d, \qquad [S,[S,[S,\mathcal{H}_{\rm mol}]]] = 0.$$

Hence, the new Hamiltonian $\tilde{\mathcal{H}}_{\rm mol} = e^S \mathcal{H} e^{-S}$ contains two new terms, effectively resulting in

$$\tilde{\mathcal{H}}_{\rm mol} = \tilde{\varepsilon} d^\dagger d + \omega_0 a^\dagger a, \tag{7.7}$$

where $\tilde{\varepsilon} = \varepsilon - \lambda^2/\omega_0$.

The method is now easy to generalize to the situation we interested in, namely, the Hamiltonian given in (7.2). We take the operator

$$S = \frac{\lambda}{\omega_0} \sum_\sigma d_\sigma^\dagger d_\sigma (a^\dagger - a), \tag{7.8}$$

and we obtain the transformed Hamiltonian

$$\tilde{\mathcal{H}}_{\text{mol}} = \sum_\sigma \tilde{\varepsilon}_\sigma d_\sigma^\dagger d_\sigma + \tilde{U} n_\uparrow n_\downarrow + \omega_0 a^\dagger a, \tag{7.9}$$

where we now have $\tilde{\varepsilon}_\sigma = \varepsilon_\sigma - \lambda^2/\omega_0$ and $\tilde{U} = U - 2\lambda^2/\omega_0$.

It is, thus, clear that the canonical transformation provides a decoupling of the electronic and vibronic degrees of freedom, at least in the atomic limit model. The question that arise is what happens with the electron and vibron operators. This is something we need to figure out since we want to couple our molecule to the electrodes, or leads. Therefore, in applying the canonical transformation to the individual operators, we obtain

$$\tilde{d}_\sigma = d_\sigma \mathcal{X}, \qquad \mathcal{X} = e^{-(\lambda/\omega_0)(a^\dagger - a)}, \qquad \tilde{a} = a - \frac{\lambda}{\omega_0}\sum_\sigma d_\sigma^\dagger d_\sigma,$$

$$\tilde{d}_\sigma^\dagger = d_\sigma^\dagger \mathcal{X}^\dagger, \qquad \mathcal{X}^\dagger = e^{(\lambda/\omega_0)(a^\dagger - a)}, \qquad \tilde{a}^\dagger = a^\dagger - \frac{\lambda}{\omega_0}\sum_\sigma d_\sigma^\dagger d_\sigma,$$

which operators satisfy $\tilde{n}_\sigma = \tilde{d}_\sigma^\dagger \tilde{d}_\sigma = d_\sigma^\dagger \mathcal{X}^\dagger d_\sigma \mathcal{X} = n_\sigma \mathcal{X}^\dagger \mathcal{X} = n_\sigma$, and $\tilde{a}^\dagger \tilde{a} = a^\dagger a - (\lambda/\omega_0)\sum_\sigma n_\sigma (a^\dagger + a) + (\lambda/\omega_0)^2 \sum_{\sigma\sigma'} n_\sigma n_{\sigma'}$. Inserting these new operators into (7.2), of course, also results in (7.9).

Thinking in terms of the Green function for the levels in the molecule, we find from the definition that the Green function $G_{\sigma\sigma'}(t,t') = (-i)\langle \mathrm{T} d_\sigma(t) d_{\sigma'}^\dagger(t')\rangle$ becomes

$$\begin{aligned}
G_{\sigma\sigma'}(t,t') &= (-i)\langle \mathrm{T} d_\sigma(t) d_{\sigma'}^\dagger(t')\rangle = (-i)\langle \mathrm{T} d_\sigma(t) d_{\sigma'}^\dagger(t') e^{-S} e^{S}\rangle \\
&= (-i)\langle \mathrm{T} e^{S} d_\sigma(t) d_{\sigma'}^\dagger(t') e^{-S}\rangle = (-i)\langle \mathrm{T} \tilde{d}_\sigma(t) \tilde{d}_{\sigma'}^\dagger(t')\rangle \\
&= (-i)\langle \mathrm{T} d_\sigma(t) \mathcal{X}(t) d_{\sigma'}^\dagger(t') \mathcal{X}^\dagger(t')\rangle \\
&= (-i)\langle \mathrm{T} d_\sigma(t) d_{\sigma'}^\dagger(t')\rangle_{\text{el}} \langle \mathcal{X}(t)\mathcal{X}^\dagger(t')\rangle_{\text{vib}} \\
&= \tilde{G}_{\sigma\sigma'}(t,t') \langle \mathcal{X}(t)\mathcal{X}^\dagger(t')\rangle_{\text{vib}},
\end{aligned} \tag{7.10}$$

where the subscripts refer to averaging over the electronic and vibronic degrees of freedom, respectively. The meaning of the above equalities is that we can calculate the properties of the resonant level in the molecule as we have done before, but in the Hamiltonian system (7.9), and then finally multiply it by the average $\langle \mathcal{X}(t)\mathcal{X}^\dagger(t')\rangle_{\text{vib}}$, where $\mathcal{X}(t) = e^{i\omega_0 a^\dagger a t}\mathcal{X}e^{-i\omega_0 a^\dagger a t}$. This average can, on the other hand, be exactly calculated. We have to be a bit cautious when using the introduced procedure though, since this procedure works very well when the coupling between the electrons and vibrons is weak. There will be an influence from the electron-vibron coupling on the tunneling, which we cannot neglect in general. However, doing first things first, we will return to these issues later.

For a calculation of the average $\langle \mathcal{X}(t)\mathcal{X}^\dagger(t')\rangle_{\text{vib}}$, we introduce the function

$$F(t,t') = \langle \mathcal{X}(t)\mathcal{X}^\dagger(t')\rangle_{\text{vib}} = \frac{\sum_{n=0}^{\infty}\langle n|e^{-\beta\omega_0 a^\dagger a}\mathcal{X}(t)\mathcal{X}(t')|n\rangle}{\sum_{n=0}^{\infty}\langle n|e^{-\beta\omega_0 a^\dagger a}|n\rangle}, \tag{7.11}$$

where $|n\rangle = (a^\dagger)^n|0\rangle/\sqrt{n!}$ denotes the bosonic state with n excitations. Here, the denominator equals $\sum_n e^{-\beta\omega_0 n} = (1 - e^{-\beta\omega_0})^{-1} = e^{\beta\omega_0}N_p$, where $N_p = (e^{\beta\omega_0} - 1)^{-1}$. The time-development of the operators $\mathcal{X}$ can be rewritten from its fundamental expression to read

$$\mathcal{X}(t) = e^{i\omega_0 a^\dagger a t}\mathcal{X}e^{-i\omega_0 a^\dagger a t} = e^{-(\lambda/\omega_0)^2/2}e^{i\omega_0 a^\dagger a t}e^{-\lambda a^\dagger/\omega_0}e^{\lambda a/\omega_0}e^{-i\omega_0 a^\dagger a t}, \tag{7.12}$$

where we have used Feynman's theorem for disentangling of operators, i.e. $e^{A+B} = e^A e^B e^{-[A,B]/2}$ if both A and B commutes with $[A,B]$ [17]. Then, between the third and fourth exponential we insert $e^{-i\omega_0 a^\dagger a t}e^{i\omega_0 a^\dagger a t}$, and, moreover, using that

$$\begin{aligned}
&e^{i\omega_0 a^\dagger a t}e^{-\lambda a/\omega_0}e^{-i\omega_0 a^\dagger a t} \\
&\quad = \sum_{n=0}^{\infty}\frac{(-\lambda/\omega_0)^n}{n!}\{a^n + i\omega_0 t[a^\dagger a, a^n] + \cdots\} \\
&\quad = \sum_{n=0}^{\infty}\frac{(-\lambda/\omega_0)^n}{n!}a^n e^{-i\omega_0 n t} = \exp\left(-\frac{\lambda}{\omega_0}ae^{-i\omega_0 t}\right),
\end{aligned} \tag{7.13}$$

we find

$$\begin{aligned}
\mathcal{X}(t) &= e^{-(\lambda/\omega_0)^2/2}\exp\left(-\frac{\lambda}{\omega_0}a^\dagger e^{i\omega_0 t}\right)\exp\left(\frac{\lambda}{\omega_0}ae^{-i\omega_0 t}\right) \\
&= e^{-(\lambda/\omega_0)^2}\exp\left(-\frac{\lambda}{\omega_0}\left[a^\dagger e^{i\omega_0 t} - ae^{-i\omega_0 t}\right]\right).
\end{aligned} \tag{7.14}$$

The average $F(t,t')$ can now be written

$$\begin{aligned}
F(t,t') = {}& \frac{e^{-\beta\omega_0}}{N_p}\sum_{n=0}^{\infty}e^{-\beta\omega_0 n}e^{-(\lambda/\omega_0)^2} \\
&\times \langle n|e^{-\lambda a^\dagger e^{i\omega_0 t}/\omega_0}e^{\lambda ae^{-i\omega_0 t}/\omega_0}e^{\lambda a^\dagger e^{-i\omega_0 t'}/\omega_0}e^{-\lambda ae^{i\omega_0 t'}/\omega_0}|n\rangle.
\end{aligned} \tag{7.15}$$

The next step is to arrange the operators such that all destruction operators stand to the right in the average. We begin by considering the product

$$e^{\lambda ae^{-i\omega_0 t}/\omega_0}e^{\lambda a^\dagger e^{-i\omega_0 t'}/\omega_0} = e^{\lambda a^\dagger e^{-i\omega_0 t'}/\omega_0}\left[e^{-\lambda a^\dagger e^{-i\omega_0 t'}/\omega_0}e^{\lambda ae^{-i\omega_0 t}/\omega_0}e^{\lambda a^\dagger e^{-i\omega_0 t'}/\omega_0}\right]. \tag{7.16}$$

The expression in brackets is straightforwardly calculated using the expansions introduced above. The result is

$$e^{\lambda ae^{-i\omega_0 t}/\omega_0}e^{\lambda a^\dagger e^{-i\omega_0 t'}/\omega_0} = e^{-(\lambda/\omega_0)^2 e^{-i\omega_0(t-t')}}e^{\lambda a^\dagger e^{-i\omega_0 t'}/\omega_0}e^{\lambda ae^{-i\omega_0 t}/\omega_0}. \tag{7.17}$$

Hence, the average can be written as

$$F(t,t') = \exp\left\{-\left(\frac{\lambda}{\omega_0}\right)^2 \left[1 - e^{-i\omega_0(t-t')}\right]\right\} \frac{e^{-\beta\omega_0}}{N_p} \times \sum_{n=0}^{\infty} e^{-\beta\omega_0 n} \langle n| e^{\lambda a^\dagger [e^{i\omega_0 t'} - e^{i\omega_0 t}]/\omega_0} e^{-\lambda a [e^{-i\omega_0 t'} - e^{-i\omega_0 t}]/\omega_0} |n\rangle. \quad (7.18)$$

The final step in the calculation of the average is taken by acting with the exponential $e^{-\phi(t)a}$ on the states $|n\rangle$. Here, it is advisable to notice that

$$a|n\rangle = \sqrt{n}|n-1\rangle, \quad a^2|n\rangle = \sqrt{n(n-1)}|n-2\rangle, \ldots, a^m|n\rangle = \sqrt{\frac{n!}{(n-m)!}}|n-m\rangle,$$

such that $|n-m\rangle = 0$ for all $m > n$. Summing up the series results in

$$e^{-\phi(t)a}|n\rangle = \sum_{m=0}^{n} (-1)^m \frac{\phi^m(t)}{m!} \sqrt{\frac{n!}{(n-m)!}} |n-m\rangle, \quad (7.19a)$$

$$\langle n| e^{\phi^*(t)a^\dagger} = \langle n-m| \sum_{m=0}^{n} \frac{[\phi^*(t)]^m}{m!} \sqrt{\frac{n!}{(n-m)!}}. \quad (7.19b)$$

By orthogonality of the states we, thus, obtain

$$\begin{aligned}
&\langle n| e^{\lambda a^\dagger [e^{i\omega_0 t'} - e^{i\omega_0 t}]/\omega_0} e^{-\lambda a [e^{-i\omega_0 t'} - e^{-i\omega_0 t}]/\omega_0} |n\rangle \\
&= \sum_{m=0}^{n} \frac{(-1)^m}{m!} \frac{n!}{m!(n-m)!} \left(\frac{\lambda}{\omega_0} \left|e^{-i\omega_0 t'} - e^{-i\omega_0 t}\right|\right)^{2m} \\
&= \mathcal{L}_n\left(\left[\frac{\lambda}{\omega_0} \left|e^{-i\omega_0 t'} - e^{-i\omega_0 t}\right|\right]^2\right), \quad (7.20)
\end{aligned}$$

where $\mathcal{L}_n(x) = \mathcal{L}_n^0(x)$ is the Laguerre polynomial. In e.g. [18] one finds that the sum $\sum_{n=0}^{\infty} \mathcal{L}_n(x) y^n = (1-y)^{-1} \exp[(xy)/(y-1)]$. By identifying y by $e^{-\beta\omega_0}$ we can write $(1-y)^{-1} \exp[(xy)/(y-1)] = N_p e^{\beta\omega_0 - xN_p}$, which leads to that we can finally write the average $F(t,t')$ according to

$$\begin{aligned}
F(t,t') &= \exp\left\{-\left(\frac{\lambda}{\omega_0}\right)^2 \left[1 - e^{-i\omega_0(t-t')} + \left|1 - e^{-i\omega_0(t-t')}\right|^2 N_p\right]\right\} \\
&= \exp\left\{-\left(\frac{\lambda}{\omega_0}\right)^2 \left[(1 - e^{-i\omega_0(t-t')})(1+N_p) + (1 - e^{i\omega_0(t-t')})N_p\right]\right\}.
\end{aligned} \quad (7.21)$$

The Green function for the resonant level in the molecular quantum dot coupled to the vibrational mode ω_0, thus, becomes

$$G_{\sigma\sigma'}(t,t') = \tilde{G}_{\sigma\sigma'}(t,t') \exp\left\{-\left(\frac{\lambda}{\omega_0}\right)^2 \left[(1-e^{-i\omega_0(t-t')})(1+N_p) + (1-e^{i\omega_0(t-t')})N_p\right]\right\}. \tag{7.22}$$

Let us now assume stationary conditions, so that we can Fourier transform into frequency space. Doing this for the retarded Green function and assuming that $\tilde{G}_{\sigma\sigma'}(\omega) = \int \tilde{G}_{\sigma\sigma'}(t,t')e^{i\omega\tau}dt'$, where $\tau = t - t'$, give

$$\begin{aligned} G^r_{\sigma\sigma'}(\omega) &= \int \tilde{G}^r_{\sigma\sigma'}(t,t') \exp\left\{-\left(\frac{\lambda}{\omega_0}\right)^2 \left[(1-e^{-i\omega_0\tau})(1+N_p) + (1-e^{i\omega_0\tau})N_p\right]\right\} \\ &= e^{-(\lambda\sqrt{1+2N_p}/\omega_0)^2} \sum_n I_n\left(2\left[\frac{\lambda}{\omega_0}\right]^2 \sqrt{N_p(1+N_p)}\right) \\ &\quad \times e^{n\beta\omega_0/2} \tilde{G}^r_{\sigma\sigma'}(\omega - n\omega_0), \end{aligned} \tag{7.23}$$

where $I_n(x)$ is the modified Bessel function. Hence, if the electronic Green function is given by e.g. $\tilde{G}^r_{\sigma\sigma'}(\omega) = \delta_{\sigma\sigma'}/(\omega - \varepsilon_\sigma + i\delta)$, $\delta > 0$, the full Green function which includes both the electronic and vibronic degrees of freedom becomes

$$G^r_{\sigma\sigma'}(\omega) = \delta_{\sigma\sigma'} e^{-(\lambda\sqrt{1+2N_p}/\omega_0)^2} \sum_n \frac{I_n(2[\frac{\lambda}{\omega_0}]^2\sqrt{N_p(1+N_p)})}{\omega - (\varepsilon_\sigma + n\omega_0) + i\delta} e^{n\beta\omega_0/2}, \tag{7.24}$$

which clearly illustrates the full effect of the coupling between the resonant electronic level and the vibrational mode in that there appear a series of side-peaks ($n \neq 0$) around the resonant level ($n = 0$). This transformation of the single resonant level into a multitude of levels is illustrated in Fig. 7.2.

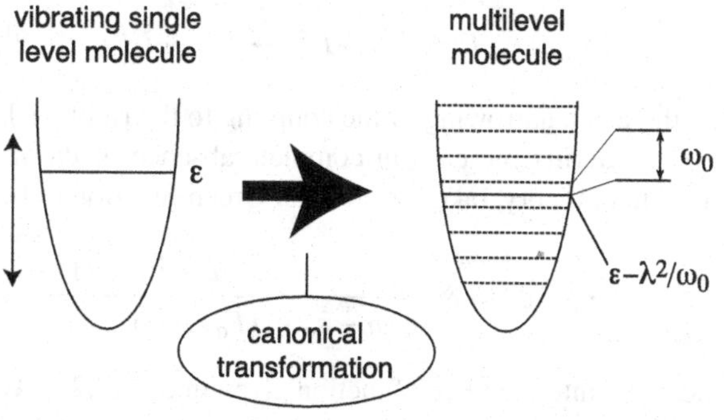

Fig. 7.2 Before and after the canonical transformation, where the molecule before the canonical transformation is represented by a single level structure which is moving back and forth, whereas the molecule has turned into a multilevel structure after the transformation

Recall that the molecular quantum dot is supposed to be coupled to electrodes, or leads, and therefore we include these into the picture at this moment, cf. (7.1). The canonical transformation that we applied to the molecular part of the Hamiltonian also has to be applied to the remainder of the model, giving

$$\tilde{\mathcal{H}} = \sum_{\mathbf{k}\sigma\in L,R} \varepsilon_{\mathbf{k}\sigma} c^\dagger_{\mathbf{k}\sigma} c_{\mathbf{k}\sigma} + \sum_\sigma \tilde{\varepsilon}_\sigma d^\dagger_\sigma d_\sigma + \tilde{U} n_\uparrow n_\downarrow + \omega_0 a^\dagger a$$
$$+ \sum_{\mathbf{k}\sigma} (v_{\mathbf{k}\sigma} c^\dagger_{\mathbf{k}\sigma} d_\sigma \mathcal{X} + H.c.). \tag{7.25}$$

Thus, the diagonalization of the molecular part of the model simply transferred the complexity of the coupling between the electronic and vibronic degrees of freedom into the tunneling term. The problem is only that we wanted to avoid the coupling between the electrons and vibron at all, which is possible under certain conditions. Hewson and Newns [19] investigated this issue for equilibrium situations and found that the electron-vibron coupling becomes important to include whenever

$$2\frac{\lambda^2}{\omega_0} > \Gamma, \qquad \omega_0 > \left|\varepsilon_\sigma - \frac{\lambda^2}{\omega_0}\right|, \tag{7.26a}$$

$$\frac{\Gamma}{2\omega_0} e^{-(\lambda/\omega_0)^2} < 1 < \left(\frac{\lambda}{\omega_0}\right)^2, \tag{7.26b}$$

where $\Gamma = 2\pi \sum_{\mathbf{k}\sigma} |v_{\mathbf{k}\sigma}|^2 \delta(\omega - \varepsilon_{\mathbf{k}\sigma})$. The first inequality means that the electron-vibron coupling is strong in comparison with the tunneling, whereas the second inequality says that the vibron frequency should be large compared to the electronic energy (relative to the Fermi level). The two last inequalities mean that vibronic satellite peaks are separated from one another and the number of vibrons involved in the processes is appreciable, respectively. Thus, if these conditions are not met in the system under investigation, one can neglect the influence from the electron-vibron coupling in the tunneling Hamiltonian.

In non-equilibrium the situation may be different since, e.g. the electron level might be close to the chemical potential of the left lead but not of the right. If this is the case, then the total tunneling coupling strength [11]

$$\Gamma = \Gamma^L + \Gamma^R \to \Gamma^L + \Gamma^R e^{-(\lambda/\omega_0)(1+2N_p)}, \tag{7.27}$$

i.e. there is a narrowing of the coupling to the right lead.

Considering the case in complete absence of the narrowing effect one can, in mean field theory, take the electron Green function to be

$$\tilde{G}^r_{\sigma\sigma'}(\omega) = \delta_{\sigma\sigma'} \frac{\omega - \tilde{\varepsilon}_\sigma - \langle 1 - n_{\bar{\sigma}}\rangle \tilde{U}}{(\omega - \tilde{\varepsilon}_\sigma + i\Gamma_\sigma/2)(\omega - \tilde{\varepsilon}_\sigma - \tilde{U}) + i\langle n_{\bar{\sigma}}\rangle \tilde{U}\Gamma_\sigma/2},$$

and put it into the Green function given in e.g. (7.23). We obtain

$$G^r_{\sigma\sigma'}(\omega) = \delta_{\sigma\sigma'} e^{-(\lambda\sqrt{1+2N_p}/\omega_0)^2} \sum_n I_n(z) e^{n\beta\omega_0/2}$$

$$\times \frac{\omega - \tilde{\varepsilon}_\sigma - n\omega_0 - \langle 1 - n_{\bar{\sigma}} \rangle \tilde{U}}{(\omega - \tilde{\varepsilon}_\sigma - n\omega_0 + i\Gamma_\sigma/2)(\omega - \tilde{\varepsilon}_\sigma - n\omega_0 - \tilde{U}) + i\langle n_{\bar{\sigma}} \rangle \tilde{U}\Gamma_\sigma/2}, \tag{7.28}$$

with $z = 2(\lambda/\omega_0)^2\sqrt{N_p(1+N_p)}$. In the limit $U/\Gamma_\sigma \gg 1$ we have approximately

$$\tilde{G}^r_{\sigma\sigma'}(\omega) \approx \delta_{\sigma\sigma'}\left(\frac{\langle 1 - n_{\bar{\sigma}} \rangle}{\omega - \tilde{\varepsilon}_\sigma + i\Gamma_\sigma/4} + \frac{\langle n_{\bar{\sigma}} \rangle}{\omega - \tilde{\varepsilon}_\sigma - \tilde{U} + i\Gamma_\sigma/4}\right),$$

i.e. one resonance at $\tilde{\varepsilon}_\sigma$ and one at $\tilde{\varepsilon}_\sigma + \tilde{U}$, both with a width of roughly $\Gamma_\sigma/2$, which in the vibrating molecular quantum dot results in

$$G^r_{\sigma\sigma'}(\omega) = \delta_{\sigma\sigma'} e^{-(\lambda\sqrt{1+2N_p}/\omega_0)^2} \sum_n I_n(z) e^{n\beta\omega_0/2}\left(\frac{\langle 1 - n_{\bar{\sigma}} \rangle}{\omega - \tilde{\varepsilon}_\sigma - n\omega_0 + i\Gamma_\sigma/4}\right.$$
$$\left. + \frac{\langle n_{\bar{\sigma}} \rangle}{\omega - \tilde{\varepsilon}_\sigma - n\omega_0 - \tilde{U} + i\Gamma_\sigma/4}\right). \tag{7.29}$$

In this expression one clearly sees that there appear vibrational side-peaks around both resonant levels, as one would expect.

In the above discussion we have completely neglected the influence of the electron-vibron coupling on the Green function self-energy. Later in this chapter, we will return to this issue and also address a way to include such effects in the Green function.

7.2.2 *Electron Spin Resonance*

We consider an example of an application of the small polaron formation model in connection to the ESR set-up for tunnel current systems. In this set-up, the spin of a local defect is measured by variations in the charge, or spin, current as the frequency of a time-dependent magnetic field is varied.

Here, one may think of a resonator which is oscillating with the frequency ω_0 and where the mechanical oscillations are weakly coupled to the molecular electrons—we denote the coupling strength by λ. We have in mind a single molecular level ε_0 coupled to external baths, or leads. The level is spin-split by the external magnetic field B_0, such that $\varepsilon_\downarrow - \varepsilon_\uparrow = \omega_r = g\mu_B B_0$, where g and μ_B is the gyromagnetic ratio and Bohr magneton, respectively. In addition, the spins are coupled to each other by a rotating magnetic field $B_1(\cos\omega_1 t, \sin\omega_1 t)$ which is applied perpendicular to B_0. We assume that $2g\mu_B B_1$ is much less than both ω_0 and ω_r. In principle, we employ the model $\mathcal{H}=\mathcal{H}_c+\mathcal{H}_d+\mathcal{H}_T$ [16], where

$$\mathcal{H}_d=\sum_\sigma\left[\varepsilon_\sigma+\lambda(a^\dagger+a)+\frac{U}{2}n_{\bar\sigma}\right]n_\sigma-g\mu_B B_1\left[d_\uparrow^\dagger d_\downarrow r^{i\omega_1 t}+d_\downarrow^\dagger d_\uparrow r^{-i\omega_1 t}\right]+\omega_0 a^\dagger a, \tag{7.30}$$

$\mathcal{H}_c=\sum_{\mathbf{k}\sigma}\varepsilon_\mathbf{k}c_{\mathbf{k}\sigma}^\dagger c_{\mathbf{k}\sigma}$, and $\mathcal{H}_T=\sum_{\mathbf{k}\sigma}v_\mathbf{k}c_{\mathbf{k}\sigma}^\dagger d_\sigma+H.c.$, describe the electrons in the molecule, the reservoir, and their mutual tunneling interactions, respectively.

The second term in $\mathcal{H}_d$ is non-diagonal in terms of the Fermi operators and, moreover, it is time-dependent. First, we remove the time-dependence in the model by transforming the system into the rotating reference frame of the magnetic field. This can be achieved through the unitary transformation

$$\mathcal{H}_\text{rf}=e^{S_\text{rf}}\mathcal{H}e^{-S_\text{rf}}+i\left(\frac{\partial}{\partial t}e^{S_\text{rf}}\right)e^{-S_\text{rf}}, \tag{7.31a}$$

$$S_\text{rf}=i\frac{\omega_1 t}{2}\left[n_\downarrow-n_\uparrow+\sum_\mathbf{k}(n_{\mathbf{k}\downarrow}-n_{\mathbf{k}\uparrow})\right]. \tag{7.31b}$$

Again using the expansion $e^S A e^{-S}=A+[S,A]+\frac{1}{2!}[S,[S,A]]+\cdots$ we obtain, for instance,

$$\begin{aligned}e^{S_\text{rf}}d_\uparrow^\dagger d_\downarrow e^{i\omega_1 t}e^{-S\text{rf}}&=\left(d_\uparrow^\dagger d_\downarrow+i\frac{\omega_1 t}{2}[n_\downarrow-n_\uparrow,d_\uparrow^\dagger d_\downarrow]\right.\\&\quad\left.+\left[i\frac{\omega_1 t}{2}\right]^2\frac{1}{2!}[n_\downarrow-n_\uparrow,[n_\downarrow-n_\uparrow,d_\uparrow^\dagger d_\downarrow]]+\cdots\right)e^{i\omega_1 t}\\&=d_\uparrow^\dagger d_\downarrow\left(1+i\frac{\omega_1 t}{2}(-2)+\left[i\frac{\omega_1 t}{2}\right]^2\frac{(-2)^2}{2!}+\cdots\right)e^{i\omega_1 t}\\&=d_\uparrow^\dagger d_\downarrow e^{-i\omega_1 t}e^{i\omega_1 t}=d_\uparrow^\dagger d_\downarrow.\end{aligned} \tag{7.32}$$

The overall effect of the transformation results in

$$\begin{aligned}\mathcal{H}_\text{rf}&=\sum_{\mathbf{k}\sigma}\varepsilon_{\mathbf{k}\sigma}^\text{rf}n_{\mathbf{k}\sigma}+\omega_0 a^\dagger a+\mathcal{H}_T\\&\quad+\sum_\sigma\left(\varepsilon_\sigma^\text{rf}+\lambda(a^\dagger+a)+\frac{U}{2}n_{\bar\sigma}\right)n_\sigma-g\mu_B B_1(d_\uparrow^\dagger d_\downarrow+H.c.),\end{aligned} \tag{7.33}$$

where $\varepsilon_\sigma^\text{rf}=\varepsilon_\sigma+\sigma\omega_1/2$ and $\varepsilon_{\mathbf{k}\sigma}^\text{rf}=\varepsilon_\mathbf{k}+\sigma\omega_1/2$.

Clearly, removing the time-dependence in the model comes with the cost that we introduce a spin-splitting of the electronic energies. The spin-splitting of the electrons in the reservoir originates from the magnetic pumping field acting on the localized electrons which, in turn, hybridize with the de-localized electrons in the reservoir. The magnetic pumping propagates energy from the molecule to the reservoir and generates the spin chemical potentials $\mu_\sigma=-\sigma\omega_1/2$ in the reservoir (remember that all energies are given relative to the equilibrium chemical potential μ).

In this sense, the frequency ω_1 of the oscillating field can be viewed as a (spin) bias applied to the system. Despite this imbalance between the spin channels the charge chemical potential is, nevertheless, $\mu = (\mu_\uparrow + \mu_\downarrow)/2 = 0$.

Although the system as a whole remains in equilibrium, the one-photon imbalance between the spin-channels generates a non-equilibrium condition for the two spin projections of the electrons. A spin $\downarrow$ electron the in the reservoir at the energy $\varepsilon^{\text{rf}}_{\mathbf{k}\downarrow}$ can tunnel into the local molecular level $\varepsilon^{\text{rf}}_\downarrow$. The rotating magnetic field does, in turn, flip the spin of the electron and puts it into the level $\varepsilon^{\text{rf}}_\uparrow$, from which it may tunnel into the reservoir state $\varepsilon^{\text{rf}}_{\mathbf{k}\uparrow}$. Repeated occurrence of such tunneling and spin-flip events build up a stationary current between the two spin-channels in the reservoir.

The next step is to decouple the vibronic and electronic degrees of freedom, for which we employ the procedure of previous section, i.e. introducing $\tilde{\mathcal{H}} = e^{S_{\text{ph}}}\mathcal{H}_{\text{rf}}e^{-S_{\text{ph}}}$, $S_{\text{ph}} = (\lambda/\omega_0)(a^\dagger - a)\sum_\sigma n_\sigma$. We then obtain the transformed energy levels $\tilde{\varepsilon}_\sigma = \varepsilon^{\text{rf}}_\sigma - \lambda^2/\omega_0$ and charging energy $\tilde{U} = U - 2\lambda^2/\omega_0$, while the tunneling Hamiltonian is shifted into $\tilde{\mathcal{H}}_T = \sum_{\mathbf{k}\sigma} v_{\mathbf{k}} c^\dagger_{\mathbf{k}\sigma} d_\sigma \mathcal{X} + H.c.$ Here, we make use of the assumptions of small currents and weak coupling between the vibronic and electronic levels and, thus, neglect the effect of the vibrons on the tunneling, i.e. let $\mathcal{X} \to \langle \mathcal{X} \rangle_{\text{ph}}$ and let $v_{\mathbf{k}}\langle \mathcal{X} \rangle_{\text{ph}} \to v_{\mathbf{k}}$ in the tunneling Hamiltonian.

In the atomic limit ($\tilde{U} = 0$) and in absence of the vibron-electron coupling ($\lambda = 0$), the molecule is reduced to a simple driven two-level system, which is characterized by a coherent weight transfer between the two spin states, i.e. Rabi oscillations. This weight transfer has the resonance frequency $\omega_1 = \omega_r$. Further, the Rabi frequency of the spin oscillations is given by $\Omega = \sqrt{\Delta^2 + 4(g\mu_B B_1)^2}$, where $\Delta = \omega_1 - \omega_r$ denotes the detuning from the resonance.

We notice that the electronic states of the molecule still are represented in a non-diagonal fashion, both in that there is a spin-flip term and there is a charging term in the Hamiltonian. This can be taken care of by different means, and first we approach the problem by a semi-diagonalization of the molecular electron states through the transformation [20]

$$\begin{pmatrix} d_\uparrow \\ d_\downarrow \end{pmatrix} = \mathbf{u} \begin{pmatrix} c_\uparrow \\ c_\downarrow \end{pmatrix}, \quad \mathbf{u} = \begin{pmatrix} \cos\phi & -\sin\phi \\ \sin\phi & \cos\phi \end{pmatrix}, \tag{7.34}$$

where $\tan\phi = 2g\mu_B B_1/(\Omega - \Delta)$. In this new basis for the molecular electron states the molecular Hamiltonian becomes

$$\tilde{\mathcal{H}}_d = \sum_\sigma E_\sigma c^\dagger_\sigma c_\sigma + \tilde{U} c^\dagger_\uparrow c_\uparrow c^\dagger_\downarrow c_\downarrow, \tag{7.35}$$

with $E_\sigma = (\tilde{\varepsilon}_\uparrow + \tilde{\varepsilon}_\downarrow - \sigma\Omega)/2 = \varepsilon_0 - \lambda^2/\omega_0 - \sigma\Omega/2$.

The current we are to consider is the flow of electrons of a given spin projection σ in the reservoir, through the molecule and back into the reservoir in a different spin state. Thus, we cannot simply study the full charge current $I = -e\partial_t \sum_{\mathbf{k}\sigma}\langle n_{\mathbf{k}\sigma}\rangle$, since this current must vanish. The current we are looking for is, however, the spin current $I_\sigma = -e\partial_t \sum_{\mathbf{k}}\langle n_{\mathbf{k}\sigma}\rangle$, and following the procedure introduce in Chap. 4, we find that this current can be written as

$$I_\sigma = \frac{ie}{\hbar}\mathrm{tr}\int \Gamma_\sigma\big\{f_\sigma(\omega)\mathbf{G}^>(\omega) + [1-f_\sigma(\omega)]\mathbf{G}^<(\omega)\big\}d\omega. \tag{7.36}$$

Here, the coupling matrix $\Gamma_\sigma = \mathbf{u}_\sigma \Gamma$, with $\Gamma = 2\pi\sum_{\mathbf{k}}|v_{\mathbf{k}}|^2\delta(\omega-\varepsilon_{\mathbf{k}})$, and

$$\mathbf{u}_\uparrow = \sigma^y \mathbf{u}_\downarrow \sigma^y, \quad \mathbf{u}_\downarrow = \begin{pmatrix} \sin^2\phi & \sin\phi\cos\phi \\ \sin\phi\cos\phi & \cos^2\phi \end{pmatrix}. \tag{7.37}$$

The y-component of the Pauli matrix vector is represented by σ^y, whereas $f_\sigma(\omega) = f(\omega-\mu_\sigma)$ is the Fermi function for the spin σ channel. The Green function matrix is defined by $\mathbf{G}^{</>} = \{G^{</>}_{\sigma\sigma'}\}_{\sigma\sigma'}$, and is calculated using $\mathbf{G}^{</>} = \mathbf{G}^r\Sigma^{</>}\mathbf{G}^a$, where e.g. the retarded Green function $G^r_{\sigma\sigma'}(t,t') = (-i)\theta(t-t')\langle\{c_\sigma(t), c^\dagger_{\sigma'}(t')\}\rangle$. The canonical electron-vibron decoupling procedure casts the Green function into the form $G^r_{\sigma\sigma'}(t,t') = \tilde{G}^r_{\sigma\sigma}(t,t')\langle\mathcal{X}(t)\mathcal{X}(t')\rangle_{\mathrm{ph}}$.

Calculating the electronic Green function in the mean field approximation, we find that $\tilde{G}^r_{\sigma\bar{\sigma}}(\omega) = 0$ and $\tilde{G}^r_{\sigma\sigma}(\omega) = \tilde{G}^r_\sigma(\omega)$ with

$$\tilde{G}^r_\sigma(\omega) = \frac{\omega - E_\sigma - \langle 1 - n_{\bar{\sigma}}\rangle\tilde{U}}{(\omega - E_\sigma + i\Gamma/2)(\omega - E_\sigma - \tilde{U}) + i\langle n_{\bar{\sigma}}\rangle\tilde{U}\Gamma/2}, \tag{7.38}$$

and $\langle n_\sigma\rangle = (-i)\int\tilde{G}^<_\sigma(\omega)d\omega/(2\pi)$. Consequently, we have our total Green function as given in (7.28), or

$$G^r_\sigma(\omega) = e^{-(\lambda\sqrt{1+2N_p}/\omega_0)^2}\sum_n I_n(z)e^{n\beta\omega_0/2} \times \frac{\omega - E_\sigma - n\omega_0 - \langle 1-n_{\bar{\sigma}}\rangle\tilde{U}}{(\omega - E_\sigma - n\omega_0 + i\Gamma_\sigma/2)(\omega - E_\sigma - n\omega_0 - \tilde{U}) + i\langle n_{\bar{\sigma}}\rangle\tilde{U}\Gamma_\sigma/2}. \tag{7.28$'$}$$

Under the assumption that the electron-vibron coupling is weak it is justified to neglect the contributions to the self-energy arising due to the electron-vibron interactions. The lesser and greater self-energies can, therefore, be approximated as

$$\Sigma^<(\omega) = if_\uparrow(\omega)\Gamma_\uparrow + if_\downarrow(\omega)\Gamma_\downarrow, \tag{7.39a}$$

$$\Sigma^>(\omega) = -i\big[1 - f_\uparrow(\omega)\big]\Gamma_\uparrow - i\big[1 - f_\downarrow(\omega)\big]\Gamma_\downarrow. \tag{7.39b}$$

One should be aware that those forms of the lesser and greater self-energies are somewhat simplified, since the self-energy which is related to the electron correlations have been omitted. Qualitatively though, the effect we are studying here does not depend on the particular form of the self-energy, but rather on the coupling between different states. Hence, the introduced approximation is justified. Inserting these self-energies into the spin σ current, we obtain

$$I_\sigma = \frac{e}{h}\Gamma^2 \int \mathcal{T}_s(\omega)\big[f_\sigma(\omega) - f_{\bar\sigma}(\omega)\big]d\omega, \tag{7.40a}$$

$$\mathcal{T}_s(\omega) = \big|G^r_\uparrow(\omega) - G^r_\downarrow(\omega)\big|^2 \sin^2\phi\cos^2\phi. \tag{7.40b}$$

Here, we notice that the transmission $\mathcal{T}_s$ is spin-independent which, of course, is expected in the stationary regime. The total spin current is then $I_s = \sum_\sigma \sigma^z_{\sigma\sigma} I_\sigma = 2I_\uparrow$, since $I_\downarrow = -I_\uparrow$.

The form of the transmission coefficient suggests an interpretation of the spin current as interference between the wave functions that propagate back and forth between the reservoir and the molecule. In the present context of electron-vibron coupled system, such an interpretation is especially appealing. In order to make a simple argument of this interpretation we make things as simple as possible and consider the case of vanishing effective charging energy, i.e. $\tilde{U} = 0 \Leftrightarrow U = 2\lambda^2/\omega_0$. The first factor in the transmission can be written

$$\begin{aligned}
&\big|G^r_\uparrow(\omega) - G^r_\downarrow(\omega)\big|^2 \\
&= \left|\sum_n \frac{\Omega I_n(z)e^{n\beta\omega_0/2}}{(\omega-\omega^r_\uparrow - n\omega_0)(\omega - \omega^r_\downarrow - n\omega_0)}\right|^2 \\
&= \left|\frac{\Omega}{(\omega-\omega^r_\uparrow)(\omega-\omega^r_\downarrow)} + \sum_{n\neq 0} \frac{\Omega I_n(z)e^{n\beta\omega_0/2}}{(\omega-\omega^r_\uparrow - n\omega_0)(\omega - \omega^r_\downarrow - n\omega_0)}\right|^2,
\end{aligned} \tag{7.41}$$

where $\omega^{r/a}_\sigma = E_\sigma \mp i\Gamma/2$. The transmission, and hence the current, peaks at the resonance frequency $\omega_1 = \omega_r$, giving rise to the ESR peak. At resonance we also have $\Omega = 2g\mu_B B_1$ and $E_\sigma = \varepsilon_0 - \lambda^2/\omega_0 - \sigma g\mu_B B_1$, which means that $|E_\uparrow - E_\downarrow|$ is minimal with respect to the rotating frequency, see Fig. 7.3(a). In the present form of the spin current, the ESR peak is generated by the first term ($n = 0$) in the expansion of $\mathcal{T}$, which is also the only term that remains in absence of the electron-vibron coupling.

The sum over $n \neq 0$ arises because of the electron-vibron coupling and from this sum we pick up additional features in the spin current, which consequently is an effect from presence of the many conductance channels. The schematically drawn electronic structure for the molecule in this set-up is displayed in Fig. 7.3(b), from which we see that something interesting may occur at frequencies $\omega_1 = \omega_r + n\omega_0$, $n \neq 0$. Then, the assumption $2g\mu_B B_1 \ll |n|\omega_0$ leads to that $\Omega = \sqrt{(n\omega_0)^2 + 4(g\mu_B B_1)^2} \approx |n|\omega_0(1 + 2[g\mu_B B_1/(n\omega_0)]^2)$.

We proceed by studying the first two terms in the expansion $|G^r_\uparrow - G^r_\downarrow|^2$, that is, terms with $n = 0, 1$,

$$\begin{aligned}
&\Omega^2\big[I_0^2(z)|\tilde{G}^r_\uparrow(\omega)\tilde{G}^r_\downarrow(\omega)|^2 + I_1^2(z)|\tilde{G}^r_\uparrow(\omega-\omega_0)\tilde{G}^r_\downarrow(\omega-\omega_0)|^2 e^{\beta\omega_0} \\
&+ 2\,\mathrm{Re}\, I_0(z)I_1(z)\tilde{G}^r_\uparrow(\omega)\tilde{G}^r_\downarrow(\omega)\tilde{G}^r_\uparrow(\omega-\omega_0)\tilde{G}^r_\downarrow(\omega-\omega_0)e^{\beta\omega_0/2}\big].
\end{aligned} \tag{7.42}$$

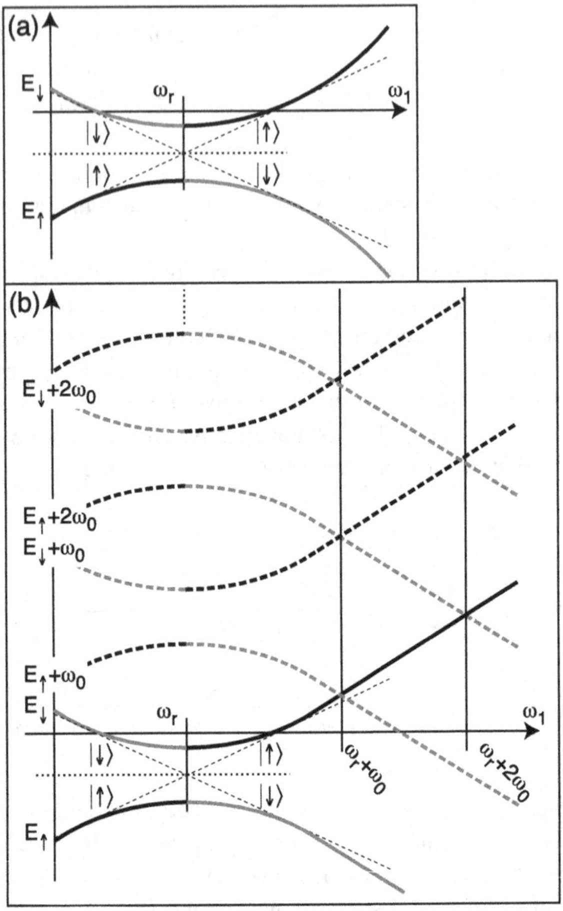

Fig. 7.3 Schematic electronic structure of the molecule in (**a**) absence and in (**b**) presence of electron-vibron coupling. At vanishing time-dependent magnetic field, the electron levels are separated by ω_r, while at finite B_1 and resonance frequency $\omega_1 = \omega_r$, the levels are separated by $2g\mu_B B_1 < \omega_r$. and the levels diverge for increasing frequencies ω_1. (**a**) There is only one frequency at which the levels converge. (**b**) There are additional frequencies where the molecular levels converge, or, cross. At those crossings, levels with different vibrational excitations are degenerate

The first terms add positively to the transmission and they peak at $\omega = E_\sigma$ and $\omega = E_\sigma + \omega_0$, respectively. The last term is more interesting though. It is proportional to

$$-\operatorname{Re}\frac{\Omega}{\omega_0 + i\Gamma}\left(\frac{1}{(\omega - E_\uparrow - \omega_0 - i\Gamma/2)(\omega - E_\downarrow + i\Gamma/2)} - \frac{1}{(\omega - E_\downarrow - \omega_0 - i\Gamma/2)(\omega - E_\uparrow + i\Gamma/2)}\right), \tag{7.43}$$

and it is negligible at pumping frequencies $\omega_1 = \omega_r$ since then $E_\uparrow \approx E_\downarrow$, which leads to that the two contributions cancel. At pumping frequencies $\omega_1 \to \omega_r + \omega_0$, however, we have that $\Omega \to \sqrt{\omega_0^2 + 4(g\mu_B B_1)^2} \approx \omega_0$, which leads to that

$E_\uparrow + \omega_0 \approx E_\downarrow$. Consequently, the above expression reduces to approximately

$$-\frac{1}{1+(\Gamma/\omega_0)^2}\frac{1}{(\omega-E_\downarrow)^2+(\Gamma/2)^2}, \tag{7.44}$$

since the second term is negligible. This expression peaks at $\omega = E_\downarrow \approx E_\uparrow + \omega_0$ and adds negatively to the total transmission $\mathcal{T}_s$. We can estimate the impact of this negative contribution to the transmission through the ratios between the third and first terms, and third and second terms, in (7.42) at the frequency $\omega_1 \approx \omega_r + \omega_0$. We obtain

$$\left|\frac{I_1(z)}{2I_0(z)}\right| e^{\beta\omega_0/2}\mathcal{L}(\omega_0), \qquad \left|\frac{I_0(z)}{2I_1(z)}\right| e^{-\beta\omega_0/2}\mathcal{L}(\omega_0), \tag{7.45}$$

respectively, where $\mathcal{L}(\omega_0) = \omega_0^2/[1+(\omega_0/\Gamma)^2] = \Gamma^2/[1+(\Gamma/\omega_0)^2]$. This shows that the transmission is significantly reduced when the detuning Δ equals the first vibrational side band. By including all remaining contributions to the transmission, i.e. summing over all n, we obtain analogous reductions in the transmission at the pumping frequencies $\omega_1 = \omega_r + n\omega_0$. The argument is certainly also true in presence of a finite effective charging energy, hence, the result is quite general.

When calculating the physical properties of the system we have to perform self-consistent calculations of (7.38) with respect to the occupation numbers $\langle n_\sigma \rangle$. Here, for instance, we are interested in the (equilibrium) spin current flowing through the molecular level. In Fig. 7.4(a) we plot the spin current as function of the pumping frequency ω_1, and the result clearly illustrates the main ESR peak at $\omega_1 = \omega_r$ as well as the vibrational anti-resonances at $\omega_1 = \omega_r + n\omega_0$, $n \neq 0$.

A peculiar feature in the spin current is that it decreases for increasing coupling λ at $\tilde{U} = 0$. This can understood as an effect of that the electron density is distributed among an increasing number of vibrational side-bands for increasing λ. Hence, for a given spin bias ω_1 there is less electron density available for conductance within the bias window as the coupling λ increases. Analogously, there is an increasing spin current for increasing correlation charging energy, which is a result of focusing the electron density in the spin bias window as $\tilde{U}$ grows.

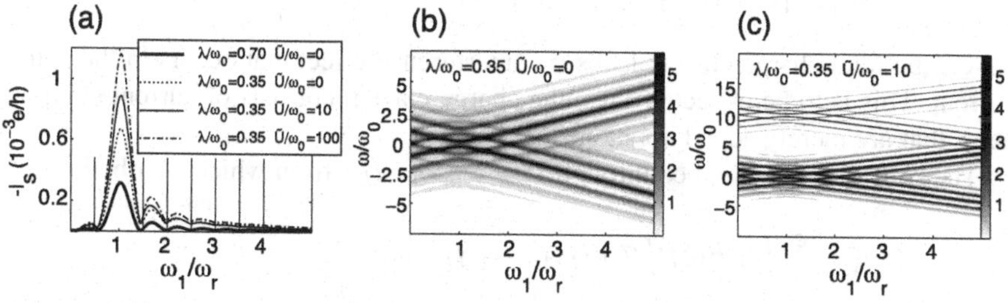

Fig. 7.4 (**a**) Dependence of the equilibrium spin current on the electron-vibron coupling λ and effective charging energy $\tilde{U}$. Local molecular DOS for $\lambda/\omega_0 = 0.35$, and $\tilde{U} = 0$ (**b**) and $\tilde{U}/\omega_0 = 10$ (**c**). Here, $\omega_r = 2$, $g\mu_B B_1 = 0.2315$, $\Gamma = 4\sqrt{5}/25$, and $k_B T = 10$, in units of ω_0

7.2.2.1 Two-Terminal System and Non-Equilibrium Conditions

The above framework is straightforwardly generalized to systems with two leads and non-equilibrium conditions. We define the bias voltage $V = (\mu_L - \mu_R)/e$ and in each lead we have the spin imbalance such that $\mu_\chi = (\mu_{\chi\uparrow} + \mu_{\chi\downarrow})/2$, where $\chi = L, R$. Using the same procedure as above, we derive the current $I_{L\sigma}$ for the spin σ current flowing from the left lead into the molecule, which results in

$$I_{L\sigma} = \frac{e}{h}\int \Gamma^L\big\{\Gamma^R\mathcal{T}_c(\omega)\big(f_{L\sigma}(\omega) - f_{R\sigma}(\omega)\big) + \Gamma^L\mathcal{T}_s(\omega)\big(f_{L\sigma}(\omega) - f_{L\bar{\sigma}}(\omega)\big) \\ + \Gamma^R\mathcal{T}_s(\omega)\big(f_{L\sigma}(\omega) - f_{R\bar{\sigma}}(\omega)\big)\big\}d\omega. \tag{7.46}$$

Here, we have defined the transmission coefficient $\mathcal{T}_c = |G^r_\uparrow \cos^2\phi + G^r_\downarrow \sin^2\phi|^2$, whereas $f_{\chi\sigma} = f(\omega - \mu_{\chi\sigma})$. The expression for the current is obtained by noticing that

$$\Sigma^< = i\sum_{\chi\sigma} f_{\chi\sigma}\Gamma^\chi_\sigma, \qquad \Sigma^> = -i\sum_{\chi\sigma}(1 - f_{\chi\sigma})\Gamma^\chi_\sigma. \tag{7.47}$$

The first contribution in (7.46) can be identified as the usual charge current through molecule in analogy to the result in [21], the second contribution is analogous to the one discussed above, whereas the third contribution describes the spin current between the leads.

The total charge current flowing through the molecule is defined as $I_c = \sum_\sigma I_{L\sigma}$, for which we obtain

$$I_c = \frac{e}{h}\sum_\sigma \int \Gamma^L\Gamma^R\big(\mathcal{T}_s(\omega) + \mathcal{T}_c(\omega)\big)\big(f_{L\sigma}(\omega) - f_{R\sigma}(\omega)\big)d\omega, \tag{7.48}$$

which is just the sum of the different transmissions between the leads. We notice here that

$$\begin{aligned}\mathcal{T}_s + \mathcal{T}_c &= |G^r_\uparrow|\cos^4\phi + |G^r_\downarrow|^2\sin^4\phi + 2\,\mathrm{Re}(G^r_\uparrow G^a_\downarrow + G^r_\downarrow G^a_\uparrow)\cos^2\phi\sin^2\phi \\ &\quad + \big(|G^r_\uparrow| + |G^r_\downarrow|^2 - 2\,\mathrm{Re}(G^r_\uparrow G^a_\downarrow + G^r_\downarrow G^a_\uparrow)\big)\cos^2\phi\sin^2\phi \\ &= |G^r_\uparrow|\cos^2\phi + |G^r_\downarrow|^2\sin^2\phi,\end{aligned} \tag{7.49}$$

hence, the total charge current lacks the interference effect that occurs in the spin-current. This is to be expected since the charge current consists of current components that are merely added upon one another.

The spin current is defined through $I_s = \sum_\sigma \sigma^z_{\sigma\sigma} I_{L\sigma}$ from which we obtain

$$\begin{aligned}I_s = \frac{e}{h}\int \Gamma^L\big(&2\Gamma^L\mathcal{T}_s\big(f_{L\uparrow}(\omega) - f_{L\downarrow}(\omega)\big) \\ &+ \Gamma^R\mathcal{T}_s\big(f_{L\uparrow}(\omega) - f_{R\downarrow}(\omega) + f_{R\uparrow}(\omega) - f_{L\downarrow}(\omega)\big) \\ &+ \Gamma^R\mathcal{T}_c\big(f_{L\uparrow}(\omega) - f_{R\uparrow}(\omega) + f_{R\downarrow}(\omega) - f_{L\downarrow}(\omega)\big)\big)d\omega.\end{aligned} \tag{7.50}$$

The first of the contributions to the spin-current, i.e. proportional to $\Gamma^L \mathcal{T}_s$, has the same origin as the spin-current discussed in the single terminal case, and describes the spin-current between the spin-channels within one lead. The second contribution has an analogous origin but with the difference that this spin-current flows between the spin-channels in *different* leads. The last contribution arise due to the spin-imbalance in the charge current which is caused by the spin-bias acting on the leads. In contrast to the first two contributions, this last one may vanish under equilibrium conditions such that $\mu_L - \mu_R = 0$ although there might be a spin-bias applied on the system. If the spin-bias acts on the leads such that $\omega_1 = \mu_{\chi\uparrow} - \mu_{\chi\downarrow}$, then $\mu_{L\sigma} - \mu_{R\sigma} = 0$ which leads to that $f_{L\sigma} - f_{R\sigma} = 0$ and, hence, the spin-current vanishes. If the spin-bias, on the other hand, is such that $\omega_1 = \mu_{L\uparrow} - \mu_{L\downarrow} = \mu_{R\downarrow} - \mu_{R\uparrow}$, then there would be a spin-$\uparrow$ current flowing between the leads, say, from the left to the right lead. Simultaneously, there is a spin-$\downarrow$ current flowing from the right to the left lead. Summing up those two current, this leads to a finite contribution to the total spin-current.

Typical examples of the non-equilibrium spin-current is plotted in Fig. 7.5 for a few different values of the bias voltage. The characteristics of the spin-current can be explained by the following: For low pumping frequencies ω_1, the transport that is assisted by spin-flips in the molecular structure dominates the spin-current, c.f. first and second contributions to I_s in (7.50). Thus, the main ESR peak is visible as well as are the vibrational anti-resonances in analogy to the equilibrium case. For increasing frequencies ω_1 ones sees that the spin-current drifts off from the equilibrium value. This characteristics is expected since for increasing ω_1, the potential barrier for spin-flips increases which leads to that the transport that is assisted by the ac magnetic field becomes suppressed. The current that is not assisted by the spin-flips, i.e. the last contribution in (7.50), increase for increasing ω_1 due to the non-equilibrium conditions. As ω_1 grows, the distance between the spin-chemical potentials $\mu_{\chi\uparrow}$ and $\mu_{\chi\downarrow}$ increases, as well as the distance between the molecular level spin-projections $E_\uparrow$ and $E_\downarrow$. The spin-$\uparrow$ current flowing between the leads via the molecular energy $E_\uparrow$ grows since the loss due to spin-flips decreases. The same arguments holds for the spin-$\downarrow$ channel. Eventually, the current from the last contribution in (7.50) saturates in accordance with usual transport through a molecular level.

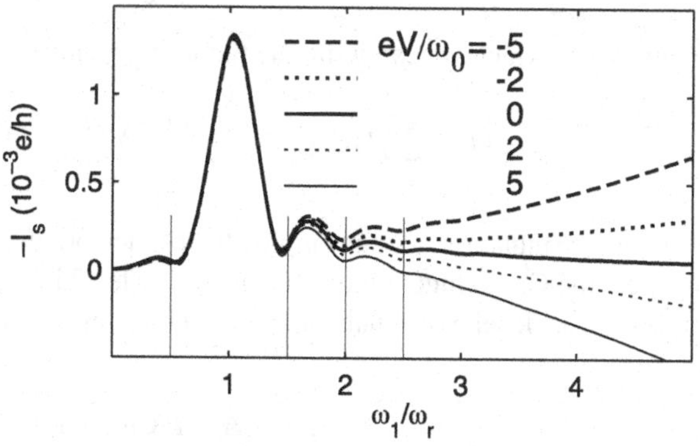

Fig. 7.5 Bias voltage dependence of the spin-current for $\lambda/\omega_0 = 0.35$ and $\tilde{U} = 0$. Other parameters are as in Fig. 7.4

As we have seen in the previous discussion, the lesser and greater self-energies of the molecular Green function have been incorrectly treated. Although it is not impossible to improve on the description of the present picture, such that the lesser and greater self-energies are appropriately accounted for within the given approximation, it is found not to be straight forward. Therefore, if we are to improve on the electronic side of the problem, one choice is to describe the electronic degrees of freedom through Hubbard operators. This is our task in the succeeding section.

7.2.3 Converting the Electronic Operators to Many-Body Operators

The lesser and greater self-energies for the Green function given in (7.28′) are simplified since they do not include effects from the electron-electron interactions occurring within the molecular quantum dot. This unsatisfactory property can be remedied e.g. by transforming the electronic operators into Hubbard operators, i.e. letting the operators c_σ which are diagonal in spin space be expanded as $c_\sigma = X^{0\sigma} + \sigma X^{\bar{\sigma}2}$. We then have the transformed Hamiltonians

$$\tilde{\mathcal{H}}_d = \sum_{p=0,\sigma,2} \mathbb{E}_p h^p, \tag{7.51a}$$

$$\begin{aligned}\tilde{\mathcal{H}}_T = \sum_{\mathbf{k}\sigma} v_{\mathbf{k}} c^\dagger_{\mathbf{k}\sigma} \big(&[X^{0\sigma} + \sigma X^{\bar{\sigma}2}] \cos\phi \\ &- \sigma[X^{0\bar{\sigma}} + \bar{\sigma} X^{\sigma 2}] \sin\phi \big) \mathcal{X} + H.c.,\end{aligned} \tag{7.51b}$$

where $\mathbb{E}_p \in \{0, E_\sigma, \sum_\sigma E_\sigma + \tilde{U}\} = \{0, \varepsilon_0 - \lambda^2/\omega_0 - \sigma\Omega/2, 2\varepsilon_0 + U - 4\lambda^2/\omega_0\}$. We introduce the coupling vectors

$$v_{\mathbf{k}\uparrow} = v_{\mathbf{k}}(\cos\phi, -\sin\phi, \cos\phi, \sin\phi), \tag{7.52a}$$

$$v_{\mathbf{k}\downarrow} = v_{\mathbf{k}}(\sin\phi, \cos\phi, \sin\phi, \cos\phi), \tag{7.52b}$$

by means of which we can write the tunneling Hamiltonian

$$\tilde{\mathcal{H}}_T = \sum_{\mathbf{k}\sigma} v_{\mathbf{k}\sigma} c^\dagger_{\mathbf{k}\sigma} (X^{0\uparrow}, X^{0\downarrow}, X^{\downarrow 2}, X^{\uparrow 2})^T \mathcal{X} + H.c. \tag{7.53}$$

With this formulation, we can employ the framework that we introduced in Sect. 6.3. We treat the Green function for the molecular level in the Hubbard-I-approximation omitting the level renormalization, in which approximation we have the equation

$$(i\omega - \Delta - \mathbf{PV})\tilde{\mathbf{G}} = \mathbf{P}, \tag{7.54}$$

with

$$\mathbf{V} = \sum_\sigma \mathbf{V}_\sigma = \sum_{\mathbf{k}\sigma} \frac{|v_\mathbf{k}|^2}{i\omega - \varepsilon_\mathbf{k}} \mathbf{U}_\sigma, \tag{7.55a}$$

$$\mathbf{U}_\uparrow = \begin{pmatrix} \mathbf{u}_\uparrow & \mathbf{u}_\uparrow \sigma^z \\ \sigma^z \mathbf{u}_\uparrow & \sigma^z \mathbf{u}_\uparrow \sigma^z \end{pmatrix}, \qquad \mathbf{U}_\downarrow = \begin{pmatrix} \mathbf{u}_\downarrow & \mathbf{u}_\downarrow \sigma^z \\ \sigma^z \mathbf{u}_\downarrow & \sigma^z \mathbf{u}_\downarrow \sigma^z \end{pmatrix}, \tag{7.55b}$$

where the matrices $\mathbf{u}_\sigma$ are defined in (7.37). Using this description, we find that the lesser and greater forms can be written as $\tilde{\mathbf{G}}^{</>} = \tilde{\mathbf{G}}^r \mathbf{V}^{</>} \tilde{\mathbf{G}}^a$, where the correct forms of $\mathbf{V}^{</>}(\omega) = (\pm i) \sum_\sigma \Gamma \mathbf{u}_\sigma f_\sigma(\pm\omega)$ are obtained even in presence of finite Coulomb interactions, c.f. the lesser and greater self-energies for the Green function given in (7.28′). Thanks to the product of $\mathbf{P}$ and $\mathbf{V}$ in the self-energy, the effects from the electronic correlations are included into the full form of the Green function. For more elaborate forms of the self-energy we can, of course, consult the theoretical formalism in Chap. 5.

The retarded and advanced forms of the Green functions can be readily obtained by inverting the equation for $\tilde{\mathbf{G}}$ above. We find

$$G_n^{r/a}(\omega) = \frac{P_n}{\omega - \Delta_n - \Sigma_n^{r/a} - \sum_{m \neq n} \frac{\Sigma_{nm}^{r/a} \Sigma_{mn}^{r/a}}{\omega - \Delta_m - \Sigma_m^{r/a}}}, \tag{7.56a}$$

$$G_{nm}^{r/a}(\omega) = -\frac{\Sigma_{nm}^{r/a}}{\omega - \Delta_n - \Sigma_n^{r/a}} G_m^{r/a}, \tag{7.56b}$$

where $G_n \equiv G_{nn}$, whereas the self-energies $\Sigma_{mn}^{r/a}$ are given by

$$\Sigma_{nm}^{r/a} = \sum_l P_{nl} V_{lm}^{r/a}, \quad m \neq n, \ n, m = 1, 2, \tag{7.56c}$$

$$\Sigma_n^{r/a} \equiv \Sigma_{nn}^{r/a}; \tag{7.56d}$$

where we have used that the equation of motion can be turned into a block structure in which the blocks are 2×2-matrices.

Having solved the Green function for the molecular level and, in addition, provided the corresponding non-equilibrium Green functions with the correct self-energies, we can transform the problem back into the framework of the previous section, by the transformation $G_\sigma(t,t') = G_{0\sigma}(t,t') + \sigma[G_{0\sigma 2\bar{\sigma}}(t,t') + G_{\bar{\sigma}2\sigma 0}(t,t')] + G_{\bar{\sigma}2}(t,t')$. Nevertheless, while we have solved the non-equilibrium physics of the electronic level more appropriately, there is still no improvement on the influence from the electron-vibron coupling.

7.3 Beyond Weak Electron-Vibron Coupling

In all of the previous discussions, we have assumed that the coupling between the electronic and vibronic degrees of freedom is weak, and that the effects of the coupling does not influence the self-energy of the electronic Green function. It has been shown by several authors, however, that even in the weak coupling limit one cannot neglect the influence from the vibronic modes on the electronic structure, see e.g. [12, 22]. Here, we will address a more generalized situation in which the polaron effects are also taken into account for in the description of the molecular levels.

7.3.1 Intermediate and Strong Electron-Vibron Coupling Regime

We begin by a simple extension of the previous discussion, see Sect. 7.2.1, in which we consider the Green functions $\tilde{G}_{\sigma\sigma'}(t,t') = (-i)\langle \mathrm{T} d_\sigma(t) d^\dagger_{\sigma'}(t')\rangle$ and $K(t,t') = \langle \mathrm{T}\mathcal{X}(t)\mathcal{X}^\dagger(t')\rangle$, such that $G_{\sigma\sigma'}(t,t') = \tilde{G}_{\sigma\sigma'}(t,t')K(t,t')$. We here, follow the line-out proposed in [22], which provides a self-consistent set of equations for $\tilde{G}$ and K and is valid in the intermediate to strong electron-vibron coupling regimes. The approach can, furthermore, be viewed as a strong coupling analog of the self-consistent Born approximation.

The Green functions are calculated in the Hamiltonian system (7.25). For simplicity, however, we omit the correlation term $Un_\uparrow n_\downarrow$ such that the two-electron Green function does not have to be considered.

We have the equation of motion for $\tilde{G}$ given by

$$(i\partial_t - \tilde{\varepsilon}_\sigma)\tilde{G}_{\sigma\sigma'}(t,t') = \delta_{\sigma\sigma'}\delta(t-t') + \sum_{\mathbf{k}} v^*_{\mathbf{k}\sigma}(-i)\langle \mathrm{T}(\mathcal{X}^\dagger c_{\mathbf{k}\sigma})(t) d^\dagger_{\sigma'}(t')\rangle, \quad (7.57)$$

which leads to

$$\tilde{G}_{\sigma\sigma'}(t,t') = \delta_{\sigma\sigma'}\tilde{g}_\sigma(t,t') + \int_C \tilde{g}_\sigma(t,\tau)\sum_{\mathbf{k}} v^*_{\mathbf{k}\sigma}(-i)\langle \mathrm{T}(\mathcal{X}^\dagger c_{\mathbf{k}\sigma})(\tau) d^\dagger_{\sigma'}(t')\rangle d\tau, \quad (7.58)$$

where $(i\partial_t - \varepsilon_\sigma)g_\sigma(t,t') = \delta(t-t')$. Acting from the right on $\tilde{G}$ with the operator $-i\partial_{t'} - \varepsilon_\sigma$, which satisfies $g_\sigma(t,t')(-i\partial_{t'} - \varepsilon_\sigma) = \delta(t-t')$, we obtain

$$\tilde{G}_{\sigma\sigma'}(t,t') = \delta_{\sigma\sigma'}\tilde{g}_\sigma(t,t') + \delta_{\sigma\sigma'}\int_C \tilde{g}_\sigma(t,\tau)\Sigma_\sigma(\tau,\tau')\tilde{g}_\sigma(\tau',t')d\tau d\tau', \quad (7.59a)$$

where the self-energy

$$\Sigma_\sigma(t,t') = \sum_{\mathbf{k}} |v_{\mathbf{k}\sigma}|^2 g_{\mathbf{k}\sigma}(t,t')K(t',t). \quad (7.59b)$$

Obviously, the presence of the vibronic modes also gives an influence on the Green function $\tilde{G}$, and not only on G as in the weakly coupled regime.

The vibron Green function K is treated somewhat differently. Consider the expansion

$$\langle \mathrm{T}\mathcal{X}(t)\mathcal{X}^{\dagger}(t')\rangle = \sum_{n,m=0}^{\infty} \frac{(-i\lambda)^n}{n!}\frac{(i\lambda)^m}{m!}\langle \mathrm{T}P^n(t)P^m(t')\rangle, \tag{7.60}$$

where the vibron momentum operator $P = (-i)(a - a^{\dagger})$. It is convenient to introduce the cumulant $\phi_p(t,t')$ of order p, in terms of which, the vibron Green function can be rewritten into

$$\begin{aligned}
&\langle \mathrm{T}\mathcal{X}(t)\mathcal{X}^{\dagger}(t')\rangle \\
&= \exp\left\{\sum_{p=1}^{\infty}\frac{\lambda^p}{p!}\phi_p(t,t')\right\} \\
&= 1 + \sum_{p=1}^{\infty}\frac{\lambda^p}{p!}\phi_p(t,t') + \frac{1}{2}\sum_{pq}\frac{\lambda^{p+q}}{p!q!}\phi_p(t,t')\phi_q(t,t') + \cdots .
\end{aligned} \tag{7.61}$$

Retaining terms up to second order in λ and equating same orders in λ in the two above expressions, we obtain the relations

$$\phi_1(t,t') = i\langle P(t)\rangle - i\langle P(t')\rangle, \tag{7.62a}$$

$$\phi_2(t,t') + \phi_1^2(t,t') = 2\langle \mathrm{T}P(t)P(t')\rangle - \langle P^2(t)\rangle - \langle P^2(t')\rangle. \tag{7.62b}$$

In the stationary regime, the averages $\langle P^n(t)\rangle = \langle P^n(t')\rangle = \langle P^n\rangle$, which leads to that $\phi_1(t,t') = 0$, whereas

$$\phi_2(t,t') = 2\langle \mathrm{T}P(t)P(t')\rangle - 2\langle P^2\rangle. \tag{7.63}$$

Making use of those results in the cumulant expansion of K, yields the following form:

$$K(t,t') = \exp\{i\lambda^2[D(t,t') + i\langle P^2\rangle]\}, \tag{7.64}$$

where $D(t,t') = (-i)\langle \mathrm{T}P(t)P(t')\rangle$ is the phonon momentum Green function, whereas $\langle P^2(t)\rangle = iD^<(t,t) = iD^>(t,t)$.

The next step is to consider the equation of motion for D, that is,

$$i\partial_t D(t,t') = (-i)\omega_0(-i)\langle \mathrm{T}Q(t)P(t')\rangle, \tag{7.65}$$

where $Q = a + a^{\dagger}$ is the phonon displacement operator. Here, we have used that $[P, a^{\dagger}a] = -iQ$ and that P commutes with $\mathcal{X}$. This operator obeys the equation of motion

$$i\partial_t Q(t) = i\omega_0 P(t) - 2\lambda\sum_{\mathbf{k}\sigma}(v_{\mathbf{k}\sigma}c^{\dagger}_{\mathbf{k}\sigma}d_{\sigma}\mathcal{X} - H.c.), \tag{7.66}$$

since $[Q, a^\dagger a] = iP$. We moreover have that $[Q, P^n] = nP^{n-1}$ which implies that

$$[Q, \mathcal{X}] = \sum_{n=0}^{\infty} \frac{(i\lambda)^n}{n!}[Q, P^n] = \cdots = -2\lambda\mathcal{X}. \tag{7.67}$$

Putting together the two equations for D and $(-i)\langle \mathrm{T}Q(t)P(t')\rangle$ leads to that we can write

$$\begin{aligned}\frac{1}{2\omega_0}\big((i\partial_t)^2 - \omega_0^2\big)D(t,t') = \delta(t-t') + i\lambda\sum_{\mathbf{k}\sigma}\big(&v_{\mathbf{k}\sigma}(-i)\langle \mathrm{T}(c_{\mathbf{k}\sigma}^\dagger d_\sigma \mathcal{X})(t)P(t')\rangle \\ &- v_{\mathbf{k}\sigma}^*(-i)\langle \mathrm{T}(\mathcal{X}^\dagger d_\sigma^\dagger c_{\mathbf{k}\sigma})(t)P(t')\rangle\big),\end{aligned} \tag{7.68}$$

where we have neglected contributions that mix different processes, using the so-called *non-crossing approximation* (NCA) [23]. Applying the differentiation operator $D_0^{-1} = [(i\partial_t)^2 - \omega_0^2]/(2\omega_0)$ from the right, analogously as the case of the electronic Green function, we obtain the equation

$$D(t,t') = D_0(t,t') + \int_C D_0(t,\tau)\Pi(\tau,\tau')D_0(\tau',t')d\tau d\tau', \tag{7.69a}$$

where the self-energy

$$\begin{aligned}\Pi(t,t') = -i\lambda^2\sum_{\mathbf{k}\sigma}|v_{\mathbf{k}\sigma}|^2\big(&g_{\mathbf{k}\sigma}(t',t)\tilde{G}_\sigma(t,t')K(t,t') \\ &+ g_{\mathbf{k}\sigma}(t,t')\tilde{G}_\sigma(t',t)K(t',t)\big).\end{aligned} \tag{7.69b}$$

In order to improve on the description of the electronic structure, we replace the rightmost (bare) propagators in both (7.59a) and (7.69a), by their dressed counterparts. In this way we obtain Dyson-like equations for both the electronic and vibronic Green functions, which have to be self-consistently solved.

The next step is to determine the lesser and greater forms of $\tilde{G}$ and K. In order to do this calculation, we assume time-independent conditions, such that all propagators depend on the time-difference $t - t'$. Thanks to that both $\tilde{G}$ and D are given in terms of Dyson-like equations, we find that

$$\tilde{G}_\sigma^{</>}(\omega) = \tilde{G}_\sigma^r(\omega)\Sigma_\sigma^{</>}(\omega)\tilde{G}_\sigma^a(\omega), \tag{7.70a}$$

$$D^{</>}(\omega) = D^r(\omega)\Pi^{</>}(\omega)D^a(\omega), \tag{7.70b}$$

where the electronic lesser and greater self-energies are expressed as

$$\Sigma_\sigma^{</>}(\omega) = V_\sigma^{</>}(\omega)K^{>/<}(\omega) = (\pm i)\sum_\chi f_\chi(\pm\omega)\Gamma_\sigma^\chi K^{>/<}(\omega), \tag{7.71a}$$

whereas the corresponding vibronic self-energies are written according to

$$\Pi^{</>}(t,t') = -i\lambda^2 \sum_{\sigma} \big(V_{\sigma}^{>}(t',t) G_{\sigma}^{<}(t,t') + V_{\sigma}^{<}(t',t) G_{\sigma}^{>}(t,t')\big) K^{</>}(t,t'). \tag{7.71b}$$

Here, we have used the tunneling propagator $V_{\sigma}(i\omega) = \sum_{\mathbf{k}} |v_{\mathbf{k}\sigma}|^2 g_{\mathbf{k}\sigma}(i\omega)$, and $g_{\mathbf{k}\sigma}(i\omega) = 1/(i\omega - \varepsilon_{\mathbf{k}\sigma})$. Finally, from the definition of K we obtain

$$K^{>}(t,t') \equiv \langle \mathcal{X}(t)\mathcal{X}^{\dagger}(t')\rangle = \exp\{i\lambda^2 [D^{>}(t,t') - D^{>}(t,t)]\}, \tag{7.72a}$$

$$K^{<}(t,t') \equiv \langle \mathcal{X}^{\dagger}(t')\mathcal{X}(t)\rangle = \exp\{i\lambda^2 [D^{<}(t,t') - D^{<}(t,t)]\}. \tag{7.72b}$$

Before the description is complete, we also need to find expressions for the retarded and advanced Green functions. In principle, those can be obtained through the definitions e.g. $G^{r/a}(t,t') = \pm\theta(\pm t \mp t')[G^{>}(t,t') - G^{<}(t,t')]$. We may, on the other hand, just well make use of the equations for the Green functions, which in the time-independent regime give

$$\tilde{G}_{\sigma}^{r/a}(\omega) = \frac{1}{\omega - \tilde{\varepsilon}_{\sigma} - \frac{K^{r/a}(\omega)}{\omega - \tilde{\varepsilon}_{\sigma} \pm i\delta}}, \tag{7.73a}$$

$$D^{r/a}(\omega) = \frac{2\omega_0}{\omega^2 - \omega_0^2 - 2\omega_0 \Pi^{r/a}(\omega)}, \tag{7.73b}$$

where e.g.

$$\begin{aligned} \Pi^{r}(t,t') &= \theta(t-t')[\Pi^{>}(t,t') - \Pi^{<}(t,t')] \\ &= -i\lambda^2 \theta(t-t')\big(V^{>}(t',t)\tilde{G}_{\sigma}^{<}(t,t') + V^{<}(t',t)\tilde{G}_{\sigma}^{>}(t,t')\big) \\ &\quad \times \big(K^{>}(t,t') - K^{<}(t,t')\big). \end{aligned} \tag{7.74}$$

Using this procedure, we obtain a self-consistent scheme for solving the electronic structure of the vibrating molecule. In this scheme, it is obvious that the coupling between the electronic and vibronic degrees of freedom are affecting the properties of both the electronic and vibronic Green functions, such that the vibronic modes are also influencing the electron level broadening.

7.4 Transforming It All to Many-Body Operators

We saw in Sect. 7.2.1 that there are reasons to consider other ways to deal with the electron-vibron coupled system, and an important reason is that we want to be able to handle the strongly coupled cases. In the literature there is a huge number of approaches, especially within non-equilibrium theory since most methods have proven useful in different regimes and for different purposes. Here, the aim is to connect to the many-body operator formalism and elucidate its abilities within this part of the theoretical formalism. The method as such, is based on the previous ideas of using the many-body states of the central unit, e.g. molecular quantum dot. These ideas have subsequently been extended by Galperin et al. [24] to also include the vibronic side-levels into the many-body states.

The idea is straightforward from the previous introduction of the many-body operators. As an example, we again take the model introduced in (7.1), here repeated for convenience

$$\mathcal{H} = \sum_{\mathbf{k}\sigma\in L,R} \varepsilon_{\mathbf{k}\sigma} c^\dagger_{\mathbf{k}\sigma} c_{\mathbf{k}\sigma} + \sum_\sigma \varepsilon_\sigma d^\dagger_\sigma d_\sigma + U n_\uparrow n_\downarrow + \omega_0 a^\dagger a$$
$$+ \lambda \sum_\sigma d^\dagger_\sigma d_\sigma (a^\dagger + a) + \sum_{\mathbf{k}\sigma} (v_{\mathbf{k}\sigma} c^\dagger_{\mathbf{k}\sigma} d_\sigma + H.c.), \qquad (7.1)$$

which is canonically transformed into the form

$$\tilde{\mathcal{H}} = \sum_{\mathbf{k}\sigma\in L,R} \varepsilon_{\mathbf{k}\sigma} c^\dagger_{\mathbf{k}\sigma} c_{\mathbf{k}\sigma} + \sum_\sigma \tilde{\varepsilon}_\sigma d^\dagger_\sigma d_\sigma + \tilde{U} n_\uparrow n_\downarrow + \omega_0 a^\dagger a$$
$$+ \sum_{\mathbf{k}\sigma} (v_{\mathbf{k}\sigma} c^\dagger_{\mathbf{k}\sigma} d_\sigma \mathcal{X} + H.c.). \qquad (7.25)$$

In absence of the coupling to the vibrational mode, the local electronic states are $|p\rangle$, $p \in \{0, \uparrow, \downarrow, 2\}$. These eigenstates are now extended to the set $|p, \mu\rangle$, where μ denotes the vibronic excitation of the electronic state labeled by p. The local electron operators are, in this basis, given by the expansion

$$d_\sigma = \sum_{pq} \sum_{\mu\nu} \langle p, \mu | d_\sigma | q, \nu\rangle X^{p\mu, q\nu}. \qquad (7.75)$$

The electron operators act only on the electron degree of freedom in the state $|p, n\rangle$, hence $\sum_q d_\sigma |q, \nu\rangle = d_\sigma [|\sigma, \nu\rangle + |2, \nu\rangle] = |0, \nu\rangle + \sigma |\bar{\sigma}, \nu\rangle$, which leads to

$$d_\sigma = \sum_{\mu\nu} [\langle 0, \mu | 0, \nu\rangle\, X^{0\mu, \sigma\nu} + \sigma\, \langle \bar{\sigma}, \mu | \bar{\sigma}, \nu\rangle\, X^{\bar{\sigma}\mu, 2\nu}] = \sum_\mu [X^{0\mu, \sigma\mu} + \sigma X^{\bar{\sigma}\mu, 2\mu}]. \qquad (7.76)$$

The vibron operators, on the other hand, only act on the bosonic degree of freedom, and due to the expansion

$$a^m = \sum_{pq} \sum_{\mu\nu} = \langle p, \mu | a^m | q, \nu\rangle X^{p\mu, q\nu}$$
$$= \sum_{pq} \sum_{\mu\nu} \sqrt{\frac{\nu!}{(\nu - m)!}} \langle p, \mu | q, \nu - m\rangle\, X^{p\mu, q\nu}$$
$$= \sum_{p\mu} \sqrt{\frac{(\mu + m)!}{\mu!}} X^{p\mu, p\mu+m}, \qquad (7.77)$$

we obtain the combination

$$d_\sigma a^m = \sum_\mu \sqrt{\frac{(\mu+m)!}{\mu!}}[X^{0\mu,\sigma\mu+m} + \sigma X^{\bar{\sigma}\mu,2\mu+m}]. \tag{7.78}$$

Further, by making use of the identity $e^{-\lambda(a^\dagger - a)/\omega_0} = e^{-(\lambda/\omega_0)^2/2} e^{-\lambda a^\dagger/\omega_0} e^{\lambda a/\omega_0}$, we find the matrix elements

$$\langle p,\mu | d_\sigma \mathcal{X} | q,\nu\rangle = e^{-(\lambda/\omega_0)^2/2}\left(\frac{\lambda}{\omega_0}\right)^{\nu-\mu}\sqrt{\frac{\mu!}{\nu!}}\mathcal{L}_\mu^{\nu-\mu}\left(\left[\frac{\lambda}{\omega_0}\right]^2\right), \tag{7.79}$$

where we, without loss of generality, have assumed that $\mu < \nu$. The tunneling Hamiltonian accordingly becomes

$$\tilde{\mathcal{H}}_T = \sum_{\mathbf{k}\sigma} v_{\mathbf{k}\sigma} c^\dagger_{\mathbf{k}\sigma} d_\sigma + H.c. = \sum_{\mathbf{k}\sigma a} v_{\mathbf{k}\sigma a} c^\dagger_{\mathbf{k}\sigma} X^a + H.c., \tag{7.80a}$$

$$v_{\mathbf{k}\sigma a} = \langle p,\mu | d_\sigma \mathcal{X} | q,\nu\rangle, \quad a = (p,\mu)(q,\nu), \tag{7.80b}$$

where the transition index a carries both the electronic and vibronic degrees of freedom.

The many-body eigenstates were introduced in order to make the molecular part of the Hamiltonian diagonal, i.e.

$$\tilde{\mathcal{H}}_{\text{mol}} = \sum_\sigma \tilde{\varepsilon}_\sigma d^\dagger_\sigma d_\sigma + \tilde{U} n_\uparrow n_\downarrow + \omega_0 a^\dagger a = \sum_{p\mu} E_{p\mu} h^{p\mu} = \sum_\xi E_\xi h^\xi, \tag{7.81}$$

where $E_{p\mu} = E_p + \mu\omega_0$, with $E_0 = 0$, $E_\sigma = \tilde{\varepsilon}_\sigma$, and $E_2 = \tilde{\varepsilon}_\uparrow + \tilde{\varepsilon}_\downarrow + \tilde{U}$. Here, also the transition index $\xi = (p,\mu)$ denotes the diagonal transition. We, thus, have control of the full Hamiltonian for the vibrating molecular system in the eigenstate representation and we can, therefore, also use the framework developed in Chap. 5. It is interesting to notice that the many-body representation very clearly describes the vibrating molecular structure through an infinite set of eigenstates $|\xi\rangle = |p,\mu\rangle$, and that the tunneling current through the molecule is mediated by transitions between those (infinitely many) eigenstates. In this sense, the concept of opening conduction channels due to the vibrations in the structure becomes even more conceivable. Through the eigenstate description we also avoid the discussion of inelastic processes since, strictly speaking, there can only be elastic (direct) processes between different eigenstates. There can, however, occur indirect inelastic transitions between the eigenstates if those are assisted by some corresponding inelastic process in the lead-molecule system.

In principle, we can perform as high order diagrammatic expansion of the Green function defined in terms of the many-body operators X^a as we would find necessary, using the technique introduced in Chap. 5. It is, however, interesting to notice that already at the level of the Hubbard-I-approximation, the resulting molecular electronic structure is largely influenced by the vibronic modes. In particular, this manifest through non-uniform widths of the vibronic side-peaks, i.e. different vibronic side-peaks acquire different width due to the different transition matrix elements, or Franck-Condon factors. This is in sharp contrast to the picture obtained in the small polaron picture. From a physical point of view, this latter scenario is expected to be the most sensible one, since there is no principle mechanism that supports a uniform broadening of the vibronic side-peaks. Such a negative statement cannot be taken as a proof of principle. It is, on the other hand, easy to see that one would expect a larger probability for transitions between nearby states, than between states that largely differ in energy. From other symmetry reasons, it would also be quite obvious that transitions between states with more alike symmetry should be more likely, than between states with less symmetry in common.

References

1. Jaklevic, R.C., Lambe, J.: Phys. Rev. Lett. **17**, 1139 (1966)
2. Park, H., Park, J., Lim, A.K.L., Anderson, E.H., Alivisatos, A.P., McEuen, P.L.: Nature **407**, 57 (2000)
3. Smit, R.H.M., Noat, Y., Untiedt, C., Lang, N.D., van Hemert, M.C., van Ruitenbeek, J.M.: Nature **419**, 906 (2002)
4. Wang, W., Lee, T., Kretzschmar, I., Reed, M.A.: Nano Lett. **4**, 643 (2004)
5. Yu, L.H., Keane, Z.K., Ciszek, J.W., Cheng, L., Stewart, M.P., Tour, J.M., Natelson, D.: Phys. Rev. Lett. **93**, 266802 (2004)
6. Hahn, J.R., Lee, H.J., Ho, W.: Phys. Rev. Lett. **85**, 1914 (2000)
7. Gawronski, H., Melhorn, M., Morgenstern, K.: Science **319**, 930 (2008)
8. Persson, B.N.J., Baratoff, A.: Phys. Rev. Lett. **59**, 339 (1987)
9. Tikhodeev, S., Natario, M., Makoshi, K., Mii, T., Ueba, H.: Surf. Sci. **493**, 63 (2001)
10. Mii, T., Tikhodeev, S., Ueba, H.: Surf. Sci. **502–503**, 26 (2002)
11. Lundin, U., McKenzie, R.: Phys. Rev. B **66**, 075303 (2002)
12. Mitra, A., Aleiner, I., Millis, A.J.: Phys. Rev. B **69**, 245302 (2004)
13. Zhu, J.-X., Balatsky, A.V.: Phys. Rev. B **67**, 165326 (2003)
14. Flensberg, K.: Phys. Rev. B **68**, 205323 (2003)
15. Galperin, M., Ratner, M.A., Nitzan, A.: Nano Lett. **5**, 125 (2005)
16. Fransson, J., Zhu, J.-X.: Phys. Rev. B **78**, 113307 (2008)
17. Feynman, R.P.: Phys. Rev. **84**, 108 (1951)
18. Gradshteyn, I.S., Ryzhik, I.M.: Table of Integrals, Series, and Products, 6th edn. Academic Press, San Diego (2001)
19. Hewson, A.C., Newns, D.M.: J. Phys. C **13**, 4477 (1980)
20. Zhang, P., Xue, O.-K., Xie, X.C.: Phys. Rev. Lett. **91**, 196602 (2003)
21. Meir, Y., Wingreen, N.S.: Phys. Rev. Lett. **68**, 2512 (1992)
22. Galperin, M., Nitzan, A., Ratner, M.A.: Phys. Rev. B **73**, 045314 (2006)
23. Bickers, N.E.: Rev. Mod. Phys. **59**, 845 (1987)
24. Galperin, M., Nitzan, A., Ratner, M.A.: Phys. Rev. B **78**, 125320 (2008)

Chapter 8
Nanomechanical Oscillator

Abstract Coupling of electronic and vibrational degrees of freedom in systems with superconducting leads further enriches the physics in tunneling systems. We consider back-action effects of the current on the mechanical motion of a subsystem and show that the mechanical oscillations can be externally excited by the supercurrent. On the other hand, the nanomechanical motion leaves off-prints in the supercurrent which may lead to e.g. additional Shapiro steps in the I–V characteristics. We finally investigate the dynamics of a vibrating quantum dot embedded in a Josephson junction, and we find that new time-scales emerge. These emergent time-scales are intimately related to the energies of the single-electron transitions occurring in the quantum dot. We derive a phase diagram for the dynamics of the quantum dot occupation. In particular, we find one regime in which the occupation numbers are constants of the motion and in which the mechanical dynamics is, thus, set by the Josephson frequency.

8.1 Introduction

Coupling between electronic and vibronic degrees of freedom introduces, as we saw in the previous chapter, additional complexity to the characteristics of the nanoscale system. Unlike static defects, vibrational modes posses internal degrees of freedom which may give rise to additional peaks and dips in the differential conductance of molecular systems. Incorporating superconducting electronics in combination within nanomechanical set-up, inevitably enrichess both e.g. Josephson currents as well as the nanomechanical motion. A prominent example of superconducting electronics in combination with nanomechanics is a Cooper pair shuttle [1], in which a superconducting grain, or island, has an oscillating motion inside a Josephson junction. By varying the superconducting phases in the superconducting leads and island, the magnitude of the dc current can be controlled. Superconducting transport through a vibrating molecular structure in the subgap regime can be interpreted in terms of a ladder picture of inelastic multiple Andreev reflection processes [2]. Yet other studies of the Josephson effect in superconducting junctions coupled to mechanical oscillators show the possibility to introduce additional Shapiro steps in the

J. Fransson, *Non-Equilibrium Nano-Physics*,
Lecture Notes in Physics 809,
DOI 10.1007/978-90-481-9210-6_8, © Springer Science+Business Media B.V. 2010

I–V characteristics whose positions are determined by the frequency of the vibrations [3, 4].

In this chapter we discuss a few non-equilibrium characteristic of the Josephson effect in superconducting junctions coupled to mechanical degrees of freedom. For single and double Josephson junctions, we derive the time-dependent supercurrent from which we define a Hamiltonian which is inserted into the model of the nanomechanical motion. The resulting motion is put back into the supercurrent. Through this procedure we can study both the current driven nanomechanical motion as well as the effects of the electron-vibron coupling on the supercurrent. As a final step, we also discuss these ideas in a vibrating single level quantum dot embedded in a Josephson junction.

8.2 Vibrating Josephson Junction

The physical system we will discuss in this part, comprise a superconducting electrode (SCR) of mass m_c attached to a cantilever with a spring constant k_c, see Fig. 8.1, cf. [3]. The movable electrode is located at some distance from an infinitely massive counterelectrode (SCL). The Hamiltonian describing this system is given by

$$\mathcal{H} = \mathcal{H}_L + \mathcal{H}_R + \mathcal{H}_T, \tag{8.1a}$$

$$\mathcal{H}_\chi = \sum_{\mathbf{k}\sigma} \varepsilon_{\mathbf{k}} c^\dagger_{\mathbf{k}\sigma} c_{\mathbf{k}\sigma} + \sum_{\mathbf{k}} [\Delta_\chi c^\dagger_{\mathbf{k}\uparrow} c^\dagger_{-\mathbf{k}\downarrow} + H.c.], \quad \chi = L, R, \tag{8.1b}$$

$$\mathcal{H}_T = \sum_{\mathbf{pq}\sigma} T_{\mathbf{pq}} c^\dagger_{\mathbf{p}\sigma} c_{\mathbf{q}\sigma} + H.c., \quad \mathbf{p} \in L,\ \mathbf{q} \in R. \tag{8.1c}$$

Here, the leads are described by the BCS Hamiltonians $\mathcal{H}_\chi$, where $\varepsilon_{\mathbf{k}}$ is the single-particle energies of the conduction electrons, whereas Δ_χ is the superconducting pair potential, or gap function, in the leads. We assume that the superconductors are of a spin-singlet s-wave pairing symmetry, and we consider the Josephson junction at zero temperature. The latter assumption implies that the derived theory is applicable for temperatures much lower than the relevant energy scales related

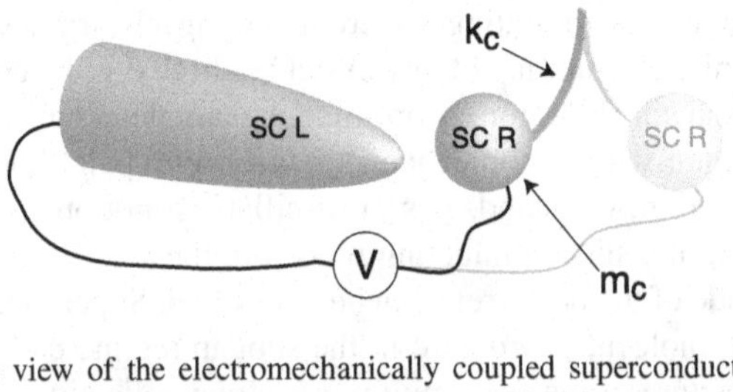

Fig. 8.1 Schematic view of the electromechanically coupled superconducting tunnel junction. The superconducting electrode attached to the cantilever (SCR) is modeled as a harmonic oscillator with spring constant k_c and mass m_c, located at some distance from an infinitely massive superconducting counterelectrode (SCL). The device is biased with a voltage V

to the nanomechanical motion, ω_0, and Josephson frequency, ω_J, i.e. $T \ll \omega_0, \omega_J$. Specifically, typical frequencies for the nanomechanical oscillations can be in the range $\omega_0 \sim 10^8$–10^9 Hz, which requires that $T \sim 10$–100 mK.

The tunneling between the superconductors is given at the rate $T_{\mathbf{pq}}$. The nanomechanical motion of the oscillating electrode is included in this tunneling rate through

$$T_{\mathbf{pq}} = T_{\mathbf{pq}}^{(0)}[1+\alpha q], \tag{8.2}$$

where α describes the coupling between the electronic tunneling and vibrational motion. The quantity q is the displacement operator for the oscillator, such that the equilibrium point ($q=0$) of the oscillating electrode is located within the junction at a position that does not correspond to a special point of symmetry between the leads. This linear approximation of the motion captures the modulation of the tunnel barrier triggered by the oscillations.

The energy associated with the frequency of the nanomechanical oscillations,

$$\omega_0 = \sqrt{\frac{k_c}{m_c}} \sim 10^{-6}\text{–}10^{-1}\ \text{eV}, \tag{8.3}$$

is much smaller than the typical electronic energy on the order of 1 eV. Hence, the mechanical oscillations is very slow compared to the time scale of the electronic processes which allows us to apply the Born-Oppenheimer approximation to treat the electronic degrees of freedom. In this approximation, the moving island is regarded to be static at every instantaneous location. Under those conditions we study the dynamics of the mechanical oscillator as affected by the tunneling electrons.

8.2.1 The Supercurrent

As usual, we consider the number of electrons N_L in the left lead and study the rate of change of electrons within the model. We have the total current ($\hbar = 1$)

$$I(t) = 2e\,\mathrm{Re}\int_{-\infty}^{t} \left(\langle [A(t), A^{\dagger}(t')]\rangle e^{ieV(t-t')} + \langle [A(t), A(t')]\rangle e^{ieV(t+t')}\right)dt', \tag{8.4}$$

where $A(t) = \sum_{\mathbf{pq}} T_{\mathbf{pq}}(t) c_{\mathbf{p}\sigma}^{\dagger}(t) c_{\mathbf{q}\sigma}(t)$, whereas the time-dependent operators $c_{\mathbf{k}\sigma}(t) = e^{i\mathcal{K}_\chi t} c_{\mathbf{k}\sigma} e^{-i\mathcal{K}_\chi t}$. The first term is associated with conventional single-electron tunneling which does not contribute to the supercurrent. In the following we shall only be concerned with the supercurrent, which is described by the second term, i.e.

$$I_S(t) = 2e\,\mathrm{Re}\int_{-\infty}^{t} \langle [A(t), A(t')]\rangle e^{ieV(t+t')}dt'. \tag{8.5}$$

The average contains two creation (destruction) operators from the left (right) lead, that is, it describes the tunneling of two electrons from the right to the left lead which is the transfer of a Cooper pair through the junction. The current thus, includes the anomalous averages e.g. $\mathcal{F}^{\dagger>}_{\mathbf{p}\sigma\mathbf{p}'\sigma'}(t,t') = (-i)\langle c^\dagger_{\mathbf{p}\sigma}(t)c^\dagger_{\mathbf{p}'\sigma'}(t')\rangle$ and $\mathcal{F}^{>}_{\mathbf{q}\sigma\mathbf{q}'\sigma'}(t,t') = (-i)\langle c_{\mathbf{q}\sigma}(t)c_{\mathbf{q}'\sigma'}(t')\rangle$.

We proceed by making a Bogoliubov-Valatin transformation $c_{\mathbf{k}\sigma} = u_{\mathbf{k}}\gamma_{\mathbf{k}\sigma} - \sigma v^*_{\mathbf{k}}\gamma^\dagger_{\mathbf{k}\bar{\sigma}}$, where the subscripts $\sigma = \uparrow, \downarrow$, whereas the factor $\sigma = \pm 1$. Using this transformation we have introduced the quasi-particle energies $E_{\mathbf{k}} = \sqrt{(\varepsilon_{\mathbf{p}} - \mu_\chi)^2 + |\Delta_\chi|^2}$, and the coherence factors

$$u_{\mathbf{k}} = \sqrt{\frac{1}{2}\left(1 + \frac{\varepsilon_{\mathbf{k}} - \mu_\chi}{E_{\mathbf{k}}}\right)}, \qquad v_{\mathbf{k}} = \sqrt{\frac{1}{2}\left(1 - \frac{\varepsilon_{\mathbf{k}} - \mu_\chi}{E_{\mathbf{k}}}\right)}, \tag{8.6}$$

such that $|u_{\mathbf{k}}|^2 + |v_{\mathbf{k}}|^2 = 1$ and $u^*_{\mathbf{k}}v_{\mathbf{k}} = |\Delta_\chi|e^{i\phi_\chi}/(2E_{\mathbf{k}})$, where ϕ_χ is the phase associated with the superconductor χ. In terms of those new operators, $\gamma_{\mathbf{p}\sigma}, \gamma^\dagger_{\mathbf{p}\sigma}$, we will pair the operators γ and $\gamma^\dagger$ with one another to form a quasi-particle description of the transport. We obtain e.g.

$$\begin{aligned}
\mathcal{F}^{>}_{\mathbf{k}\sigma\mathbf{k}'\sigma'}(t,t') &= (-i)\langle c_{\mathbf{k}\sigma}(t)c_{\mathbf{k}'\sigma'}(t')\rangle \\
&= (-i)\langle (u_{\mathbf{k}}\gamma_{\mathbf{k}\sigma} - \sigma v^*_{\mathbf{k}}\gamma^\dagger_{\mathbf{k}\bar{\sigma}})(t)(u_{\mathbf{k}'}\gamma_{\mathbf{k}'\sigma'} - \sigma' v^*_{\mathbf{k}'}\gamma^\dagger_{\mathbf{k}'\bar{\sigma}'})(t')\rangle \\
&= i\sigma'\delta_{\sigma\bar{\sigma}'}u_{\mathbf{k}}v^*_{\mathbf{k}'}\langle\gamma_{\mathbf{k}\sigma}(t)\gamma^\dagger_{\mathbf{k}'\bar{\sigma}'}(t')\rangle + i\sigma\delta_{\bar{\sigma}\sigma'}v^*_{\mathbf{k}}u_{\mathbf{k}'}\langle\gamma^\dagger_{\mathbf{k}\bar{\sigma}}(t)\gamma_{\mathbf{k}'\sigma'}(t')\rangle \\
&= i\delta_{\mathbf{k}\mathbf{k}'}\sigma\delta_{\bar{\sigma}\sigma'}u_{\mathbf{k}}v^*_{\mathbf{k}'}\left[f(E_{\mathbf{k}})e^{iE_{\mathbf{k}}(t-t')} - f(-E_{\mathbf{k}})e^{-iE_{\mathbf{k}}(t-t')}\right],
\end{aligned} \tag{8.7}$$

where the last line is obtained by observing that $\langle\bar{\gamma}^\dagger_{\mathbf{k}\bar{\sigma}}(t)\gamma_{\mathbf{k}'\sigma'}(t')\rangle = f(E_{\mathbf{k}}) \times \exp[iE_{\mathbf{k}}(t-t')]$ defines the occupation number for the quasi-particle at the energy $E_{\mathbf{k}}$ and momentum $\mathbf{k}$. For short, we use the notation $\mathcal{F}_{\mathbf{k}\sigma\sigma'} = \delta_{\mathbf{k}\mathbf{k}'}\mathcal{F}_{\mathbf{k}\sigma\mathbf{k}'\sigma'}$. In summary, we have the four different anomalous Green functions

$$\mathcal{F}^{<}_{\mathbf{k}\sigma\sigma'}(t,t') = i\delta_{\sigma'\bar{\sigma}}\sigma u_{\mathbf{k}}v^*_{\mathbf{k}}\left[f(E_{\mathbf{k}})e^{-iE_{\mathbf{k}}(t-t')} - f(-E_{\mathbf{k}})e^{iE_{\mathbf{k}}(t-t')}\right], \tag{8.8a}$$

$$\mathcal{F}^{>}_{\mathbf{k}\sigma\sigma'}(t,t') = i\delta_{\sigma'\bar{\sigma}}\sigma u_{\mathbf{k}}v^*_{\mathbf{k}}\left[f(E_{\mathbf{k}})e^{iE_{\mathbf{k}}(t-t')} - f(-E_{\mathbf{k}})e^{-iE_{\mathbf{k}}(t-t')}\right], \tag{8.8b}$$

$$\mathcal{F}^{\dagger<}_{\mathbf{k}\sigma\sigma'}(t,t') = i\delta_{\sigma'\bar{\sigma}}\sigma u^*_{\mathbf{k}}v_{\mathbf{k}}\left[f(-E_{\mathbf{k}})e^{iE_{\mathbf{k}}(t-t')} - f(E_{\mathbf{k}})e^{-iE_{\mathbf{k}}(t-t')}\right], \tag{8.8c}$$

$$\mathcal{F}^{\dagger>}_{\mathbf{k}\sigma\sigma'}(t,t') = i\delta_{\sigma'\bar{\sigma}}\sigma u^*_{\mathbf{k}}v_{\mathbf{k}}\left[f(-E_{\mathbf{k}})e^{-iE_{\mathbf{k}}(t-t')} - f(E_{\mathbf{k}})e^{iE_{\mathbf{k}}(t-t')}\right]. \tag{8.8d}$$

At zero temperature, which is assumed here, the Fermi function $f(E_{\mathbf{k}}) = 0$. The current can, thus, be reformulated into ($\tau = t - t'$, $\omega_J = 2eV$, and $\phi = \phi_L - \phi_R$)

$$\begin{aligned}
I_S(t) = 2e\,\mathrm{Re}\sum_{\mathbf{p}\mathbf{q}\sigma}\int_{-\infty}^{t} & T_{\mathbf{p}\mathbf{q}}(t)T_{\mathbf{p}\mathbf{q}}(t')\big(\mathcal{F}^{\dagger>}_{\mathbf{p}\sigma\bar{\sigma}}(t,t')\mathcal{F}^{<}_{\mathbf{q}\bar{\sigma}\sigma}(t',t) \\
& - \mathcal{F}^{\dagger<}_{\mathbf{p}\sigma\bar{\sigma}}(t,t')\mathcal{F}^{>}_{\mathbf{q}\bar{\sigma}\sigma}(t',t)\big)e^{-ieV\tau}dt'e^{i\omega_J t}
\end{aligned}$$

$$= 2e\,\mathrm{Re}\sum_{\mathbf{pq}\sigma}\frac{|\Delta_L||\Delta_R|}{4E_{\mathbf{p}}E_{\mathbf{q}}}$$

$$\times \int_{-\infty}^{t} T_{\mathbf{pq}}(t)T_{\mathbf{pq}}(t')\big(e^{-i(E_{\mathbf{p}}+E_{\mathbf{q}})\tau} - e^{i(E_{\mathbf{p}}+E_{\mathbf{q}})\tau}\big)e^{-ieV\tau}dt'e^{i(\omega_J t+\phi)}. \quad (8.9)$$

Were it not for the unknown time-dependence in the tunneling parameter, this time-integral would be solvable. However, thanks to the discussion above, we are allowed to use the Born-Oppenheimer approximation and treat the nanomechanical motion as very slow in comparison with the electronic processes, i.e. $E_{\mathbf{k}} \gg \omega_J, \omega_0$. Hence, we can make the approximation $q(t') \approx q(t) - \tau\dot{q}(t)$. By introducing the parameters

$$J_S(eV) = e\sum_{\mathbf{pq}}|T^{(0)}_{\mathbf{pq}}|^2\frac{|\Delta_L||\Delta_R|}{E_{\mathbf{p}}E_{\mathbf{q}}}\left(\frac{1}{eV+E_{\mathbf{p}}+E_{\mathbf{q}}} - \frac{1}{eV-E_{\mathbf{p}}-E_{\mathbf{q}}}\right), \quad (8.10)$$

which describes the amplitude of the supercurrent in absence of the vibrational mode, and

$$\Gamma_S(eV) = e\sum_{\mathbf{pq}}|T^{(0)}_{\mathbf{pq}}|^2\frac{|\Delta_L||\Delta_R|}{E_{\mathbf{p}}E_{\mathbf{q}}}\left(\frac{1}{(eV+E_{\mathbf{p}}+E_{\mathbf{q}})^2} - \frac{1}{(eV-E_{\mathbf{p}}-E_{\mathbf{q}})^2}\right), \quad (8.11)$$

which reflects the modified amplitude of the supercurrent due to the vibrational mode, we can finally write the Josephson current according to

$$I_S(t) = J_S[1+\alpha q]^2\sin(\omega_J t+\phi) - \Gamma_S[1+\alpha q]\alpha\dot{q}\cos(\omega_J t+\phi), \quad (8.12)$$

where $\dot{q} = \partial_t q$. Making an order of magnitude estimate of those parameters one finds for small bias voltages that

$$\omega_J\frac{\Gamma_S}{J_S} \sim \left(\frac{eV}{|\Delta|}\right)^2 \ll 1. \quad (8.13)$$

8.2.2 *Motion of the Oscillating Electrode*

It is clear from (8.12) that the current very strongly depends on the position of the oscillating electrode. Hence, in order to achieve a comprehensive picture of the effects from the movable electrode on the supercurrent we need a model for the nanomechanical motion. We can achieve this picture by constructing a Hamiltonian $\mathcal{H}_J$ for the Josephson junction using the requirement that $2e\partial\mathcal{H}_J/\partial\phi = I_S$. We, thus, find

$$\mathcal{H}_J = E_J[1+\alpha q]^2[1-\cos(\omega_J t+\phi)] - \frac{\Gamma_S}{2e}[1+\alpha q]\alpha\dot{q}\sin(\omega_J t+\phi), \quad (8.14)$$

where $E_J = J_S/(2e)$. This Hamiltonian captures the Josephson back-action effect, or in other words, the influence from the electronic degrees of freedom on the nanomechanical motion. The entire mechanical motion will now be modeled by the classical Hamiltonian

$$\mathcal{H}_{\text{osc}} = \frac{p^2}{2m_c} + \frac{k_c q^2}{2} + \mathcal{H}_J. \tag{8.15}$$

Using the Hamilton equations of motion, $\dot{q} = \partial\mathcal{H}_{\text{osc}}/\partial p$ and $\dot{p} = -\partial\mathcal{H}_{\text{osc}}/\partial q$, we find the equation of motion for the coordinate variable q according to

$$m_c\ddot{q} + [\gamma_N + \gamma_S(t)]\dot{q} + k_c q = F(t). \tag{8.16}$$

Evidently, the nanomechanical motion of the movable electrode describes a driven and damped harmonic oscillation, here with the time-dependent damping

$$\gamma_S(t) = -\alpha^2 \frac{\Gamma_S}{e}\sin(\omega_J t + \phi), \tag{8.17}$$

while the damping γ_N is related to e.g. the mechanical friction, and driving force

$$F(t) = -2\alpha E_J(1 + \alpha q)\left(1 - \left[1 + \frac{\omega_J \Gamma_S}{4eE_J}\right]\cos(\omega_J t + \phi)\right). \tag{8.18}$$

The net effect of the Josephson back-action on the nanomechanical dynamics is found to be two-fold. First, the coupling to the Josephson current generates a modification of the mechanical stiffness—the quantity $2\alpha^2 E_J$ can be removed from the driving force and added to the spring constant, hence, adding a so-called Josephson stiffness to the effective spring constant. Second, the oscillatory contribution to the driving force and the time-dependent part of the damping lead to a coherent back-action.

It is an important fact to observe that the damping γ_S originates from the coupling between the tunneling electrons and the mechanical motion of the system, and one can separate out two particular aspects. One of those aspects is that γ_S depends on the bias voltage and vanishes for vanishing bias voltage, since $\Gamma_S \to 0$ as $eV \to 0$. The second important aspect is that γ_S is a periodic function of time, with the Josephson frequency, for finite bias voltages. There is no analogous effect in a normal metal junction due to the absence of a quasi-particle energy gap, hence, the time-dependent damping is unique for the Josephson system.

We study the mechanical motion under low bias voltages, such that Γ_S is much smaller than J_S. By also assuming weak electron-vibron coupling α, the main physics is captured by neglecting terms that are proportional to Γ_S, $\alpha\Gamma_S$ and α^2, in the equation of motion for q. The movable electrode, then, obeys the motion

$$q(t) = q_0\sin(\widetilde{\omega}_0 t + \delta_0)e^{-\gamma_N t/2m_c} - 2\frac{\alpha E_J}{k_c}(1 - H(t;\omega_J,\phi)). \tag{8.19}$$

Here, $\widetilde{\omega}_0 = \sqrt{\omega_0^2 - (\gamma_N/2m)^2}$ is the eigenfrequency of the damped oscillations, whereas the function

$$\begin{aligned} H(t;\omega,\phi) &= \frac{1-(\omega/\omega_0)^2}{[1-(\omega/\omega_0)^2]^2+(\gamma_N\omega/k_c)^2} \\ &\quad\times\left[\cos(\omega t+\phi)+\frac{\gamma_N}{k_c}\frac{\omega}{1-(\omega/\omega_0)^2}\sin(\omega t+\phi)\right] \\ &= \frac{1-(\omega/\omega_0)^2}{[1-(\omega/\omega_0)^2]^2+(\gamma_N\omega/k_c)^2}\sqrt{1+\left(\frac{\gamma_N}{k_c}\frac{\omega}{1-(\omega/\omega_0)^2}\right)^2} \\ &\quad\times\sin(\omega t+\tilde{\phi}), \end{aligned} \tag{8.20}$$

where $\tilde{\phi} = \phi + \arctan\{(k_c/\gamma_N)[1-(\omega/\omega_0)^2]/\omega\}$. Here also, the parameters q_0 and δ_0 are to be determined by the initial conditions.

Tuning the Josephson frequency $\omega_J \to \omega_0$ leads to that $H(\omega_J) \to (k_c/\gamma_N\omega_0) \times \sin(\omega_0 t + \phi)$, and to the motion

$$q(t) = q_0 \sin(\widetilde{\omega}_0 t + \delta_0)e^{-\gamma_N t/2m_c} - 2\frac{\alpha E_J}{\gamma_N\omega_0}\sin(\omega_0 t+\phi). \tag{8.21}$$

For a weakly damped system ($\gamma_N \to 0$), the motion appears to occur with an uncontrolled amplitude at the frequency ω_0, something which is a typical resonant characteristics. In the strongly damped case, on the other hand, the oscillatory motion with frequency $\widetilde{\omega}_0$ rapidly decays whereas the part oscillating at the frequency ω_0 persists indefinitely with a small amplitude.

Substituting this motion into the expression for the current, (8.12), under the same approximation regarding Γ_S, however, retaining contributions up to α^2, leads to the Josephson current

$$\begin{aligned} \frac{I_S(t)}{J_S} &= \left(1-4\frac{\alpha_0^2}{K}\right)\sin(\omega_J t+\phi) + \Bigg(2\alpha_0\sin(\widetilde{\omega}_0 t+\delta_0)e^{-\gamma_N t/2m_c} \\ &\quad + \alpha_0^2\sin^2(\widetilde{\omega}_0 t+\delta_0)e^{-\gamma_N t/m_c} + 4\frac{\alpha_0^2}{K}H(t;\omega_J,\phi)\Bigg)\sin(\omega_J t+\phi), \end{aligned} \tag{8.22}$$

where $\alpha_0 = q_0\alpha$ and $K = k_c q_0^2/E_J$. The Josephson current is obviously not only modulated by the Josephson frequency ω_J but also by the (effective) vibrational frequency $\widetilde{\omega}_0$. In particular, for weakly damped oscillations ($\gamma_N \to 0$), one finds contributions to the Josephson current which are proportional to $\sin(\widetilde{\omega}_0 t+\delta_0)\sin(\omega_J t+\phi)$, from the second term in (8.22), which gives the frequencies $\omega_J \pm \omega_0$, and $\sin^2(\widetilde{\omega}_0 t+\delta_0)\cos(\omega_J t+\phi) \sim [1-\cos 2(\widetilde{\omega}_0 t+\delta_0)]\sin(\omega_J t+\phi)$, from the third term in (8.22), giving the frequencies $\omega_J \pm 2\omega_0$. From the last term in (8.22) we finally find a contributions which is proportional to $\cos(\omega_J t+\phi)\sin(\omega_J t+\phi)$, which gives a frequency $2\omega_J$.

Thus, apart from the fundamental Josephson frequency ω_J, which is provided by e.g. the first term in (8.22), the coupling between the tunneling electrons and the mechanical motion provides modulations of the Josephson current by the doubled Josephson frequency $2\omega_J$ and, in addition, with the frequencies $\omega_J \pm \omega_0$ and $\omega_J \pm 2\omega_0$. Especially, tuning the bias voltage such that either $\omega_J = \omega_0$ or $\omega_J = 2\omega_0$ leads to dc components, Shapiro steps, in the current.

8.3 Vibrating Superconducting Island in a Josephson Junction

As we saw above, the coupling between the tunneling electrons and mechanical motion leads to additional complexity of both the mechanical motion and Josephson current in the system. Without the electro-mechanical coupling, the Josephson current is simply described by $I_S(t) = J_S \sin(\omega_J t + \phi)$, and this contribution is retained in the electro-mechanically coupled system, cf. (8.22). In the electro-mechanically system, however, the Josephson current assumes the functional form $I_S(t) = \tilde{J}_s(t)\sin(\omega_J t + \phi) + \tilde{\Gamma}_S(t)\cos(\omega_J t + \phi)$, where the parameters $\tilde{J}_S$ and $\tilde{\Gamma}_S$ depend on the motion of the movable electrode in a non-trivial fashion, see e.g. (8.22). Here, we proceed to build on our intuition by adding complexity to the set-up in that we introduce a second Josephson junction. We, furthermore, let the island between the two Josephson junctions be suspended on a spring, or cantilever, see Fig. 8.2.

In the spirit of the previous example, we model the present system with the Hamiltonian

$$\mathcal{H} = \mathcal{H}_L + \mathcal{H}_I + \mathcal{H}_R + \mathcal{H}_T, \tag{8.23a}$$

$$\mathcal{H}_\chi = \sum_{\mathbf{n}\sigma} \varepsilon_\mathbf{n} c^\dagger_{\mathbf{n}\sigma} c_{\mathbf{n}\sigma} + \sum_\mathbf{n} \left[\Delta_\chi c^\dagger_{\mathbf{n}\uparrow} c^\dagger_{-\mathbf{n}\downarrow} + H.c. \right], \quad \chi = L, I, R, \tag{8.23b}$$

$$\mathcal{H}_T = \sum_{\mathbf{pk}\sigma} T_{\mathbf{pk}} c^\dagger_{\mathbf{p}\sigma} c_{\mathbf{k}\sigma} + \sum_{\mathbf{qk}\sigma} T_{\mathbf{qk}} c^\dagger_{\mathbf{q}\sigma} c_{\mathbf{k}\sigma} + H.c. \quad \mathbf{p} \in L,\ \mathbf{k} \in I,\ \mathbf{q} \in R, \tag{8.23c}$$

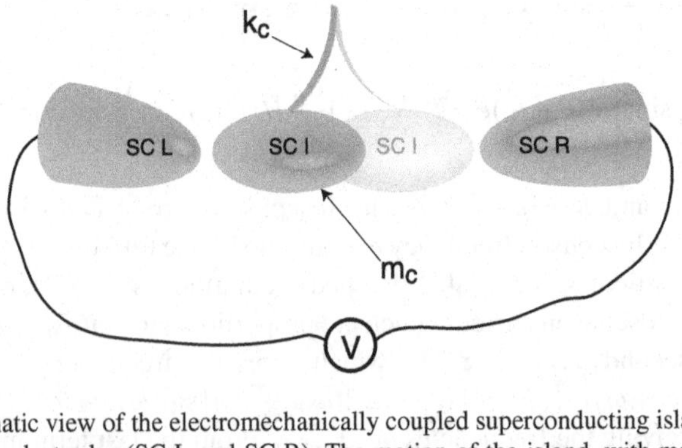

Fig. 8.2 Schematic view of the electromechanically coupled superconducting island (SC I) to the superconducting electrodes (SC L and SC R). The motion of the island, with mass m_c, which is suspended on e.g. a cantilever and located between the infinitely massive superconducting leads, is modeled as a harmonic oscillator with spring constant k_c. The device is biased with a voltage V

where $T_{\mathbf{pk}} = T_{\mathbf{pk}}^{(0)}(1+\alpha_L q)$ and $T_{\mathbf{qk}} = T_{\mathbf{qk}}^{(0)}(1+\alpha_R q)$, with the couplings α_L and α_R to the left and right leads. Performing an analogous derivation as was done for the single junction system, we obtain the supercurrent $I_S^\chi(t)$ between the lead $\chi = L, R$ and the island according to

$$I_S^\chi(t) = J_S^\chi(\omega_\chi)[1+\alpha_\chi q]^2 \sin(\omega_J t+\phi_\chi) - \Gamma_S^\chi(\omega_\chi)[1+\alpha_\chi q]\alpha_\chi \dot{q}\cos(\omega_J t+\phi_\chi), \tag{8.24}$$

where the Josephson frequency $\omega_\chi = 2(\mu_\chi - \mu_I)$ and ϕ_χ is the phase difference between the lead χ and the island. Moreover, the amplitude of the Josephson current in absence of electromechanical coupling is given by

$$J_S^\chi(\mu_\chi) = e \sum_{\mathbf{n}\in\chi,\mathbf{k}} |T_{\mathbf{nk}}^{(0)}|^2 \frac{|\Delta_\chi||\Delta_I|}{E_{\mathbf{n}}E_{\mathbf{k}}}\left(\frac{1}{\mu_\chi + E_{\mathbf{n}} + E_{\mathbf{k}}} - \frac{1}{\mu_\chi - E_{\mathbf{n}} - E_{\mathbf{k}}}\right), \tag{8.25}$$

where Δ_χ and Δ_I is the superconducting gap in lead χ and the island, respectively, whereas the quasi-particle energies $E_{\mathbf{n}} = \sqrt{(\varepsilon_{\mathbf{n}} - \mu_\chi)^2 + |\Delta_\chi|^2}$ and $E_{\mathbf{k}} = \sqrt{(\varepsilon_{\mathbf{k}} - \mu_I)^2 + |\Delta_I|^2}$. Here, also the second contribution to the current has the amplitude

$$\Gamma_S^\chi(\mu_\chi) = e \sum_{\mathbf{n}\in\chi,\mathbf{k}} |T_{\mathbf{nk}}^{(0)}|^2 \frac{|\Delta_\chi||\Delta_I|}{E_{\mathbf{n}}E_{\mathbf{k}}}\left(\frac{1}{(\mu_\chi + E_{\mathbf{n}} + E_{\mathbf{k}})^2} - \frac{1}{(\mu_\chi - E_{\mathbf{n}} - E_{\mathbf{k}})^2}\right). \tag{8.26}$$

For stationary bias voltages we use that $I_S^R = -I_S^L$ such that the total supercurrent can be written $I_S = I_S^L = (I_S^L - I_S^R)/2$.

The Hamiltonian for the nanomechanical motion in which the force arising from the electro-mechanical coupling is included, is derived by requiring that $2e\partial\mathcal{H}_J/\partial\phi_\chi = I_S^\chi$. It is then found that

$$\begin{aligned}\mathcal{H}_J = \frac{1}{2}\sum_{\chi=L,R}\Big(&E_J^\chi[1+\alpha_\chi q]^2[1-\cos(\omega_\chi t+\phi_\chi)]\\ &-\frac{1}{2e}\Gamma_S^\chi[1+\alpha_\chi q]\alpha_\chi\dot{q}\sin(\omega_\chi t+\phi_\chi)\Big),\end{aligned} \tag{8.27}$$

where $E_J^\chi = J_S^\chi/(2e)$, and we find the equation of motion for the coordinate q given by an equation which is formally equal to the one given in (8.17), here with the driving force

$$F(t) = -\sum_{\chi=L,R} E_J^\chi \alpha_\chi [1+\alpha_\chi q]\left\{1 - \left(1 + \frac{\omega_\chi \Gamma_S^\chi}{4eE_J^\chi}\right)\cos(\omega_\chi t+\phi_\chi)\right\}, \tag{8.28}$$

whereas the time-dependent damping

$$\gamma_S(t) = -\frac{1}{2e}\sum_{\chi=L,R}\Gamma_S^\chi \alpha_\chi^2 \sin(\omega_\chi t+\phi_\chi). \tag{8.29}$$

In comparison with the single junction system, it appears as the only complexity added in this case is the summation over two junctions. However, the motion of the island is now affected by the supercurrent in two junctions, and the increased degrees of freedom due to this summation, increases our possibilities to control and manipulate the motion of the island and the resulting current. To begin our analysis of this feature, we perform similar approximations as above, i.e. we neglect the terms in the damping and driving force which are proportional to $\alpha_\chi \Gamma_S^\chi$ and α_χ^2. Then, the motion of the island is given by

$$q(t) = q_0 \sin(\tilde{\omega}_0 t + \delta_0) e^{-\gamma_N t/2m_c} - \sum_{\chi=L,R} \frac{\alpha_\chi E_J^\chi}{k_c} (1 - H(t;\omega_\chi,\phi_\chi)). \tag{8.30}$$

Using this motion in the expression for the Josephson current yields

$$\frac{I_S^\chi(t)}{J_S^\chi} = \bigg(1 + 2\tilde{\alpha}_\chi \sin(\tilde{\omega}_0 t + \delta_0) e^{-\gamma_N t/2m_c} + \tilde{\alpha}_\chi^2 \sin^2(\tilde{\omega}_0 t + \delta_0) e^{-\gamma_N t/m_c}$$
$$- 2\tilde{\alpha}_\chi \sum_{\chi'=L,R} \frac{\tilde{\alpha}_{\chi'}}{K_{\chi'}} [1 - H(t;\omega_{\chi'},\phi_{\chi'})]\bigg) \sin(\omega_\chi t + \phi_\chi), \tag{8.31}$$

with $\tilde{\alpha}_\chi = q_0 \alpha_\chi$ and $K_\chi = k_c q_0^2 / E_J^\chi$. Despite this current appear as innocent as the single junction current, we here have an expression with many more degrees of freedom. For instance, the appearance of two phase factors ϕ_L and ϕ_R demands that we cannot neglect them. Instead, there is an interplay between them by which we can control whether there flows a current through the junctions. The two Josephson frequencies ω_L and ω_R may, moreover, also be different if the system is asymmetric with respect to the bias voltage. Finally, the sum in the last term shows that the Josephson current in one junction depends on the current in the other, hence, one has to adjust the parameters with much more care in the double junction system.

In order to get a grasp of the richness the double junction system presents, we tune the parameters into the following regime. Assuming that $T_{\mathbf{pk}}^{(0)} = T_{\mathbf{qk}}^{(0)}$ and $|\Delta_L| = |\Delta_R|$, it is seen that $J_S^L = J_S^R = J_S$, hence, we can set $E_J^\chi = E_J$ and $K_J^\chi = K_J$. By, furthermore, assuming that $\alpha_L = -\alpha_R = \alpha$ we find that the Josephson current can be written as

$$\frac{2I_S(t)}{J_S} = \big[1 + \tilde{\alpha}^2 \sin^2(\tilde{\omega}_0 t + \delta_0) e^{-\gamma_N t/2m_c}\big]\big[\sin(\omega_L t + \phi_L) - \sin(\omega_R t + \phi_R)\big]$$
$$+ 2\tilde{\alpha}\bigg(\sin(\tilde{\omega}_0 t + \delta_0) e^{-\gamma_N t/m_c} + \frac{\tilde{\alpha}}{K_J}\big[H(t;\omega_L,\phi_L) - H(t;\omega_R,\phi_R)\big]\bigg)$$
$$\times \big[\sin(\omega_L t + \phi_L) + \sin(\omega_R t + \phi_R)\big]. \tag{8.32}$$

From this expression it is clear that under zero phase differences $\phi_\chi = 0$ and symmetrically biased system $\omega_L = -\omega_R = \omega_J$, the Josephson current becomes

$$\frac{2I_S(t)}{J_S} = 2\big[1 + \tilde{\alpha}^2 \sin^2(\omega_0 t + \delta_0) e^{-\gamma_N t/m}\big] \sin \omega_J t. \tag{8.33}$$

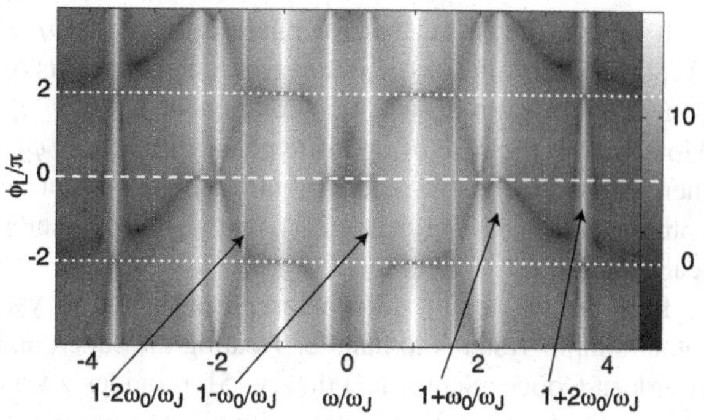

Fig. 8.3 Fourier transform of the Josephson current ($\log_{10}|2I_S(\omega)/J_S|^2$) as function of the frequency ω and the phase difference ϕ_L for undamped ($\gamma_N = 0$) and symmetric bias ($\omega_J = -\omega_R$) conditions, and fixed phase difference $\phi_R = 0$. Here, we have taken $\tilde{\alpha} = 0.1$, $K_J = 0.6$, and frequency $\omega_0 = 1.3\omega_J$

Hence, in the underdamped case ($\gamma_N \to 0$), the Josephson current comprise a dc component whenever the bias voltage matches the Shapiro step $2\tilde{\omega}_0$, analogous to the single junction system. In contrast to the single junction, however, there is no dc component at bias voltages equaling $\tilde{\omega}_0$. The electro-mechanical coupling in the system generates a dc component at each junction, but under the present conditions, those cancel each other. The net effect, thus, is that there is no dc component at $\tilde{\omega}_0$. The Josephson current under those conditions corresponds to the dashed line in Fig. 8.3, where it is readily seen that there are current components only at the frequencies ω_J, and $\omega_J \pm 2\omega_0$.

Another limiting case is to consider opposite phases, i.e. $\phi_R = \phi_L + \pi$, and we assume symmetrically biased and undamped system. Then, the Josephson current can be simplified to

$$\frac{I_S(t)}{J_S} = \left[1 + \tilde{\alpha}^2 \sin^2(\omega_0 t + \delta_0)\right] \cos\omega_J t \sin\phi_L + 4\tilde{\alpha}\sin(\omega_0 t + \delta_0)\sin\omega_J t\cos\phi_L + \frac{\tilde{\alpha}}{K_J}\sin 2\omega_J t\cos^2\phi_L. \tag{8.34}$$

Consequently, for general phase differences ϕ_L there are current components with frequencies ω_J, $2\omega_J$, $\omega_J \pm \omega_0$, and $\omega_J \pm 2\omega_0$, see Fig. 8.3. However, at $\phi_L = -\pi$, giving $\phi_R = 0$, the components at frequency ω_J and $\omega_J \pm 2\omega_0$ vanishes, something which is also illustrated in Fig. 8.3 (dotted lines).

Finally, we consider the zero bias case, i.e. $\omega_\chi = 0$. Then, the function $H(t;\omega,\phi) = \cos\phi$, and the Josephson current in (8.32) reduces to

$$\begin{aligned}\frac{2I_S(t)}{J_S} &= \left[1 + \tilde{\alpha}^2 \sin(\tilde{\omega}_0 t + \delta_0)e^{-\gamma_N t/m}\right][\sin\phi_L - \sin\phi_R] \\ &\quad + 2\tilde{\alpha}\left(\sin(\tilde{\omega}_0 t + \delta_0)e^{-\gamma_N t/2m} + \frac{\tilde{\alpha}}{K_J}[\cos\phi_L - \cos\phi_R]\right) \\ &\quad \times [\sin\phi_L + \sin\phi_R].\end{aligned} \tag{8.35}$$

Hence, for general combinations of the phase differences ϕ_χ, the zero bias Josephson current has a non-vanishing dc components as well as ac components at the frequencies $\tilde{\omega}_0$ and $2\tilde{\omega}_0$.

It is seen, however, that for phase differences $\phi_L = \phi_R$, only the contribution proportional to $\sin(\widetilde{\omega}_0 t + \delta_0)$ survives, that is, there is only one non-vanishing component to the current, and this component is time-dependent with frequency $\widetilde{\omega}_0$. Moreover, in the case of opposite phase differences $\phi_L = -\phi_R$, on the other hand, there is one dc and one ac contribution to the current, of which the time-dependent component has the frequency $2\widetilde{\omega}_0$, that is, twice the eigenfrequency of the mechanical oscillations.

Both the single and double Josephson junction systems are, in certain respect, rather simple systems to handle. Treating the superconductors in the spirit of BCS (Bardeen-Cooper-Schrieffer) theory [5], provides a very simple parametrization of the amplitudes (J_S, Γ_S) associated with the time-independent and time-dependent couplings between the superconductors. One can imagine to have very different superconducting media, and asymmetric couplings in the different junction (in the double junction system) in order to vary the circumstances. Nevertheless, we do not obtain any additional time-dependence but those of frequencies that are associated with the fundamental frequencies ω_0 and ω_J. In order to obtain something that does have other time-dependences involved, we consider in the succeeding section an oscillating single level quantum dot embedded in a Josephson junction.

8.4 Vibrating Quantum Dot Island in a Josephson Junction

Here, we imagine that the superconducting island SC I is replaced by a single level quantum dot, but we retain the mechanical degrees of freedom. Having this in mind we, thus, consider the model

$$\mathcal{H} = \sum_{\chi=L,R} \mathcal{H}_\chi + \mathcal{H}_{QD} + \mathcal{H}_T, \tag{8.36}$$

where $\mathcal{H}_\chi$ is as before, whereas the quantum dot is described by $\mathcal{H}_{QD} = \sum_\sigma [\varepsilon_0 + Un_{\bar{\sigma}}/2]n_\sigma = \sum_{p=0,\sigma,2} E_p h^p$, and the tunneling between the leads and the quantum dot by $\mathcal{H}_T = \sum_{\mathbf{k}\sigma} T_{\mathbf{k}\sigma} c^\dagger_{\mathbf{k}\sigma} d_\sigma + H.c. = \sum_{\mathbf{k}\sigma} T_{\mathbf{k}\sigma} c^\dagger_{\mathbf{k}\sigma}(X^{0\sigma} + \sigma X^{\bar{\sigma}2}) + H.c.$ The electro-mechanical coupling is, also here, taken into account through the tunneling rates

$$T_{\mathbf{k}\sigma} = T^{(0)}_{\mathbf{k}\sigma}(1 + \alpha_\chi q), \quad \mathbf{k} \in \chi = L, R. \tag{8.37}$$

In what follows, we assume spin-degenerate conditions, such that $T_{\mathbf{k}\sigma} = T_\mathbf{k}$, and we also assume that the tunneling rates slowly depends on the momentum, hence, we set $T_\mathbf{k} = T_\chi$.

Just like we did in Sect. 8.2, we derive the supercurrent flowing between the left (superconducting) lead and the quantum dot in order to obtain the formula

$$I^L_S(t) = 2e\,\mathrm{Re} \int_{-\infty}^{t} \langle [A_L(t), A_L(t')] \rangle e^{i\mu_L(t+t')} dt', \tag{8.38}$$

however, here

$$A_\chi(t)=\sum_{\mathbf{k}\sigma}T_\chi(t)c^\dagger_{\mathbf{k}\sigma}(t)d_\sigma(t)=\sum_{\mathbf{k}\sigma}T_\chi(t)c^\dagger_{\mathbf{k}\sigma}(t)\big[X^{0\sigma}(t)+\sigma X^{\bar\sigma 2}(t)\big], \tag{8.39}$$

with

$$c_{\mathbf{k}\sigma}(t)=e^{i\mathcal{K}_\chi t}c_{\mathbf{k}\sigma}e^{-i\mathcal{K}_\chi t},\quad \mathcal{K}_\chi=\mathcal{H}_\chi-\mu_\chi N_\chi, \tag{8.40a}$$

$$d_\sigma(t)=e^{i\mathcal{K}_{QD}t}d_\sigma e^{-i\mathcal{K}_{QD}t}, \tag{8.40b}$$

$$X^a(t)=e^{i\mathcal{K}_{QD}t}X^a e^{-i\mathcal{K}_{QD}t},\quad \mathcal{K}_{QD}=\mathcal{H}_{QD}. \tag{8.40c}$$

Here, we assume that the chemical potential of the quantum dot μ_{QD} equals the equilibrium chemical potential $\mu_{\text{eq}}=0$, and that it does not vary with the bias voltage.

The average inside the integration is taken care of by an analogous procedure as in the previous cases, i.e.

$$\begin{aligned}
&\langle[A_L(t),A_L(t')]\rangle\\
&=\sum_{\mathbf{p}\sigma}T_L(t)T_L(t')\big\langle c^\dagger_{\mathbf{k}\sigma}(t)d_\sigma(t)c^\dagger_{\mathbf{p}\bar\sigma}(t')d_{\bar\sigma}(t')-c^\dagger_{\mathbf{p}\bar\sigma}(t')d_{\bar\sigma}(t')c^\dagger_{\mathbf{k}\sigma}(t)d_\sigma(t)\big\rangle\\
&=-\sum_{\mathbf{p}\sigma}T_L(t)T_L(t')\big[\mathcal{F}^{\dagger>}_{\mathbf{p}\sigma\bar\sigma}(t,t')F^<_{\bar\sigma\sigma}(t',t)-\mathcal{F}^{\dagger<}_{\mathbf{p}\sigma\bar\sigma}(t,t')F^>_{\bar\sigma\sigma}(t',t)\big].
\end{aligned} \tag{8.41}$$

The anomalous Green functions for the lead, $\mathcal{F}^{\dagger</>}_{\mathbf{p}\sigma\bar\sigma}(t,t')$, are given in (8.8a)–(8.8d). The new anomalous Green functions, $F^{</>}_{\sigma\sigma'}(t,t')$, provides the dynamics of the quantum dot states, and are defined by

$$F^<_{\bar\sigma\sigma}(t',t)=i\langle d_\sigma(t)d_{\bar\sigma}(t')\rangle,\qquad F^>_{\bar\sigma\sigma}(t',t)=-i\langle d_{\bar\sigma}(t')d_\sigma(t)\rangle, \tag{8.42a}$$

$$F^{\dagger<}_{\bar\sigma\sigma}(t',t)=i\langle d^\dagger_\sigma(t)d^\dagger_{\bar\sigma}(t')\rangle,\qquad F^{\dagger>}_{\bar\sigma\sigma}(t',t)=-i\langle d^\dagger_{\bar\sigma}(t')d^\dagger_\sigma(t)\rangle, \tag{8.42b}$$

where, we for completeness, also have defined the anomalous Green functions $F^{\dagger</>}_{\sigma\sigma'}(t,t')$. Since, for instance, $d_\sigma d_{\bar\sigma}=[X^{0\sigma}+\sigma X^{\bar\sigma 2}][X^{0\bar\sigma}+\bar\sigma X^{\sigma 2}]=\bar\sigma Z^{02}$, we approximate the anomalous quantum dot Green functions by ($\tau=t-t'$)

$$F^<_{\bar\sigma\sigma}(t',t)=i\bar\sigma N_{02}e^{i\Delta_{2\sigma}\tau},\qquad F^>_{\bar\sigma\sigma}(t',t)=(-i)\sigma N_{02}e^{i\Delta_{\bar\sigma 0}\tau}, \tag{8.43a}$$

$$F^{\dagger<}_{\bar\sigma\sigma}(t',t)=i\sigma N_{20}e^{i\Delta_{\bar\sigma 0}\tau},\qquad F^{\dagger>}_{\bar\sigma\sigma}(t',t)=(-i)\bar\sigma N_{20}e^{i\Delta_{2\sigma}\tau}. \tag{8.43b}$$

Generally, the averages $N_{02}=\langle Z^{02}\rangle=\langle Z^{20}\rangle^*$ is time-dependent, and below we shall go into how this time-dependence is treated within this context. We notice, however, that N_{02} describes the rate at which transitions between the doubly occupied and empty states occur within the quantum dot, that is, the coherent tunneling of two electrons out of the quantum dot to one of the leads.

Recalling that we are performing the study at zero temperature, we find that the supercurrent can be written as

$$I_S^L(t) = 2e\,\mathrm{Re}\sum_{\mathbf{p}\sigma}\frac{|\Delta_L|}{2E_{\mathbf{p}}}\int_{-\infty}^{t} T_L(t)T_L(t')N_{02}(t')$$
$$\times\left(e^{i(E_{\mathbf{p}}+\Delta_{\bar{\sigma}0})\tau}-e^{-i(E_{\mathbf{p}}-\Delta_{2\sigma})\tau}\right)e^{-i\omega_L\tau/2}dt'e^{i(\omega_L t+\phi_L)}, \quad (8.44)$$

where $\omega_\chi = 2(\mu_\chi - \mu_{QD})$, $\chi = L, R$. When it comes to the time-dependence of the tunneling rates, we can again employ the Born-Oppenheimer approximation, that is, $T_{\mathbf{k}}(t') \approx T_{\mathbf{k}}(t) - \tau\dot{T}_{\mathbf{p}}(t)$. In doing so, we define the time-independent parameters

$$J_S^\chi(\omega_\chi;\omega) = e\sum_{\mathbf{k}\sigma}\frac{|\Delta_\chi||T_\chi^{(0)}|^2}{E_{\mathbf{k}}}$$
$$\times\left(\frac{1}{\omega-\omega_\chi/2+E_{\mathbf{k}}+\Delta_{\bar{\sigma}0}}-\frac{1}{\omega-\omega_\chi/2-E_{\mathbf{k}}+\Delta_{2\sigma}}\right), \quad (8.45)$$

$$\Gamma_S^\chi(\omega_\chi;\omega) = e\sum_{\mathbf{k}\sigma}\frac{|\Delta_\chi||T_\chi^{(0)}|^2}{E_{\mathbf{k}}}$$
$$\times\left(\frac{1}{(\omega-\omega_\chi/2+E_{\mathbf{k}}+\Delta_{\bar{\sigma}0})^2}-\frac{1}{(\omega-\omega_\chi/2-E_{\mathbf{k}}+\Delta_{2\sigma})^2}\right), \quad (8.46)$$

which allows us to write the Josephson current as

$$I_S^L(t) = -\mathrm{Re}\int N_{02}(t')\big(J_S^L(\omega_L;\omega)(1+\alpha_L q)^2\sin(\omega_L t+\phi_L)$$
$$-\Gamma_S^L(\omega_L;\omega)\alpha_L(1+\alpha_L q)\dot{q}\cos(\omega_L t+\phi_L)\big)e^{-i\omega\tau}dt'\frac{d\omega}{2\pi}. \quad (8.47)$$

Notice that this expression is the quantum dot analogy to the one given in the superconducting island case, cf. (8.24). Also, notice that this expression includes a time-dependence in the average N_{02}, which cannot be neglected, in general. Nonetheless, in situations when the time-dependence of N_{02} is negligible, the time-integration yields $2\pi\delta(\omega)$, and the behavior of the quantum dot motion, as well as the Josephson current, is described in very much the same way as the superconducting island.

Generally, the time-dependence of the average N_{02} is very much governed by the electronic processes between the leads and the quantum dot. Fortunately, we can study its time-dependence through the equation of motion for N_{02} within the model. We find that

$$(i\partial_t - \Delta_{20})N_{02}(t) = -\sum_{\mathbf{k}\sigma}\sigma T_\chi\langle[X^{0\sigma}+\sigma X^{\bar{\sigma}2}]c_{\mathbf{k}\bar{\sigma}}\rangle e^{-i\omega_\chi t/2}. \quad (8.48)$$

The correlation function is treated by means of perturbation theory, e.g. using the first order expansion $\langle \mathcal{A}(t)\rangle \approx (-i)\int \theta(\tau)\langle[\mathcal{A}(t), \mathcal{H}(t')]\rangle dt'$. To this order, we then obtain

$$\langle([X^{0\sigma} + \sigma X^{\bar{\sigma}2}]c_{\mathbf{k}\bar{\sigma}})(t)\rangle = i\int_{-\infty}^{t} T_\chi(t')\big([G^{>}_{0\sigma}(t,t') + G^{>}_{\bar{\sigma}2}(t,t')]\mathcal{F}^{<}_{\mathbf{k}\sigma\bar{\sigma}}(t',t) - [G^{<}_{0\sigma}(t,t') + G^{<}_{\bar{\sigma}2}(t,t')]\mathcal{F}^{>}_{\mathbf{k}\sigma\bar{\sigma}}(t',t)\big)e^{-i\omega_\chi t'/2}dt', \tag{8.49}$$

where we have neglected terms proportional to N_{02} (and N_{20}) since such terms leads to a higher order of approximation than we are interested in here. The lesser and greater quantum dot Green functions are here approximated by

$$G^{<}_{0\sigma}(t,t') = iN_\sigma e^{-i\Delta_{\sigma 0}\tau}, \qquad G^{>}_{0\sigma}(t,t') = (-i)N_0 e^{-i\Delta_{\sigma 0}\tau}, \tag{8.50a}$$

$$G^{<}_{\bar{\sigma}2}(t,t') = iN_2 e^{-i\Delta_{2\bar{\sigma}}\tau}, \qquad G^{>}_{\bar{\sigma}2}(t,t') = (-i)N_{\bar{\sigma}} e^{-i\Delta_{2\bar{\sigma}}\tau}. \tag{8.50b}$$

We, then, obtain the equation of motion

$$(i\partial_t - \Delta_{20})N_{02}(t) \approx -\sum_\chi \int \big((1+\alpha_\chi q)^2 U_\chi(\omega_\chi; t', \omega) - i\alpha_\chi(1+\alpha_\chi q)\dot{q}V_\chi(\omega_\chi; t', \omega)\big)e^{-i\omega\tau}\frac{d\omega}{2\pi}dt' e^{-i(\omega_\chi t + \phi_\chi)}, \tag{8.51}$$

where

$$U_\chi(\omega_\chi; t, \omega) = \sum_{\mathbf{k}\sigma} \frac{|\Delta_\chi||T^{(0)}_\chi|^2}{2E_\mathbf{k}}\bigg(\frac{N_0(t)}{\omega + \omega_\chi/2 - E_\mathbf{k} - \Delta_{\sigma 0}} + \frac{N_\sigma(t)}{\omega + \omega_\chi/2 + E_\mathbf{k} - \Delta_{\sigma 0}} + \frac{N_{\bar{\sigma}}(t)}{\omega + \omega_\chi/2 - E_\mathbf{k} - \Delta_{2\bar{\sigma}}} + \frac{N_2(t)}{\omega + \omega_\chi/2 + E_\mathbf{k} - \Delta_{2\bar{\sigma}}}\bigg), \tag{8.52a}$$

$$V_\chi(\omega_\chi; t, \omega) = \sum_{\mathbf{k}\sigma} \frac{|\Delta_\chi||T^{(0)}_\chi|^2}{2E_\mathbf{k}}\bigg(\frac{N_0(t)}{(\omega + \omega_\chi/2 - E_\mathbf{k} - \Delta_{\sigma 0})^2} + \frac{N_\sigma(t)}{(\omega + \omega_\chi/2 + E_\mathbf{k} - \Delta_{\sigma 0})^2} + \frac{N_{\bar{\sigma}}(t)}{(\omega + \omega_\chi/2 - E_\mathbf{k} - \Delta_{2\bar{\sigma}})^2} + \frac{N_2(t)}{(\omega + \omega_\chi/2 + E_\mathbf{k} - \Delta_{2\bar{\sigma}})^2}\bigg). \tag{8.52b}$$

An order of magnitude estimate for the relative ratio between the parameters J_S^χ, Γ_S^χ, U_χ, and V_χ, gives

$$\omega_\chi \frac{\Gamma_S^\chi}{J_S^\chi} \sim \left|\frac{eV}{\Delta_\chi}\right|^2, \qquad \omega_\chi \frac{V_\chi}{U_\chi} \sim \frac{1+|eV/\Delta_\chi|^2}{1-|eV/\Delta_\chi|^2}, \qquad \frac{U_\chi}{J_S^\chi} \sim \frac{eV}{|\Delta_\chi|}. \tag{8.53}$$

In the case that the quantum dot occupation numbers N_p, $p = 0, \sigma, 2$, are time-independent, the equation for N_{02} can be integrated and we obtain

$$\begin{aligned} N_{02}(t) \approx \sum_\chi \big[& \big((1+\alpha_\chi q)^2 U'_\chi(\omega_\chi) - \alpha_\chi[(1+\alpha_\chi q)\dot{q} + \alpha_\chi \ddot{q}]V'_\chi(\omega_\chi) \\ & - 2\alpha_\chi^2 \dot{q}^2 U'''_\chi(\omega_\chi)\big)\cos(\omega_\chi t + \phi_\chi) - \alpha_\chi \dot{q}\big((1+\alpha_\chi q)V'_\chi(\omega_\chi) \\ & + 2(1+\alpha_\chi q)U''_\chi(\omega_\chi) - 2\alpha_\chi \ddot{q} V'''_\chi(\omega_\chi)\big)\sin(\omega_\chi t + \phi_\chi)\big] \end{aligned} \tag{8.54}$$

where we have introduced the notation $U'_\chi(\omega_\chi) = U_\chi(\omega_\chi)/(\Delta_{20} - \omega_\chi)$, $U''_\chi(\omega_\chi) = U_\chi(\omega_\chi)/(\Delta_{20} - \omega_\chi)^2$, and $U'''_\chi(\omega_\chi) = U_\chi(\omega_\chi)/(\Delta_{20} - \omega_\chi)^3$, and analogously for V'_χ, V''_χ, and V'''_χ. Under those conditions, N_p being constants of motion, the time-dependence imposed on N_{02} has a period which is directly determined by the Josephson frequencies ω_χ, $\chi = L, R$. Thus, there is no new time-dependence introduced.

We describe the more general motion by also treating the occupation numbers N_p as time-dependent quantities, with rates of change

$$\partial_t N_0 = 2\,\mathrm{Im} \sum_{\mathbf{k}\sigma} T_\chi \langle c^\dagger_{\mathbf{k}\sigma} X^{0\sigma} \rangle e^{i\omega_\chi t}, \tag{8.55a}$$

$$\partial_t N_\sigma = -2\,\mathrm{Im} \sum_{\mathbf{k}} T_\chi \big(\langle c^\dagger_{\mathbf{k}\sigma} X^{0\sigma} \rangle + \sigma \langle c^\dagger_{\mathbf{k}\bar{\sigma}} X^{\sigma 2} \rangle\big) e^{i\omega_\chi t}, \tag{8.55b}$$

$$\partial_t N_2 = -2\,\mathrm{Im} \sum_{\mathbf{k}\sigma} T_\chi \langle c^\dagger_{\mathbf{k}\sigma} X^{\bar{\sigma} 2} \rangle e^{i\omega_\chi t}. \tag{8.55c}$$

The correlation functions are expressed as

$$\begin{aligned} \langle c^\dagger_{\mathbf{k}\sigma} X^{0\sigma} \rangle e^{i\omega_\chi t} = (-i) \int_{-\infty}^{t} & T_\chi \big(\mathcal{G}^<_{\mathbf{k}\sigma}(t',t) G^>_{0\sigma}(t,t') \\ & - \mathcal{G}^>_{\mathbf{k}\sigma}(t',t) G^<_{0\sigma}(t,t')\big) e^{i\omega_\chi \tau} dt', \end{aligned} \tag{8.56a}$$

$$\begin{aligned} \langle c^\dagger_{\mathbf{k}\sigma} X^{\bar{\sigma} 2} \rangle e^{i\omega_\chi t} = (-i)\sigma \int_{-\infty}^{t} & T_\chi \big(\mathcal{G}^<_{\mathbf{k}\sigma}(t',t) G^>_{\bar{\sigma}2}(t,t') \\ & - \mathcal{G}^>_{\mathbf{k}\sigma}(t',t) G^<_{\bar{\sigma}2}(t,t')\big) e^{i\omega_\chi \tau} dt', \end{aligned} \tag{8.56b}$$

where the Green functions $\mathcal{G}^{</>}$ for the superconducting leads are given by

$$\mathcal{G}^{<}_{\mathbf{k}\sigma}(t',t) = i\big[|u_{\mathbf{k}}|^2 f(E_{\mathbf{k}})e^{iE_{\mathbf{k}}\tau} + |v_{\mathbf{k}}|^2 f(-E_{\mathbf{k}})e^{-iE_{\mathbf{k}}\tau}\big], \tag{8.57a}$$

$$\mathcal{G}^{>}_{\mathbf{k}\sigma}(t',t) = (-i)\big[|u_{\mathbf{k}}|^2 f(-E_{\mathbf{k}})e^{iE_{\mathbf{k}}\tau} + |v_{\mathbf{k}}|^2 f(E_{\mathbf{k}})e^{-iE_{\mathbf{k}}\tau}\big]. \tag{8.57b}$$

In terms of those Green functions and the ones for the QD, (8.50a)–(8.50b), we write the correlation functions as ($T \to 0$)

$$\langle c^{\dagger}_{\mathbf{k}\sigma} X^{0\sigma}\rangle e^{i\omega_\chi t} = (-i)\int_{-\infty}^{t} T_\chi\big(|v_{\mathbf{k}\sigma}|^2 e^{-iE_{\mathbf{k}}\tau} N_0 - |u_{\mathbf{k}\sigma}|^2 e^{iE_{\mathbf{k}}\tau} N_\sigma\big) \times e^{-i(\Delta_{\sigma 0}-\omega_\chi)\tau}dt', \tag{8.58a}$$

$$\langle c^{\dagger}_{\mathbf{k}\sigma} X^{\bar{\sigma}2}\rangle e^{i\omega_\chi t} = (-i)\sigma\int_{-\infty}^{t} T_\chi\big(|v_{\mathbf{k}\sigma}|^2 e^{-iE_{\mathbf{k}}\tau} N_{\bar{\sigma}} - |u_{\mathbf{k}\sigma}|^2 e^{iE_{\mathbf{k}}\tau} N_2\big) \times e^{-i(\Delta_{2\bar{\sigma}}-\omega_\chi)\tau}dt'. \tag{8.58b}$$

Inserting the former correlation function into the equation for e.g. N_0 yields

$$\partial_t N_0 = 2\,\mathrm{Im}\sum_{\chi\sigma} T_\chi(t)\int T_\chi(t')\big(\Gamma^{\chi}_{0\sigma}(\omega)N_0 - \Lambda^{\chi}_{0\sigma}(\omega)N_\sigma\big)e^{-i\omega\tau}\frac{d\omega}{2\pi}dt' \tag{8.59}$$

where we have introduced the parameters

$$\Gamma^{\chi}_{pq}(\omega) = \sum_{\mathbf{k}} \frac{|v_{\mathbf{k}}|^2}{\omega - \Delta^{\chi}_{qp} - E_{\mathbf{k}} + i\delta}, \qquad \Lambda^{\chi}_{pq}(\omega) = \sum_{\mathbf{k}} \frac{|u_{\mathbf{k}}|^2}{\omega - \Delta^{\chi}_{qp} + E_{\mathbf{k}} + i\delta}, \tag{8.60}$$

and where $\Delta^{\chi}_{qp} = \Delta_{qp} - \omega_\chi$.

The parameters Γ^{χ}_{pq} and Λ^{χ}_{pq} can be evaluated by the following procedure. We replace the **k**-summation by integration over the density of electron states N_χ in the superconducting leads χ, and we assume N_χ to vary slowly with momentum/energy. Next, we notice that since $E_{\mathbf{k}}$ is an even functions of $\varepsilon_{\mathbf{k}}$, the fraction $1/(\omega - \Delta^{\chi}_{qp} \pm E_{\mathbf{k}} + i\delta)$ is also an even function of $\varepsilon_{\mathbf{k}}$. Furthermore, by noticing that $|u_{\mathbf{k}}|^2 = [1+\varepsilon_{\mathbf{k}}/E_{\mathbf{k}}]/2$ and $|v_{\mathbf{k}}|^2 = [1-\varepsilon_{\mathbf{k}}/E_{\mathbf{k}}]/2$, and $E_{\mathbf{k}} = \sqrt{(\varepsilon_{\mathbf{k}} - \omega_\chi)^2 + |\Delta_\chi|^2}$, we can obtain and expression for the parameter e.g. Γ^{χ}_{pq} through the following calculation:

$$\Gamma^{\chi}_{pq}(\omega) = \int N_\chi \frac{[1-\varepsilon/E]/2}{\omega - \Delta^{\chi}_{qp} - E + i\delta}d\varepsilon = 2N_\chi\int_0^{\infty}\frac{1/2}{\omega - \Delta^{\chi}_{qp} - E + i\delta}d\varepsilon$$
$$= N_\chi\int_{|\Delta_\chi|}^{\infty}\frac{1}{\omega - \Delta^{\chi}_{qp} - E + i\delta}\frac{E}{\sqrt{E^2 - |\Delta_\chi|^2}}dE. \tag{8.61}$$

Under the time-independent conditions, the equation for N_0 would only contain the imaginary part of Γ^{χ}_{pq}. Although we notice that the real part of Γ^{χ}_{pq} is by no means small, the real part of the integrand diverges as $E \to |\Delta_\chi|$ which leads to a divergent integral, we only retain the imaginary part. We thus have,

$$\Gamma^{\chi}_{pq} \approx -\pi N_{\chi} \int \delta(\omega - \Delta^{\chi}_{qp} - E) \frac{\theta(E - |\Delta_{\chi}|)}{\sqrt{E^2 - |\Delta_{\chi}|^2}} E dE$$

$$= -\pi N_{\chi} \frac{\omega - \Delta^{\chi}_{qp}}{\sqrt{(\omega - \Delta^{\chi}_{qp})^2 - |\Delta_{\chi}|^2}} \theta(\omega - \Delta^{\chi}_{qp} - |\Delta_{\chi}|). \qquad (8.62)$$

In the same way we also obtain

$$\Lambda^{\chi}_{pq}(\omega) = -\pi N_{\chi} \frac{\omega - \Delta^{\chi}_{qp}}{\sqrt{(\omega - \Delta^{\chi}_{qp})^2 - |\Delta_{\chi}|^2}} \theta(-\omega + \Delta^{\chi}_{qp} - |\Delta_{\chi}|). \qquad (8.63)$$

The two parameters Γ^{χ}_{pq} and Λ^{χ}_{pq} are non-vanishing only when the energy $|\omega - \Delta^{\chi}_{qp}| > |\Delta_{\chi}|$, something that we shall make use of below.

In order to proceed, we notice that[1]

$$\int T_{\chi}(t') e^{-i\omega\tau} dt' \approx T^{(0)}_{\chi} \int (1 + \alpha_{\chi} q - \tau \alpha_{\chi} \dot{q}) e^{-i\omega\tau} d\tau$$

$$\approx 2\pi T^{(0)}_{\chi} (1 + \alpha_{\chi} q) \delta(\omega) = 2\pi T_{\chi}(t) \delta(\omega). \qquad (8.64)$$

By this approximation, we then obtain the equations of motion for the quantum dot occupation numbers given by

$$\partial_t \mathbf{N} = \sum_{\chi} (T^{(0)}_{\chi})^2 \begin{pmatrix} \sum_{\sigma} \Lambda^{\chi}_{0\sigma} & \Gamma^{\chi}_{0\uparrow} & \Gamma^{\chi}_{0\downarrow} & 0 \\ -\Lambda^{\chi}_{0\uparrow} & -\Gamma^{\chi}_{0\uparrow} + \Lambda^{\chi}_{\uparrow 2} & 0 & \Gamma^{\chi}_{\uparrow 2} \\ -\Lambda^{\chi}_{0\downarrow} & 0 & -\Gamma^{\chi}_{0\downarrow} + \Lambda^{\chi}_{\downarrow 2} & \Gamma^{\chi}_{\downarrow 2} \\ 0 & -\Lambda^{\chi}_{\uparrow 2} & -\Lambda^{\chi}_{\downarrow 2} & -\sum_{\sigma} \Gamma^{\chi}_{\sigma 2} \end{pmatrix} \mathbf{N}, \qquad (8.65)$$

where $\mathbf{N} = \{N_p\}^T_{p=0,\sigma,2}$.

There are several regimes in which the quantum dot population numbers will behave very differently. For instance, in case all transitions $\Delta^{\chi}_{\sigma 0}$ and $\Delta^{\chi}_{2\sigma}$ are located within the superconducting gaps of the leads, all the parameters Γ^{χ}_{pq} and Λ^{χ}_{pq} vanish. Hence, all the occupation numbers N_p are constants of motion, which leads to that the time-dependence of N_{02} given in (8.54) holds. This regime corresponds to region II in Fig. 8.4.

By symmetry, the cases for which $\Delta^{\chi}_{\sigma 0} < -|\Delta_{\chi}|$ and $|\Delta^{\chi}_{2\sigma}| < |\Delta_{\chi}|$, and $\Delta^{\chi}_{2\sigma} > |\Delta_{\chi}|$ and $|\Delta^{\chi}_{\sigma 0}| < |\Delta_{\chi}|$, regions I and III in Fig. 8.4, respectively, give rise to the

[1]Here, we are not entirely careful since the Fourier transform of τ is not defined. However, the Fourier cosine transform of τ must vanish since $\tau \cos\omega\tau$ is an odd function of τ, whereas $\tau \sin\omega\tau$ is even. If we replace the latter function by $2\tau^{1-\delta} \sin\omega\tau$, $\delta >$ small, and integrate over $[0, \infty)$, one finds the Fourier sine transform $2\Gamma(1-\delta)\omega^{1-\delta} \sin([1-\delta]\pi/2) \to 0$, as $\delta \to 0$, $\Gamma(x)$ is the Gamma function.

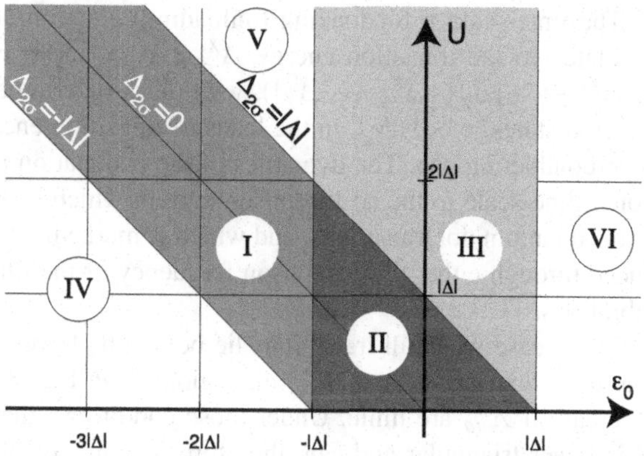

Fig. 8.4 Phase diagram for the equation of motion for **N** under the spin-degenerate conditions. In regime I (III) $\Delta^{\chi}_{\sigma 0} < -|\Delta_{\chi}$ ($\Delta^{\chi}_{2\sigma} > |\Delta_{\chi}|$) and $|\Delta^{\chi}_{2\sigma}| < |\Delta_{\chi}|$ ($|\Delta^{\chi}_{\sigma 0}| < |\Delta_{\chi}|$). In regime II, all transitions $|\Delta^{\chi}_{pq}| < |\Delta_{\chi}|$, while the transitions lie outside the superconducting gaps in the other regimes

same time-dependence of the occupation numbers, provided that we interchange N_0 and N_2 with one another. Therefore, consider e.g. region I. In this region the parameters $\Gamma^{\chi}_{pq}=0$, for all transitions $|p\rangle\langle q|$, and also $\Lambda^{\chi}_{\sigma 2}=0$ since $\Delta^{\chi}_{2\sigma} > -|\Delta_{\chi}|$, whereas $\Lambda^{\chi}_{0\sigma} \neq 0$ since $\Delta^{\chi}_{\sigma 0} < -|\Delta_{\chi}|$. Then, the equation of motion assumes the form

$$\partial_t \mathbf{N} = 2\sum_{\chi\sigma} T^2_{\chi}\Gamma^{\chi}_{0\sigma}\begin{pmatrix} 1 & 0 & 0 & 0 \\ -\delta_{\sigma\uparrow} & 0 & 0 & 0 \\ -\delta_{\sigma\downarrow} & 0 & 0 & 0 \\ 0 & 0 & 0 & 0 \end{pmatrix}\mathbf{N}, \tag{8.66}$$

that is, N_2 is a constant of motion. The time-dependence of N_0 can be determined by integrating and one finds that, up to a constant,

$$N_0(t) = \tilde{N}_0 \prod_{\chi\sigma} e^{\Lambda^{\chi}_{0\sigma}\int_{t_0}^{t}(T^{(0)}_{\chi})^2(t')dt'}, \tag{8.67a}$$

$$N_\sigma(t) = -\sum_{\chi}\Lambda^{\chi}_{0\sigma}\int_{t_0}^{t}\left(T^{(0)}_{\chi}\right)^2 N_0(t')dt', \tag{8.67b}$$

for some initial time t_0 at which $\tilde{N}_0 = N_0(t_0)$. Under the local approximation of the tunneling rate, N_0 acquires the time-evolution

$$\prod_{\chi\sigma} e^{\Lambda^{\chi}_{0\sigma}(T^{(0)}_{\chi})^3\alpha_{\chi}\dot{q}\{[1+\alpha_{\chi}q][1-\alpha_{\chi}\dot{q}(t-t_0)]+\alpha^2_{\chi}\dot{q}^2(t-t_0)^2/3\}(t-t_0)}. \tag{8.68}$$

Here, the exponent changes sign periodically with the velocity $\dot{q}$, and position q, of the quantum dot, which provides an oscillatory behavior of the occupation numbers N_0 as well as for N_σ. The physics of this behavior is that the occupation in the quantum dot periodically grows and wanes as the quantum dot moves. Such a feature may, thus, enable loading and unloading electron density on the quantum dot at different leads, or, single electron shuttling between the superconductors via the quantum dot.

The time-scale τ for loading (unloading) electron density on the quantum dot is related to the transition energy $\Delta^{\chi}_{\sigma 0}$ ($\Delta^{\chi}_{2\sigma}$), and it can be noticed that $\tau \to 0$ as $\Delta^{\chi}_{\sigma 0} \to -|\Delta_{\chi}|$ ($\Delta^{\chi}_{2\sigma} \to |\Delta_{\chi}|$), while it is determined solely by the density of electron states, $\tau \sim 1/N_{\chi}$, in the leads of transition energies far below (above) the superconducting gap. The dynamics of the occupation numbers introduce an additional time-scale to the ac Josephson current which is associated with the energies of the quantum dot transitions, and which is markedly different from the ones introduced through either the Josephson frequency or the fundamental frequency of the vibrations.

In the case when all transitions lie below the lower edges of the superconducting gaps, that is $\Delta^{\chi}_{pq} < -|\Delta_{\chi}|$ cf. region IV in Fig. 8.4, the parameters $\Gamma^{\chi}_{pq} = 0$ whereas all Λ^{χ}_{pq} are finite. Under those conditions, the matrix in the equation for **N** is lower triangular and can, therefore, be analytically solved. For instance, the time-dependence of N_0 is, up to a constant, given by (8.67a). By also integrating the equations for N_σ and N_2 we find

$$N_\sigma(t) = -\prod_{\chi} \int_{t_0}^{t} e^{\Lambda^{\chi}_{\sigma 2} \int_{t'}^{t} T^2_{\chi}(t'')dt''} \sum_{\chi'} T^2_{\chi'}(t') \Lambda^{\chi'}_{0\sigma} N_0(t')dt', \tag{8.69a}$$

$$N_2(t) = -\sum_{\chi\sigma} \Lambda^{\chi}_{\sigma 2} \int_{t_0}^{t} N_\sigma(t')dt'. \tag{8.69b}$$

Both the empty and one-electron states depend on the periodic motion and velocity of the quantum dot, as in the previous case. In addition, the occupation of the two-electron state acquires a time-dependence since it depends on the integrated time-evolution of the occupation in the other states. The time-evolution of the electron occupation in the quantum dot is, hence, related to the (four) time-scales that are associated with the transition energies $\Delta_{\sigma 0}$ and $\Delta_{2\sigma}$. The one-electron occupation N_σ in particular, strongly depend on the rates of the transitions $X^{0\sigma}$ and $X^{\sigma 2}$. The properties of regime VI are obtained by noticing that $\Gamma^{\chi}_{pq} \neq 0$ and all $\Lambda^{\chi}_{pq} = 0$, which leads to that the roles of N_0 and N_2 are interchanged.

The last regime, in which $\Delta^{\chi}_{\sigma 0} < -|\Delta_{\chi}|$ and $\Delta^{\chi}_{2\sigma} > |\Delta_{\chi}|$, region V in Fig. 8.4, provides the most complicated dynamics of the occupation numbers, since the parameters $\Lambda^{\chi}_{0\sigma} < 0$, $\Lambda^{\chi}_{\sigma 2} = 0$, $\Gamma^{\chi}_{0\sigma} = 0$, and $\Gamma^{\chi}_{\sigma 2} > 0$. The occupation number N_0 is, thus, obtained from (8.67a), whereas N_2 is obtained from an analogous expression which is found by replacing $\Lambda_{0\sigma}$ and $\tilde{N}_0$ by $\Gamma_{\sigma 2}$ and $\tilde{N}_2$, respectively. It is understood that N_0 increases while N_2 decreases, and the other way around, which is expected from conservation of probability. The one-electron occupations depend on the integrated time-evolutions of N_0 and N_2 through the expression

$$N_\sigma(t) = -\sum_{\chi} \int_{t_0}^{t} T^2_{\chi}(t')\left[\Lambda^{\chi}_{0\sigma} N_0(t') - \Gamma^{\chi}_{\sigma 2} N_2(t')\right]dt'. \tag{8.70}$$

The occupation of the one-electron states can, thus, be viewed as resulting from the imbalance between the occupation in the empty and two-electron states.

In all of the cases discussed here, one has to recall, however, that we perform the time-integrations under the assumption that $T_\chi(t') \approx T_\chi(t) - \tau\dot{T}_\chi(t)$. Moreover, in order to acquire the entire time-dependence of the occupation numbers, we also have to have information about the mechanical motion of the quantum dot itself, i.e. we need to know $q(t)$.

We understand from the discussion of the oscillating quantum dot embedded in the Josephson junction, that the new time, or energy, scale that can be introduced stem from the quantum dot transition energies Δ_{pq}. We have seen that when these transitions lie in within the superconducting gaps of the leads, the mechanical motion of the quantum dot is entirely determined by the Josephson frequencies ω_χ. If one or more transition energies lie outside the superconducting gaps, the mechanical motion is also strongly affected by the time-dependence of the occupation numbers N_p of the quantum dot.

References

1. Gorelik, L.Y., Isacsson, A., Galperin, Y.M., Shekter, R.I., Jonson, M.: Nature **411**, 454 (2001)
2. Zazunov, A., Egger, R., Mora, C., Martin, T.: Phys. Rev. B **73**, 214501 (2006)
3. Zhu, J.-X., Nussinov, Z., Balatsky, A.V.: Phys. Rev. B **73**, 064513 (2006)
4. Fransson, J., Zhu, J.-X., Balatsky, A.V.: Phys. Rev. Lett. **101**, 067202 (2008)
5. Bardeen, J., Cooper, L.N., Schrieffer, J.R.: Phys. Rev. **108**, 1175 (1957)

Chapter 9
Current-Voltage Asymmetries in Two Level Systems

Abstract We consider current-voltage asymmetries in double quantum dot systems. Such effects are here described in terms of correlation effects and we use the one-loop-approximation to find that asymmetric couplings to the left and right leads which provide asymmetric response to the bias voltage. We discuss the results in terms of the relative level spacing, detuning, and the hopping rate between the quantum dots.

9.1 Introduction

Experimental $I-V$ characteristics of e.g. quantum dot systems, carbon nanotubes, and quantum wires, often show a degree of asymmetry with respect to the bias voltage [1–3]. These observations are made on nano-devices where one part has a complex electronic structure, which is coupled to two, or more, non-complex regions, e.g. leads. The non-trivial $I-V$ characteristics becomes particularly apparent when the interacting region is asymmetrically coupled to the left and right leads.

$I-V$ asymmetries may be an effect of impurities introduced during the growth process or differences in interface roughness of the oxide layers between the contacts and the interacting region [4, 5]. Other mechanisms that may introduce asymmetries in the $I-V$, or differential conductance (dI/dV), characteristics are unintentional background charges which additionally contribute a charging energy to the interacting region [5], or higher collector barrier which enhances the charge storage in the well substantially and leads to different current amplitudes for the back- and forward biased device [6, 7].

Theoretically, it has been suggested that inelastic scattering gives different contributions in the back- and forward bias direction, something that would especially important for asymmetric structures [8, 9]. However, a full understanding of the mechanisms responsible for the observed asymmetries in the $I-V$ (dI/dV) characteristics has not yet been put forward.

J. Fransson, *Non-Equilibrium Nano-Physics*,
Lecture Notes in Physics 809,
DOI 10.1007/978-90-481-9210-6_9,

Experiments on double quantum dots coupled to metallic leads have also shown asymmetric negative differential conductance behavior in the I–V characteristics, see e.g. [10]. Here, the I–V characteristics is asymmetric in the sense that the negative differential conductance appears only in one half of the bias voltage range $(-V, V)$. Features of negative differential conductance can normally be realized in semi-conductor double- and multi-barrier structures [11, 12], which usually can be referred to as band-edge effects [13, 14]. On the other hand, sharp resonant peaks [15] related to vanishing energy distance between the levels in the two quantum dots, creating a resonant state [16], have been recorded in transport experiments on double quantum dots fabricated from semi-conductor hetero-structures. Alignment of the levels in the two quantum dots are, however, expected to create symmetric I–V characteristics with sharp resonant peaks and large peak-to-valley ratios. This is in huge contrast to the observations in [10]. Asymmetries and negative differential conductance in interacting regions coupled to metallic leads with a conduction band width of the order of eV, cannot be explained in terms of band-edge effects. Normally, one would expect an increasing current with increasing bias voltage, possibly with appearance of plateaux due to the zero-dimensional confined energy levels of the interacting region.

Here, we will seek an explanation for the I–V asymmetries and negative differential conductance in terms of electron correlations. We understand from our previous formal studies of the Green function for the interacting region that we must go to at least the one-loop-approximation in order to obtain a theory that is beyond mean-field in order to capture correlation related asymmetries.

9.2 Setting Up the Model

We will study the problem in terms of an idealized double quantum dot, cf. Fig. 9.1, each with a single level and a large intradot Coulomb repulsion $U_{A/B}$, and the quantum dots interact through the interdot Coulomb repulsion U_{AB} and hopping t. Further, the quantum dots are coupled to leads with tunneling rates $v_{\mathbf{k}\sigma}$, $\mathbf{k} \in L, R$. The energy for the system is modeled by the Hamiltonian $\mathcal{H} = \mathcal{H}_L + \mathcal{H}_R + \mathcal{H}_{DQD} + \mathcal{H}_T$, where, as usual $\mathcal{H}_\chi = \sum_{\mathbf{k}\sigma\in\chi} \varepsilon_{\mathbf{k}\sigma} c^\dagger_{\mathbf{k}\sigma} c_{\mathbf{k}\sigma}$, $\chi = L, R$, whereas

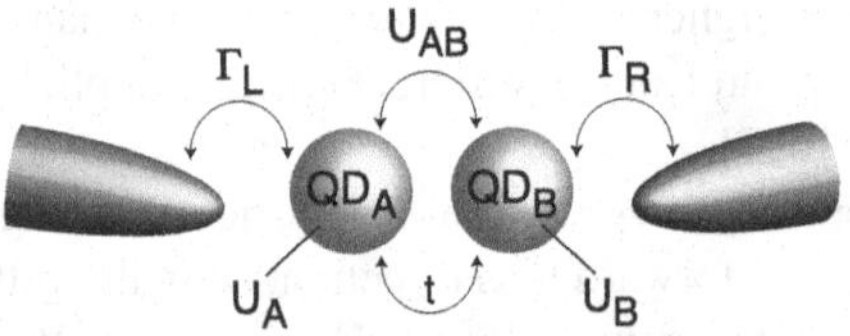

Fig. 9.1 Sketch of the double quantum dot system. The quantum dots interact through the interdot Coulomb repulsion U_{AB} and hopping t, and each quantum dot is coupled to a lead through $\Gamma^{L/R}$

$$\begin{aligned}\mathcal{H}_{DQD} &= \sum_\sigma \varepsilon_A d^\dagger_{A\sigma} d_{A\sigma} + U_A n_{A\uparrow} n_{A\downarrow} + \sum_\sigma \varepsilon_B d^\dagger_{B\sigma} d_{B\sigma} + U_B n_{B\uparrow} n_{B\downarrow} \\ &\quad + U_{AB}(n_{A\uparrow} + n_{A\downarrow})(n_{B\uparrow} + n_{B\downarrow}) + \sum_\sigma t(d^\dagger_{A\sigma} d_{B\sigma} + H.c.) \\ &= \sum_{Nn} E_{Nn} h^n_N. \end{aligned} \tag{9.1}$$

Finally, the tunneling Hamiltonian $\mathcal{H}_T = \sum_{\mathbf{k}\sigma,a} v_{\mathbf{k}\sigma a} c^\dagger_{\mathbf{k}\sigma} X^a + H.c.$, where $v_{\mathbf{k}\sigma a} = v_{\mathbf{k}\sigma}(d_\sigma)^a$, and where $d_\sigma = d_{A\sigma}$ $(d_{B\sigma})$ for $\mathbf{k} \in L$ (R).

Here, we have introduced the Hubbard operators in terms of the eigensystem $\{E_{Nn}, |N, n\rangle\}$, where N and n denote the number of electrons and state, respectively. Reference to the electron spin is not necessary, but is included for completeness. In Table 9.1 we summarize the eigenstates for the double quantum dot in the atomic limit.

In the following, we focus on the case with bias voltages less than both $U_{A/B}$ and the intradot level spacing. It is then sufficient to consider only the transitions between the empty and singly occupied states, $|0, 1\rangle$ and $|1, n\rangle$, $n = 1, \ldots, 4$, respectively. The Hamiltonian for the double quantum dot then reduces to a summation over these states only. Hence, $\mathcal{H}_T = \sum_{\mathbf{k}\sigma n} v_{\mathbf{k}\sigma n} c^\dagger_{\mathbf{k}\sigma} X^{1n}_{01} + H.c.$, where $v_{\mathbf{k}\sigma n} = v_{\mathbf{k}\sigma}(d_\sigma)^{1n}_{01}$. In reality, there may be an unknown number N of electrons in the interacting region. The conducting channels involve only one or a few of the corresponding many-body states, and therefore one can simplify the model and identify the empty state of the model with the $N - 1$ configuration.

Table 9.1 Eigenstates of the double quantum dot system. Here, we use the convention $|\sigma\rangle|\sigma'\rangle = d^\dagger_{B\sigma'} d^\dagger_{A\sigma}|0\rangle$ and $|\uparrow\downarrow\rangle = d^\dagger_\downarrow d^\dagger_\uparrow|0\rangle$. The two-electron states $|\Phi^A\rangle = |\uparrow\downarrow\rangle|0\rangle$, $|\Phi^B\rangle = |0\rangle|\uparrow\downarrow\rangle$, and $|\Phi^{AB}\rangle = [|\uparrow\rangle|\downarrow\rangle - |\downarrow\rangle|\uparrow\rangle]/\sqrt{2}$. The coefficients α_n, β_n, A_n, B_n, C_n, κ_n, and λ_n depend on the internal parameters $\varepsilon_{A/B}$, $U_{A/B}$, U_{AB}, and t

N	$\|N, n\rangle$	
0	$\|0, 1\rangle = \|0\rangle$	
1	$\|1, n\rangle = \alpha_n\|\uparrow\rangle\|0\rangle + \beta_n\|0\rangle\|\uparrow\rangle$	$n = 1, 3$
	$\|1, n\rangle = \alpha_n\|\downarrow\rangle\|0\rangle + \beta_n\|0\rangle\|\downarrow\rangle$	$n = 2, 4$
2	$\|2, 1\rangle = \|\uparrow\rangle\|\uparrow\rangle$	
	$\|2, 2\rangle = \|\downarrow\rangle\|\downarrow\rangle$	
	$\|2, 3\rangle = [\|\uparrow\rangle\|\downarrow\rangle + \|\downarrow\rangle\|\uparrow\rangle]/\sqrt{2}$	
	$\|2, n\rangle = A_n\|\Phi^A\rangle + B_n\|\Phi^B\rangle + C_n\|\Phi^{AB}\rangle$	$n = 4, 5, 6$
3	$\|3, n\rangle = \kappa_n\|\uparrow\rangle\|0\rangle + \lambda_n\|0\rangle\|\uparrow\rangle$	$n = 1, 3$
	$\|3, n\rangle = \kappa_n\|\downarrow\rangle\|0\rangle + \lambda_n\|0\rangle\|\downarrow\rangle$	$n = 2, 4$
4	$\|4, 1\rangle = \|\uparrow\downarrow\rangle\|\uparrow\downarrow\rangle$	

The eigenenergies for the one-electron states are given by

$$E_{1n} = \frac{\varepsilon_A + \varepsilon_B \pm \sqrt{\Delta\varepsilon^2 + 4t^2}}{2}, \tag{9.2}$$

where $n = 1, 2$ $(3, 4)$ correspond to the negative (positive) sign, whereas $\Delta\varepsilon = \varepsilon_A - \varepsilon_B$ is the inter-dot level off-set, or, detuning. The energy of the empty state is $E_{01} = 0$. The transition matrix elements

$$(d_{A\sigma})_{01}^{1n} = \langle 0, 1|d_{A\sigma}|1, n\rangle = \alpha_n[(\delta_{n1} + \delta_{n3})\delta_{\sigma\uparrow} + (\delta_{n2} + \delta_{n4})\delta_{\sigma\downarrow}], \tag{9.3a}$$

$$(d_{B\sigma})_{01}^{1n} = \langle 0, 1|d_{B\sigma}|1, n\rangle = \beta_n[(\delta_{n1} + \delta_{n3})\delta_{\sigma\uparrow} + (\delta_{n2} + \delta_{n4})\delta_{\sigma\downarrow}], \tag{9.3b}$$

where

$$\alpha_{1(2)} = \frac{\Delta\varepsilon - \sqrt{\Delta\varepsilon^2 + 4t^2}}{\sqrt{[\Delta\varepsilon 0 - \sqrt{\Delta\varepsilon^2 + 4t^2}]^2 + 4t^2}}, \tag{9.4a}$$

$$\beta_{1(2)} = \frac{2t}{\sqrt{[\Delta\varepsilon - \sqrt{\Delta\varepsilon^2 + 4t^2}]^2 + 4t^2}}, \tag{9.4b}$$

while $\alpha_{3(4)} = -\beta_{1(2)}$ and $\beta_{3(4)} = \alpha_{1(2)}$. Hence, $|(d_{A\sigma})_{01}^{11(2)}|^2 \neq |(d_{A\sigma})_{01}^{13(4)}|^2$ and $|(d_{B\sigma})_{01}^{11(2)}|^2 \neq |(d_{B\sigma})_{01}^{13(4)}|^2$ whenever $\varepsilon_A \neq \varepsilon_B$, as can be seen in Table 9.2 where the equilibrium properties of the double quantum dot and the transition matrix elements are listed. The properties of the matrix elements are also plotted in Fig. 9.2(a) for various hopping strengths t, whereas the drawings in Fig. 9.2(b) and (c) illustrate the system in the single-particle (b) and many-body (c) representations.

The above situation would hold in many realistic systems since the sizes of the two quantum dots are different, in general. The difference of the transition matrix elements influences the current through the system and, furthermore, one can control the properties of the system through the transition matrix elements.

Table 9.2 Equilibrium properties of the DQD given the inter-dot Coulomb repulsion $U_{AB} = 40$ meV, respectively, and the hopping $t = 0.75$ meV. The single-particle levels $\varepsilon_{A/B}$ are input parameters

	(A)	(B)	(C)
$\varepsilon_{A\sigma}$ (meV)	−3.25	−2.5	−1.75
$\varepsilon_{B\sigma}$ (meV)	−1.75	−2.5	−3.25
$E_{1\sigma}$ (meV)	−3.56	−3.25	−3.56
$E_{2\sigma}$ (meV)	−1.44	−1.75	−1.44
$\lvert(d_{A\sigma})_{01}^{01(2)}\rvert^2$	0.85	0.5	0.15
$\lvert(d_{A\sigma})_{01}^{03(4)}\rvert^2$	0.15	0.5	0.85
$\lvert(d_{B\sigma})_{01}^{01(2)}\rvert^2$	0.15	0.5	0.85
$\lvert(d_{B\sigma})_{01}^{03(4)}\rvert^2$	0.85	0.5	0.15

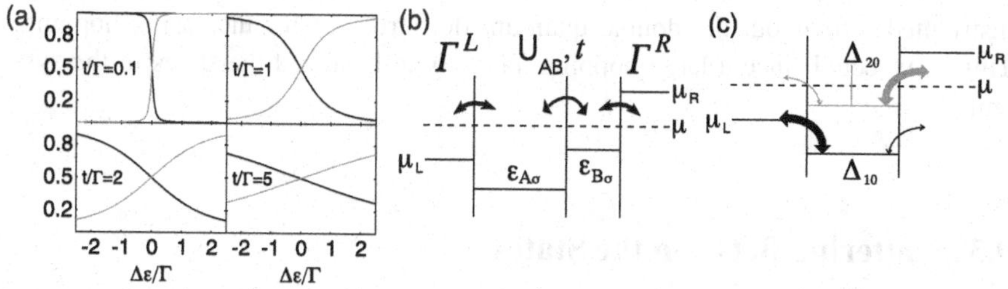

Fig. 9.2 (**a**) Transition matrix elements $|(d_{A\sigma})_{01}^{11(2)}|^2 = |(d_{B\sigma})_{01}^{13(4)}|^2$ (*black*) and $|(d_{A\sigma})_{01}^{13(4)}|^2 = |(d_{B\sigma})_{01}^{11(2)}|^2$ (*grey*) as function of the detuning $\Delta\varepsilon$ for various hopping strengths t. (**b**) Sketch of the energies in the serially coupled quantum dots. In quantum dot A (B), the bare single-electron levels $\varepsilon_{A(B)}$. (**c**) Schematic picture of the energies in the diagonal representation, where $\mu_{L(R)}$ is the chemical potential in the *left* (*right*) lead. The energies of the transitions between the one-electron states $|1, n\rangle$ and the empty state $|0, 1\rangle$ are denoted Δ_{n0}. The *arrows* illustrate the strengths of the transition probabilities between the one-electron states in the double quantum dot and the leads

The plots in Fig. 9.2(a) illustrate the dependence of $|(d_{A\sigma})_{01}^{11(2)}|^2 = |(d_{B\sigma})_{01}^{13(4)}|^2$ (black) and $|(d_{A\sigma})_{01}^{13(4)}|^2 = |(d_{B\sigma})_{01}^{11(2)}|^2$ (grey) as function of the detuning $\Delta\varepsilon$ for various hopping strengths t. It is clear that the matrix elements are equal for vanishing detuning, for all hoppings t, which is naturally understood since then e.g. $\alpha_1 = 2t/(2\cdot 2t) = 1/(2t) = \beta_1$. This is expected since the state in the two quantum dots are resonant with one another in this case. In other wordings this means that the one-electron states are equally distributed throughout the double quantum dot.

For finite detuning, one of the matrix elements approaches unity and the other to zero, as $\xi = 2t/\Delta\varepsilon \to 0$. This is easiest seen be rewriting the matrix elements in terms of the parameter ξ, giving e.g.

$$\alpha_1^2 = \frac{(\operatorname{sign}\Delta\varepsilon - \sqrt{1+\xi^2})^2}{\xi^2 + (\operatorname{sign}\Delta\varepsilon - \sqrt{1+\xi^2})^2}, \qquad \beta_1^2 = \frac{\xi^2}{\xi^2 + (\operatorname{sign}\Delta\varepsilon - \sqrt{1+\xi^2})^2}. \tag{9.5}$$

Hence, for $\Delta\varepsilon < 0$ it is obvious that $\alpha_{1(2)}^2 \to 1$ and $\beta_{1(2)}^2 \to 0$, while $\alpha_{3(4)} \to 0$ and $\beta_{3(4)} \to 1$. The physical meaning of these limits is that the state $|1, 1(2)\rangle$ has a larger weight on quantum dot A than on B, while $|1, 3(4)\rangle$ is heavier weighted on quantum dot B. In the case $\Delta\varepsilon > 0$, the limit values are interchanged, and $|1, 1(2)\rangle$ and $|1, 3(4)\rangle$ become more weighted on quantum dot B and A, respectively.

In the opposite limit, that is, $|\xi| \to \infty$, we have

$$\begin{aligned} \alpha_1^2 &= \frac{(|\xi|^{-1}\operatorname{sign}\Delta\varepsilon - \sqrt{1+\xi^{-2}})^2}{1 + (|\xi|^{-1}\operatorname{sign}\Delta\varepsilon - \sqrt{1+\xi^{-2}})^2}, \\ \beta_1^2 &= \frac{1}{1 + (|\xi|^{-1}\operatorname{sign}\Delta\varepsilon - \sqrt{1+\xi^{-2}})^2}, \end{aligned} \tag{9.6}$$

which leads to that $\alpha_n^2 \to 1/2$ and $\beta_n^2 \to 1/2$, for all n, which is indicated in Fig. 9.2(a), for increasing t. Hence, the one-electron states become more uniformly distributed throughout the double quantum dot structure for increasing hopping. This is expected since a large hopping tends to de-localized the states in the system.

9.3 Scattering Between the States

In this section we derive the non-equilibrium Green functions for the double quantum dot, and we discuss it in terms of the scattering between the states in the double quantum dot.

The charge current, I, through the systems will be calculated using the, by now well-known, formula

$$I = \frac{ie}{2h}\sum_n \int \big(\{f_L^+ \Gamma_n^L - f_R^+ \Gamma_n^R\} G_n^<(\omega) + \{f_L^- \Gamma_n^L - f_R^- \Gamma_n^R\} G_n^>(\omega)\big) d\omega, \quad (9.7)$$

where $\Gamma_n^\chi = 2\pi \sum_{\mathbf{k}\sigma\in\chi} |v_{kn\sigma}|^2 \delta(\omega - \varepsilon_{\mathbf{k}\sigma})$, $f_\chi^+ = f_\chi(\omega)$, and $f_\chi^- = f_\chi(-\omega) = 1 - f_\chi(\omega)$, whereas the Green function for the double quantum dot is defined through $G_{nm}(t,t') = (-i)\langle \mathrm{T} X_{01}^{1n}(t) X_{10}^{m1}(t')\rangle_U$. We will need the action $S = \exp[-i\int_C \mathcal{H}'(t)dt]$ with

$$\mathcal{H}'(t) = U_0(t) h_0^1 + \sum_n \big(U_n(t) h_{01}^{1n} + U_{nm}(t) Z_{11}^{nm}\big). \quad (9.8)$$

The equation of motion for the Green function is given by

$$\begin{aligned}
&\left(i\frac{\partial}{\partial t} - \Delta_n^0 - \Delta U_n(t)\right) G_{nm}(t,t') - \sum_{n'\neq n} U_{nn'}(t) G_{n'm}(t,t') \\
&\quad = \delta(t-t') P_{nm}(t') + \sum_{n'm'} \big(P_{nn'}(t^+) + R_{nn'}(t^+)\big) \\
&\quad \times \int_C V_{n'm'}(t,t'') G_{m'm}(t'',t') dt'',
\end{aligned} \quad (9.9)$$

where, we repeat that, $R_{nm}(t) = i[\delta_{nm}\delta/\delta U_0(t) + \delta/\delta U_{mn}(t)]$, and the interaction propagator $V_{nm}(t,t') = \sum_{\mathbf{k}\sigma} v_{\mathbf{k}\sigma n}^* v_{\mathbf{k}\sigma m} g_{\mathbf{k}\sigma}(t,t')$. Here, the Green function $g_{\mathbf{k}\sigma}(t,t')$ denotes an electron in the leads, and satisfies the equation $(i\partial/\partial t - \varepsilon_{\mathbf{k}\sigma}) g_{\mathbf{k}\sigma}(t,t) = \delta(t-t')$.

In the present discussion it is obvious that $G_{nm}(t,t') = \sum_{n'} D_{nn'}(t,t') P_{n'm}(t')$, such that

$$\delta G_{nm}(t,t') = \sum_{n'} \big([\delta D_{nn'}(t,t')] P_{n'm}(t') + D_{nn'}(t,t')[\delta P_{n'm}(t')]\big). \quad (9.10)$$

Recalling that $\mathbf{D}^{-1}\mathbf{D} = 1$ giving $\delta\mathbf{D} = \mathbf{D}[\delta\mathbf{D}^{-1}]\mathbf{D}$, we find that

$$\delta G_{nm}(t,t') = \sum_{n'}\Big(D_{nn'}(t,t')[\delta P_{n'm}(t')] - \sum_{\nu\nu'}\int_C D_{n\nu}(t,\tau)[\delta D_{\nu\nu'}(\tau,\tau')]D_{\nu'n'}(\tau',t')P_{n'm}(t)d\tau d\tau'\Big). \quad (9.11)$$

In the limit of zero source fields, $U_{nm}(t) \to 0$, all components of the double quantum dot Green function matrix that do not conserve either the spin or orbital moments vanish. Functional derivatives of these propagators may, as we have seen previously, be finite and will be considered. Scattering between the one-electron states $|n\rangle$ and $|m\rangle$, $n \neq m$, is included in the first order correction (first order in functional derivatives). Having these observations in mind, we find that the last term in the equation of motion can be written

$$\sum_{n'm'} R_{nn'}(t^+)\int_C V_{n'm'}(t,t'')G_{m'm}(t'',t')dt'' = i\sum_{n'\neq n}\int_C V_{n'}(t,\tau)\big(\delta_{mn}D_{n'}(\tau,t')K_{n'n}(t,t') + D_{n'}(\tau,t)G_{n'm}(t,t')\big)d\tau. \quad (9.12)$$

Here, $K_{n'n}(t,t') = (-i)\langle \mathrm{T}Z_{11}^{n'n}(t)Z_{11}^{nn'}(t')\rangle_U$ describes the scattering between the one-electron states, and satisfies the equation $(i\partial/\partial t - \Delta_{nn'})K_{n'n}(t,t') = \delta(t-t')[P_{n'}(t) - P_n(t)]$.

The first term in (9.12) plays an important role for the understanding of the scattering effects that influence the transport through the system. In particular, this term is a key part for the explanation of the I–V asymmetries as well as the asymmetric negative differential conductance, which will be discussed below. The second term provides the dressing of the transition energy arising from the kinematic interactions induced by the presence of the de-localized electrons in the leads. We, thus, identify the dressed transition energy Δ_n by

$$\Delta_n = \Delta_n^0 + i\sum_{n'\neq n}\int_C V_{n'}(t,\tau)D_{n'}(\tau,t)d\tau, \quad (9.13)$$

and the dressed end-factor $\mathbb{P}_{nm}(t,t')$ by

$$\mathbb{P}_{nm}(t,t') = \delta(t-t')P_{nm}(t) + i\delta_{mn}\sum_{n'}\int_C V_{n'}(t,\tau)D_{n'}(\tau,t')d\tau K_{n'n}(t,t'), \quad (9.14)$$

where $K_{nn}(t,t') = 0$.

We have, thus, arrived at the one-loop-approximation, which was the goal of this derivation. By restricting ourselves to the stationary regime, we can Fourier transform our result and obtain the equation of motion

$$(i\omega - \Delta_n)G_n(i\omega) = \mathbb{P}_n(i\omega) + \mathbb{P}_n(i\omega)V_n(i\omega)G_n(i\omega). \quad (9.15)$$

Although this equation is diagonal, the dressed end-factor contains, as we have seen above, scattering events between different states. Our purpose with this study is to understand the effects of the electron correlation on the I–V characteristics. Therefore, we do not include off-diagonal contributions to the Green function, although these may provide additional features to the I–V characteristics.

9.4 Asymmetric Population Numbers

The shift in the dressed transition energy Δ_n, (9.13), is caused by the kinematic interactions between electrons residing in the states $|n' \neq n\rangle$ in the double quantum dot. These interactions are possible because of the couplings to the leads, which enable tunneling of electrons between the leads and double quantum dot. Due to the fluctuating nature of the electrons, tunneling back and forth between the leads and the double quantum dot, there may be electron density in any of the states $|n\rangle$, and hence, there will be interactions between those densities. This is the physical background for the kinematic interactions in this system.

Converting the integral into real times, i.e.

$$\int_C V(t,\tau)D(\tau,t)d\tau = \int [V^r(t,\tau)D^<(\tau,t) + V^<(t,\tau)D^a(\tau,t)]d\tau,$$

introduces the Fermi function $f(\varepsilon_{\mathbf{k}\sigma})$, i.e. a restricted domain for the energy summation, which leads to an increased shift near the chemical potentials of the leads. This is illustrated Fig. 9.3(a), which shows the dressed transition energies as function of the bias voltage for the case (A) in Table 9.2. We note that the cases listed in Table 9.2 are reasonable for sizes of the quantum dots less than 10 to 100 nanometers each. It may be seen from the figure that sharp dips in Δ_n are found at specific bias voltages. If the couplings to the left and the right contact are asymmetric, that is, if the hybridization between the localized state in the double quantum dot and the states in the left lead is different from that to the right lead, then shift of the

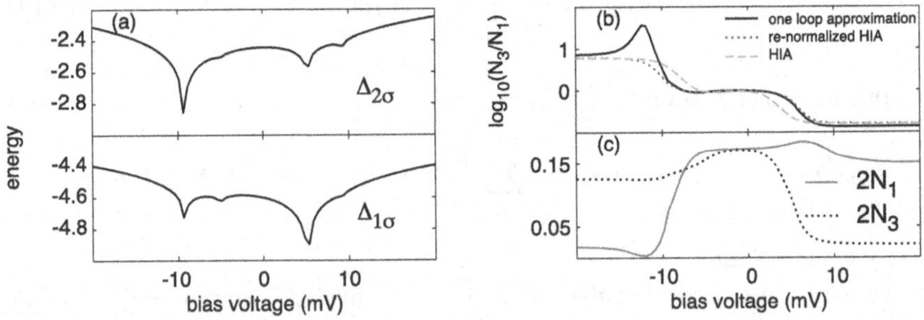

Fig. 9.3 (**a**) Dressed transition energies as function of the bias voltage for the case (A) in Table 9.2, with $\Gamma^{L/R} = 0.375$ meV, and $T = 5$ K. (**b**) Dressed population numbers $2N_1 = N_1 + N_2$ and $2N_3 = N_3 + N_4$, as function of the bias voltage for the case (A)

transition energies becomes different when the bias voltage is applied in the forward direction compared to the reverse biased system. In Fig. 9.3(a) the couplings $\Gamma^L = \Gamma^R$, however, the couplings of each transition in the double quantum dot becomes asymmetric with respect to the left and right leads, i.e. $\Gamma^L_n \neq \Gamma^R_n$, because of the asymmetry of the transitions $|0\rangle\langle n|$ in the double quantum dot.

The expression for the end-factor, (9.14), can be used to understand the basic mechanisms for the scattering between the one-electron states. First of all it should be noticed that any contribution in the sum over the states $|n\rangle$ vanishes whenever $P_n = P_m$. This happens whenever the spectral weights of the transitions $|0\rangle\langle n|$ and $|0\rangle\langle m|$ equal. Second, the leading contribution from the dressing of the end-factor has an imaginary part which peaks around Δ_n. To illustrate this with a simple model, we put $D^r_n(\omega) = (\omega - \Delta_n + i0^+)^{-1}$ and $D^<_n(\omega) = i2\pi\delta(\omega - \Delta_n)$. Then, we obtain the retarded end-factor approximately as

$$\mathbb{P}^r_{nm}(\omega) \approx P_{nm} - \delta_{mn} \sum_{n'\chi} \frac{\Gamma^\chi_{n'}}{2\pi} \frac{P_{n'} - P_n}{\omega - \Delta_n} \left(\log\left| \frac{\mu_\chi - \Delta_{n'}}{\omega - \mu_\chi - \Delta_{nn'}} \right| - i\pi f_\chi(\omega - \Delta_{nn'}) \right). \tag{9.16}$$

The real part appear to be diverging for bias voltages such that $\mu_\chi \approx \Delta_{n'}$. If we instead use the dressed locator $D^r_{n'}(\omega)$ in the above calculation, this apparent divergence is removed, hence, it is of less importance.

The physical effect from the dressing of the end-factor is an increased or decreased spectral weight for the corresponding Green function, depending on the sign of the difference $P_{n'} - P_n$. Hence, the ability for an electron to tunnel through the double quantum dot via the transition $|n\rangle\langle 0|$ is highly influenced by the scattering between the one-electron states. It should be emphasized, however, that this is a dynamical process which strongly depends on the bias voltage and, in addition, on the strength of the couplings to the left and the right leads.

The population numbers, calculated through the set of self-consistent equations such that $N_n = (-i)\int G^<_n(\omega)d\omega/(2\pi)$ and $N_0 + \sum_n N_n = 1$, are plotted in Fig. 9.3(b) and (c), where panel (c) displays the sums $N_1 + N_2 = 2N_1$ and $N_3 + N_4 = 2N_3$, for the one-electron states as function of the bias voltage for the parameters used in case (A), cf. Table 9.2. The population numbers N_1 and N_3, are almost equal in the region around equilibrium. However, it may be seen from Fig. 9.3(c) that N_3 decreases while N_1 remains more or less constant, as the bias voltage increase in the forward direction ($eV = \mu_L - \mu_R > 0$). This behavior reflects the difference of the coupling strengths for the two transition $|0\rangle\langle 1|$ and $|0\rangle\langle 3|$. The first of these two transitions couples weaker to the right lead that then other does, and hence prohibits electrons to flow through the double quantum dot via this transition. Therefore, the population remains constant around its maximum value for a large range of bias voltages.

For reverse bias voltages ($eV < 0$), the case is quite the opposite, since the first transition, $|0\rangle\langle 1|$, couples strong to the left lead whereas the second transition, $|0\rangle\langle 3|$, couples weaker, cf. Fig. 9.1(c). Hence, a larger amount of electron density can flow through $|0\rangle\langle 1|$ which leads to a decreased population in the state $|1\rangle\langle 1|$.

The dip and hump in the plots for N_1 around -11 mV and 9 mV, respectively, are both caused by the dressing of the end-factor. This fact is emphasized in Fig. 9.3(b), showing a comparison of the two mean field approximations, Hubbard-I-approximation (dashed), re-normalized Hubbard-I-approximation (dotted), and one-loop-approximation (solid). The figure illustrates logarithmic plots of the ratio N_3/N_1 as function of the bias voltage in the three approximation schemes. The dynamical effects from the dressed end-factor tend to modify the population numbers of the transitions for biases around Δ_n, as discussed above, and this modification leads to a further decrease of N_1 around -11 mV since the difference $P_3 - P_1 > 0$, whereas this difference is negative for biases around 9 mV which leads to the hump in N_1. In the mean field theories, the electrons flow through the double quantum dot directly via the transitions $|0\rangle\langle n|$, which is expected since the transverse couplings through $|n\rangle\langle n'|$ are not open in these approximations. The dynamical effects in the dressed end-factor, which arise from scattering between the one-electron states, tends to prevent the electrons in the leads to tunnel through the quantum dot via $|0\rangle\langle 1|$.

As a result of this dynamical redistribution of the spectral weight, it is expected that the current through the double quantum dot is less (larger) for biases in the range ~ -11 mV (~ 9 mV) where the population number N_1 is decreased (increased). Moreover, since both transitions between the empty and one-electron states are resonant around -11 mV, whereas only $|0\rangle\langle 3|$ is resonant around 9 mV, it is also expected that the influence of the dynamical redistribution on the resulting current should be larger in the former case than in the latter. The asymmetry of the positions of the hump and dip with respect to the bias voltage, is related to the asymmetric renormalization of the transition energies.

9.5 Current-Voltage Characteristics

We now analyze the calculated current under a few different circumstances. The important parameters in the present context are the detuning $\Delta\varepsilon = \varepsilon_A - \varepsilon_B$ and the tunneling t. These should be compared to the couplings Γ^χ between the double quantum dot and the leads. The intradot and interdot Coulomb repulsion are irrelevant as long as we restrict our discussion to bias voltages such that only transitions between the empty and one-electron states are involved in the current.

As an example of the differences between the three approximation schemes we have been considering, the I–V and dI/dV characteristics are shown in Fig. 9.4 for the parameters in case (A), cf. Table 9.2. First, one can note that the two currents in the mean-field approximations are shifted in the sense that the two transitions become resonant at lower biases in the Hubbard-I-approximation than in the re-normalized Hubbard-I-approximation. This is in agreement with the discussions in the previous section. Naturally, then the dI/dV peaks appear at lower biases in the Hubbard-I-approximation than in the re-normalized Hubbard-I-approximation. The re-normalized Hubbard-I-approximation has a slight asymmetry inherent from

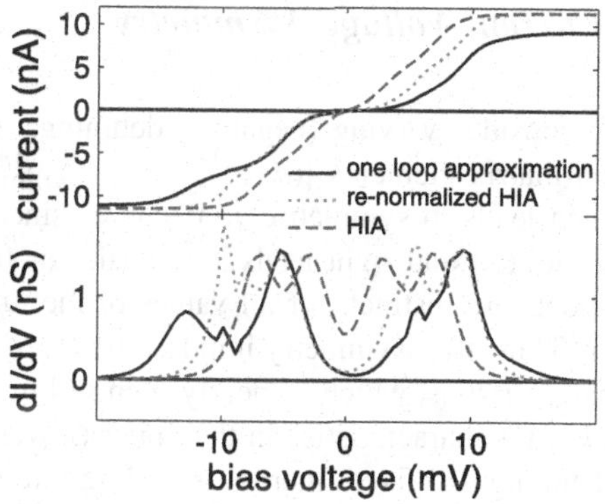

Fig. 9.4 I–V characteristics (*upper panel*) and dI/dV (*lower panel*) for the double quantum dot calculated in the one-loop-approximation (*solid*), re-normalized Hubbard-I-approximation (*dotted*), and Hubbard-I-approximation (*dashed*). Parameters are taken from case (A) in Table 9.2

the asymmetric shifts of the transition energies. The effect from this asymmetric re-normalization of the transitions is, however, negligible and would most certainly not be detectable in experiments.

In the one-loop-approximation, on the other hand, the asymmetry of the I–V and dI/dV characteristics is more apparent. Consider the reverse biased system, $eV < 0$. The large increase of the current seen in the re-normalized Hubbard-I-approximation in the range between -3 and -7 mV, remains in the one-loop-approximation, since the spectral weights are almost equal down to about -7 mV in the two cases. However, as the bias voltage approaches the range where the dynamical effects of the dressed end-factor become important, the current does not increase in a step-like fashion. The dI/dV shows a double peak with a small amplitude and minimum around -11 mV, which is where the dressed population number N_1 has its corresponding minimum, cf. Fig. 9.3(c). Similarly for forward bias voltages, the amplitude of the step around 5 mV is less in the one-loop-approximation than in the re-normalized Hubbard-I-approximation, which is understood as an effect of the scattering between the one-electron states causing a decreased probability for electrons to undergo the transition $|0\rangle\langle 3|$ in the double quantum dot. Hence, the resulting currents in the three different approximation schemes, can be understood from the discussion about the population numbers of the one-electron states. Thus, we now proceed to investigate the resulting currents through the double quantum dot in the one-loop-approximation.

The detuning of the discrete levels in the two quantum dots is of main importance in order to understand the asymmetry of the I–V characteristics of the system. When the two levels are resonant, i.e. the detuning $\Delta\varepsilon = 0$, the transition matrix elements $(d_{A\sigma})_{01}^{1n}$ and $(d_{B\sigma})_{01}^{1n}$ are equal. For negative detuning, $\Delta\varepsilon < 0$, the lower orbital in the double quantum dot couples strong/weak to the left/right lead, whereas the situation is diametrically opposite for the upper orbital. In case of positive detuning, $\Delta\varepsilon > 0$, the situation is reversed. It is thus expected that the I–V characteristics would be mirror images of one another under the detuning, which is indeed the case, cf. Fig. 8 in [17, 18].

9.5.1 *Current-Voltage Asymmetry*

We now consider varying (negative) detuning. As was previously discussed, the transition matrix elements $|(d_{A\sigma})_{01}^{1n}|^2 = |(d_{B\sigma})_{01}^{1n}|^2$ when the levels are resonant, which then leads to symmetric I–V curves since the couplings of the transitions $|0\rangle\langle n|$ to the left and right leads are equal. For finite detuning, the transition matrix elements are distinct, for all values of the tunneling rate t between the quantum dots. Thus, the asymmetry imposed on the system will provide an asymmetric I–V characteristics, which is clearly seen in Fig. 9.4(a). In Fig. 9.5(a) we plot the I–V (dI/dV) characteristics in the upper (lower) panel for increasing magnitude of the detuning and fixed interdot tunneling rate t, and the plots show the increasing asymmetry as $|\Delta\varepsilon|$ grows. The current is only slightly asymmetric for small detuning. These plots clearly illustrate the importance of the scattering between the one-electron states which is a result of the enlarged degree of asymmetric couplings to the left and right lead and follows from the increased asymmetry internally in the double quantum dot.

By instead letting the detuning be fixed and varying the interdot tunneling t, we find that the I–V characteristics follows the trends given for the transition matrix elements for varying t, cf. Fig. 9.2(a). For small tunneling rates we expect the I–V asymmetry to be significant whereas it is expected to be less pronounced for large tunneling rates. This is clear since all matrix elements tends to $1/2$ as the tunneling rate grows large, for any detuning since $\Delta\varepsilon/t \to 0$ in this limit. One would, thus, expect that the asymmetry of the I–V characteristics becomes small for large tunneling rates between the quantum dots, which is also clearly seen in Fig. 9.5(b), which displays the I–V characteristics (upper panel) for varying t. One should note that a large tunneling rate results in a large separation of the transition energies Δ_n, which eventually leads to that $\Delta_{3(4)}$, cf. Fig. 9.2(c), becomes positive for increasing t. This is the case for $t/\Gamma = 5$ (dash-dotted) in Fig. 9.5(b), showing that

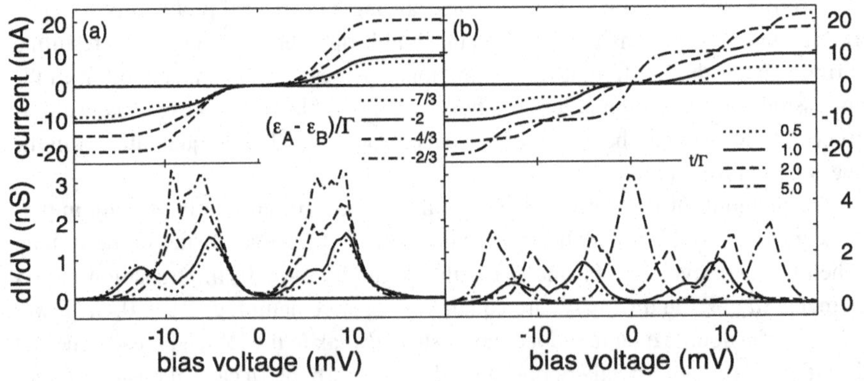

Fig. 9.5 Asymmetric I–V characteristics (*upper panel*) and dI/dV (*lower panel*) for (**a**) different detuning $\Delta\varepsilon = \varepsilon_A - \varepsilon_B < 0$ and fixed interdot tunneling $t/\Gamma = 1$, and (**b**) varying interdot tunneling t and fixed detuning $\Delta\varepsilon/\Gamma = -2$

$\Delta_{3(4)}$ lies in the vicinity of the equilibrium chemical potential μ, which gives a high conductance for low bias voltages. The plots in Fig. 9.5(b) suggest that the I–V asymmetry increases (decreases) for decreasing (increasing) tunneling rate, which is to be expected since the asymmetry internally in the double quantum dot increases (decreases) in this limit.

The plots in Fig. 9.5 show that the amplitude of the current tends to increase as the systems becomes more symmetric, i.e. $\Delta\varepsilon/t \rightarrow 0$ which leads to that $|(d_{A\sigma})_{01}^{1n}|^2$, $|(d_{B\sigma})_{01}^{1n}|^2 \rightarrow 1/2$. This behavior reflects the fact that equally strong couplings of the transitions $|0\rangle\langle n|$ to the left and right leads means that the corresponding double quantum dot orbital extends with a uniform probability amplitude throughout the structure. As the double quantum dot become strongly asymmetric, in the sense that the transition matrix elements approached 1 or 0, one finds a large probability amplitude of the orbital in one of the quantum dots and a small in the other one. The conductivity of the double quantum dot is closely related to this fact, since a strong localization of a state in one of the quantum dot yields a weak tunneling probability through the other. Hence, the overall current is reduced.

9.5.2 *Negative Differential Conductance*

Before concluding this chapter, we consider one of the more dramatic consequences of the I–V asymmetries that we have been discussing so far. From the lower panel of Fig. 9.5(a) it is clear that the separation of the peaks in the double peak structure appearing for negative voltages, increases as the magnitude of the detuning grows. Simultaneously, the conductance in the valley between the peaks approaches zero. Hence, in this picture it appears as a negative differential conductance would possibly establish if the detuning is even further increased. It seems that this increase should be accompanied with a small interdot tunneling rate t, cf. Fig. 9.5(b). Thus, fixing the tunneling rate and increasing the detuning we find that this is indeed the case, which is illustrated in Fig. 9.6(a) where the couplings $\Gamma^L/\Gamma^R = 1$.

In the previous discussion we suggested that the I–V asymmetries arise due to a significant decrease of the population number N_1, which results in that the transition $|0\rangle\langle 1|$ becomes less available for conduction. By the same token, the negative differential conductance is the result of a further decreased availability, eventually completely blocking any conduction through the double quantum dot via this transition. In the lower panel of Fig. 9.6(a) it is seen how the valley between the two conductance peaks, for negative bias voltages, evolve from being positive to negative as the detuning increases. The transition $|0\rangle\langle 1|$ begins to conduct for bias voltages slightly below its corresponding resonance value, due to the finite width of the transition. For increasingly negative values of the bias voltage, the conductance of the transition drops and, eventually, becomes more or less unavailable for conducting electrons through the double quantum dot, hence, the current drops. A further increased (negative) bias voltage results in a re-established conduction through the double quantum dot via the transition, hence the current grows again. The small

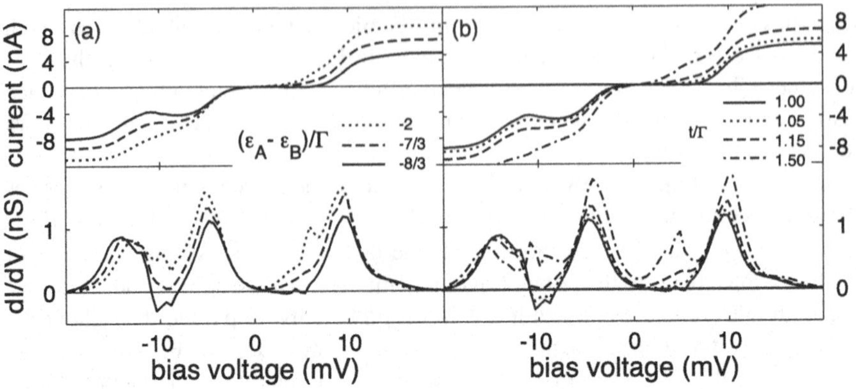

Fig. 9.6 Asymmetric I–V characteristics (*upper panel*) and dI/dV (*lower panel*) displaying negative differential conductance for (**a**) different detuning $\Delta\varepsilon = \varepsilon_A - \varepsilon_B < 0$ and fixed interdot tunneling $t/\Gamma = 1$, and (**b**) varying interdot tunneling t and fixed detuning $\Delta\varepsilon/\Gamma = -8/3$

negative differential conductance around 5 mV is most likely due to numerical errors in the numerical differentiation, since the current is vanishingly small in this region.

Above we concluded that small interdot tunneling rates tends to preserve the asymmetric properties of the double quantum dot whereas larger values of the t forces the system in to a more symmetric performance. This fact is verified for the current and is illustrates in Fig. 9.6(b), where it is seen that the negative differential conductance vanishes for increasing t.

9.6 Summary

We have seen that in the one-loop-approximation theory, scattering between the one-electron states generate regions of depleted and enhanced population of the orbital in the double quantum dot, which are asymmetrically located with respect to the bias voltage. The asymmetric behavior is a response to that the scattering between the one-electron states gives rise to a dynamical redistribution of the spectral weights. We concluded that increasing the separation between the levels in the two quantum dots and/or decreasing the interdot tunneling rate cause an increased degree of asymmetry of the I–V characteristics. For sufficiently large ratio between the detuning and the interdot tunneling rate, the asymmetric behavior may give rise to regions of negative differential conductance.

References

1. Reed, M.A., Zhou, C., Muller, C.J., Burgin, T.P., Tour, J.M.: Science **278**, 252 (1997)

2. Reichert, J., Ochs, R., Beckmann, D., Weber, H.B., Mayor, M., v. Löhneysen, H.: Phys. Rev. Lett. **88**, 176804 (2002)
3. Reichert, J., Weber, H.B., Mayor, M., v. Löhneysen, H.: Appl. Phys. Lett. **82**, 4137 (2003)
4. Leo, J., MacDonald, A.H.: Phys. Rev. Lett. **64**, 817 (1990)
5. Ferry, D.K., Goodnick, S.M.: Transport in Nanostructures. Cambridge University Press, Cambridge (1997)
6. Zaslavsky, A., Goldman, V.J., Tsui, D.C., Cunningham, J.E.: Appl. Phys. Lett. **53**, 1408 (1988)
7. Schmidt, M., Tewordt, T., Haug, R.J., von Klitzing, K., Förster, A., Lüth, H.: Solid-State Electron. **40**, 15 (1996)
8. Klimeck, G., Lake, R., Datta, S., Bryant, G.W.: Phys. Rev. B **50**, 5484 (1994)
9. Rudziński, W., Barnaś, J.: Phys. Rev. B **64**, 085318 (2001)
10. Ishibashi, K., Suzuki, M., Ida, T., Aoyagi, Y.: Appl. Phys. Lett. **79**, 1864 (2001)
11. Sollner, T.C.L.G., Goodhue, W.D., Tannenwald, P.E., Parker, C.D., Peck, D.D.: Appl. Phys. Lett. **43**, 588 (1983)
12. Sibille, A., Palmier, J.F., Wang, H., Mollot, F.: Phys. Rev. Lett. **64**, 52 (1990)
13. Tsu, R., Esaki, L.: Appl. Phys. Lett. **22**, 562 (1973)
14. Chang, L.L., Esaki, L., Tsu, R.: Appl. Phys. Lett. **24**, 593 (1974)
15. van der Vaart, N.C., Godjin, S.F., Nazarov, Y.V., Harmans, C.J.P.M., Mooij, J.E., Molenkamp, L.W., Foxon, C.T.: Phys. Rev. Lett. **74**, 4702 (1995)
16. van der Wiel, W.G., De Franceschi, S., Elzerman, J.M., Fujisawa, T., Tarucha, S., Kouwenhoven, L.P.: Rev. Mod. Phys. **75**, 1 (2003)
17. Fransson, J.: Phys. Rev. B **69**, 201304(R) (2004)
18. Fransson, J., Eriksson, O.: Phys. Rev. B **70**, 085301 (2004)

Chapter 10
Spin-Blockade

Abstract We reconsider the serially coupled double quantum dot system in the Pauli spin blockade regime and study the current and its dependence on internal and external parameters, e.g. the interdot tunneling rate and couplings to the leads. We proceed the study of the Pauli spin blockade regime by including ferromagnetic leads, in which case we find that a pure spin one state can be formed at finite bias voltages. We finally consider the prospect of obtaining an equivalent to the Pauli spin blockade phenomenon in T-shaped double quantum dots.

10.1 Pauli Spin-Blockade in Double Quantum Dots

Let us repeat some of the details from Chaps. 1 and 2 before we embark on our final analysis of the current characteristics in the Pauli spin blockade regime. The physics of the double quantum dot is captured by the Hamiltonian

$$\begin{aligned}\mathcal{H}_{DQD} = &\sum_{i=A,B}\Big(\sum_{\sigma}\varepsilon_{i\sigma}d^{\dagger}_{i\sigma}d_{i\sigma} + U_i n_{i\uparrow}n_{i\downarrow}\Big)\\ &+(U'-J/2)(n_{A\uparrow}+n_{A\downarrow})(n_{B\uparrow}+n_{B\downarrow})\\ &-2J\mathbf{s}_A\cdot\mathbf{s}_B + t\sum_{\sigma}(d^{\dagger}_{A\sigma}d_{B\sigma}+H.c.)\\ =&\sum_{Nn}E_{Nn}h^n_N, \end{aligned} \tag{10.1}$$

where the eigenstates $|N,n\rangle$ are essentially given in Table 9.1, however, now also depending on the spin-spin interaction parameter J. The leads are described simply by $\mathcal{H}_{L/R}=\sum_{\mathbf{k}\sigma\in L/R}\varepsilon_{\mathbf{k}\sigma}c^{\dagger}_{\mathbf{k}\sigma}c_{\mathbf{k}\sigma}$, and the tunneling interaction is provided by $\mathcal{H}_T=\sum_{\mathbf{k}\sigma,Nnm}v_{\mathbf{k}\sigma,Nnm}c^{\dagger}_{\mathbf{k}\sigma}X^{nm}_{NN+1}+H.c.$, where $v_{\mathbf{k}\sigma,Nnm}=v_{\mathbf{k}\sigma}(d_{i\sigma})^{nm}_{NN+1}$, with $i=A\ (B)$ if $\mathbf{k}\in L\ (R)$.

J. Fransson, *Non-Equilibrium Nano-Physics*,
Lecture Notes in Physics 809,
DOI 10.1007/978-90-481-9210-6_10, © Springer Science+Business Media B.V. 2010

In Chap. 2 we found that the equations of motion for the population numbers N_{Nn} can be written according to

$$\hbar\partial_t N_{01} = -\sum_{\chi n} \Gamma^{\chi}_{01,1n}[f^+_\chi(\Delta_{1n,01})N_{01} - f^-_\chi(\Delta_{1n,01})N_{1n}], \tag{10.2a}$$

$$\begin{aligned}\hbar\partial_t N_{Nn} = &\sum_{\chi n'}(\Gamma^{\chi}_{N-1n',Nn}[f^+_\chi(\Delta_{Nn,N-1n'})N_{N-1n'} - f^-_\chi(\Delta_{Nn,N-1n'})N_{Nn}] \\ &- \Gamma^{\chi}_{N+1n',Nn}[f^+_\chi(\Delta_{N+1n',Nn})N_{Nn} - f^-_\chi(\Delta_{N+1n',Nn})N_{N+1n'}]), \\ &N = 1, 2, 3,\end{aligned} \tag{10.2b}$$

$$\hbar\partial_t N_{41} = \sum_{\chi n} \Gamma^{\chi}_{3n,41}[f^+_\chi(\Delta_{41,3n})N_{3n} - f^-_\chi(\Delta_{41,3n})N_{41}], \tag{10.2c}$$

owing to the conditions that $\partial_t N_p = 0$ in the stationary regime, and that $\sum_p N_p = 1$ by conservation of probability, or, charge, whatever feels more convenient. Here, $\Gamma^{\chi}_{Nn,N+1n'} = 2\pi\sum_{\mathbf{k}\sigma i}|v_{\mathbf{k}\sigma}(d_{i\sigma})^{nn'}_{NN+1}|^2\delta(\Delta_{N+1n',Nn} - \varepsilon_{\mathbf{k}\sigma})$ gives the effective tunneling rate between the double quantum dot and the leads in terms of transitions between the many-body states.

10.1.1 The Pauli Spin Blockade Regime

By tuning the bias voltage into the regime $eV = \mu_L - \mu_R \in [0.1, 1]U$ and following the arguments in Sect. 2.4 we have previously seen that

$$N_{1n'} = \frac{\sum_{n=1}^5 \Gamma^R_{1n',2n}N_{2n}}{\sum_{n=1}^5 \Gamma^L_{1n',2n}}, \quad n' = 1, 2, \tag{10.3}$$

$$0 = \sum_{n'=1,2}[\Gamma^L_{1n',2n}N_{1n'} - \Gamma^R_{1n',2n}N_{2n}], \quad n = 1, \ldots, 5. \tag{10.4}$$

Spin-degenerate conditions leads to that the triplet configurations $|2, n\rangle$, $n = 1, 2, 3$, are degenerate such that $E_{2n} = E_T$ and $N_{2n} = N_T/3$, as are the one-electron states $|1, n\rangle$, $n = 1, 2$, with $E_{11} = E_{12}$ and $N_{1n} = N_1/2$. From the above equations it is then easy to see that

$$N_1 = \frac{2}{3}\frac{\Gamma^R}{\Gamma^L}\left(\frac{\alpha}{\beta}\right)^2 N_T, \tag{10.5a}$$

$$N_{2n} = \frac{1}{3}\left(\frac{L_n}{R_n}\cdot\frac{\alpha}{\beta}\right)^2 N_T, \quad n = 4, 5, \tag{10.5b}$$

where $\alpha^2 = \xi^2/[(1+\sqrt{1+\xi^2})^2 + \xi^2]$ and $\beta^2 = (1+\sqrt{1+\xi^2})^2/[(1+\sqrt{1+\xi^2})^2 + \xi^2]$, with $\xi = 2t/\Delta\varepsilon$, and where L_n and R_n are bounded and finite for all ξ. By the

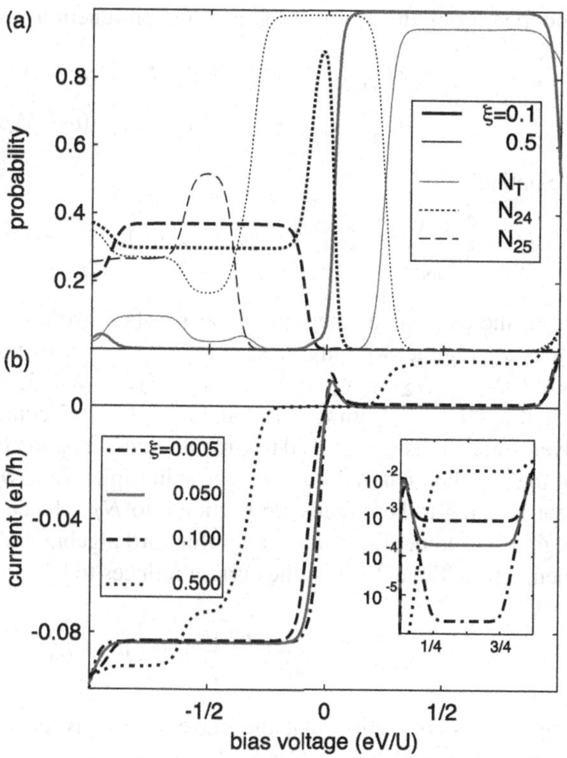

Fig. 10.1 (**a**) Two-electron population numbers N_T (*solid*), P_{24} (*dotted*), and P_{25} (*dashed*), for $\xi = 0.1$ (*bold*) and $\xi = 0.5$ (*faint*), keeping $\Delta\varepsilon$ constant, $1/\beta \sim U/10$, and $U' = U/2$. (**b**) The I–V characteristics for the double quantum dot system for various values of ξ, keeping $\Delta\varepsilon$ constant, and $\Gamma^L = \Gamma^R = \Gamma$. The inset shows the current in the Pauli spin blockade regime on a logarithmic scale

normalization condition $N_1 + N_T + N_{24} + N_{25} = 1$ we obtained

$$N_T = \left\{1 + \frac{1}{3}\left(\frac{\alpha}{\beta}\right)^2\left[2\frac{\Gamma^R}{\Gamma^L} + \sum_{n=4,5}\left(\frac{L_n}{R_n}\right)^2\right]\right\}^{-1}. \tag{10.6}$$

In this form, it is easy to see that $N_T \to 1$ as $\xi \to 0$, since $\alpha \to 0$ and $\beta \to 1$ in this limit.

The spin triplet population number, N_T (solid), is plotted in Fig. 10.1(a) along with the population numbers of the first and second spin singlet states, N_{24} (dotted) and N_{25} (dashed), respectively, for two different values of ξ. It is clear from the plots that the spin triplet state is almost fully occupied in this regime of bias voltages, $eV = \mu_L - \mu_R \in [0.1, 1]U$, for small ξ, while its population decrease for increasing ξ. It is also clear that singlet populations N_{24} and N_{25} are negligible in the Pauli spin blockade regime.

We now proceed by studying the current characteristics of the system in this regime. Writing the current according to basic relation

$$I_\chi = \frac{ie}{h}\sum_{Nnn'}\int \Gamma^\chi_{Nn,N+1n'}[f_\chi(\omega)G^>_{Nn,N+1n'}(\omega) + f_\chi(-\omega)G^<_{Nn,N+1n'}(\omega)]d\omega, \tag{10.7}$$

and identifying the lesser and greater Green functions by

$$G^{>}_{Nn,N+1n'}(\omega) = -i2\pi N_{Nn}\delta(\omega - \Delta_{N+1n',Nn}), \tag{10.8a}$$

$$G^{<}_{Nn,N+1n'}(\omega) = i2\pi N_{N+1n'}\delta(\omega - \Delta_{N+1n',Nn}), \tag{10.8b}$$

we obtain

$$I_L = \frac{e}{\hbar}\sum_{Nnn'} \Gamma^L_{Nn,N+1n'}\Big\{f_L(\Delta_{N+1n',Nn})[N_{Nn} + N_{N+1n'}] - N_{N+1n'}\Big\}. \tag{10.9}$$

From the analysis of the population numbers we conclude that, in the Pauli spin blockade regime, we only need to include the terms proportional to $N_{11} = N_{12} = N_1/2$, $N_{21} = N_{22} = N_{23} = N_T/3$, and N_{2n}, $n = 4, 5$, which leads to that only the couplings $\Gamma^L_{1n,2n'}$, with $n = 1, 2$ and $n' = 1, \ldots, 5$ contribute to the current. Moreover, since we have managed to write all non-vanishing population numbers in terms of the population number N_T for the spin triplet state, we can conclude that the current is going to be directly proportional to N_T. Hence, in the Pauli spin blockade regime, we find, after some straightforward algebra and using the transition matrix elements in Table 2.1, that the current reduces to [2]

$$I_L = \frac{2e}{3\hbar}\Gamma^R\left(\frac{\alpha}{\beta}\right)^2\left(\frac{3}{2}\beta^2 + L_4^2 + L_5^2\right)N_T. \tag{10.10}$$

This expression shows that the current strongly depends on the ratio $(\alpha/\beta)^2 = \xi^2/(1 + \sqrt{1+\xi^2})^2$ and, hence, that there is a strong suppression of the current for weakly coupled quantum dots, i.e. for $\xi \ll 1$. This property is clearly illustrated in the bias voltage range $eV/U \in [0.1, 1]$ in Fig. 10.1(b), which shows the I–V characteristics for the double quantum dot for different ξ, keeping $\Delta\varepsilon$ constant. In particular, the blockade becomes roughly 3 orders of magnitude deeper when ξ is decreased from 0.1 to 0.005. This calculated current verifies the experimental observations in [1]. Bias voltage dependent transition matrix elements generated by a 10% voltage drop between the quantum dots do not qualitatively alter this picture.

We understand that the current I_L in the Pauli spin blockade regime is not completely vanishing which is due to the finite interdot tunneling rate t. A finite interdot tunneling leads to that the population number for the spin triplet is slightly less than unity. The remaining population, i.e. $1 - N_T$, is distributed among the one-electron states $|1, n\rangle$, $n = 1, 2$ and the spin singlet states $|2, n\rangle$, $n = 4, 5$. In particular, none of the population numbers N_1, N_{24}, and N_{25}, is identically zero which implies that there occurs transitions between those states by letting electrons in and out from the double quantum dot to the leads, i.e. there is a transport of electrons between the leads which is another way to say that there is charge current flowing in the system. It is, furthermore, clear that the leakage current grows quadratically with ξ since $(\alpha/\beta)^2 = \xi^2/[1 + \sqrt{1+\xi}]^2$, while N_T decays by $[1 + (\alpha/\beta)^2/3]^{-1} \approx 1 - (\alpha/\beta)^2/3$.

The above analysis of the population numbers is made on the assumption that $\mu = 0$. Shifting the position of μ in experiments correspond to application of a gate

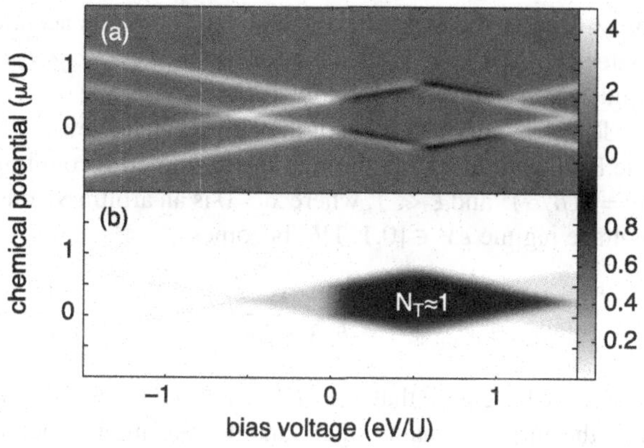

Fig. 10.2 (**a**) Differential conductance dI/dV for the double quantum dot for varying positions of the chemical potential $\mu \in [-1, 2]U$. (**b**) Contour plot of the spin triplet population number N_T as function of the bias voltage and the position μ. The truncated diamond in the bias voltage range $[0, 1.5]U$ marks the region with $N_T \approx 1$. Here, $\xi = 0.05$, while the other parameters are as in Fig. 10.1

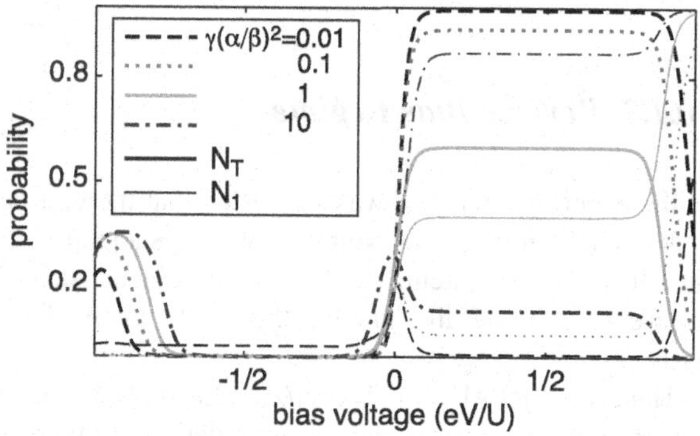

Fig. 10.3 Spin-triplet population number N_T (*bold*) for different asymmetries $\gamma = \Gamma^R/\Gamma^L$ of the couplings to the leads. The sum of the one-electron population numbers N_1 (*faint*) are plotted as reference. Here, $\xi = 0.05$, whereas other parameters are as in Fig. 10.1

voltage across the system, by means of which the levels in the quantum dot can be changed relative to μ. By shifting μ, we find that the Pauli spin blockade regime extends over a range of gate voltages, where $N_T \approx 1$ and the current through the double quantum dot is small. This is illustrated in Fig. 10.2, where the differential conductance and N_T are plotted in panels (a) and (b), respectively, as functions of the bias voltage and the chemical potential. In panel (b), the dark truncated diamond readily shows the expected domain of the Pauli spin blockade regime. The corresponding conductance in panel (a) displays regions of negative differential conductance shown through the dark ridges along the borders of the diamond.

The population number N_T strongly depends on the ratio $(\alpha/\beta)^2$ as we have seen above. It also depends on the couplings $\Gamma^{L/R}$ to the leads through the ratio Γ^R/Γ^L, cf. (10.6). Hence, assuming proportional couplings such that $\Gamma^R/\Gamma^L = \gamma \geq 0$, and letting $\gamma \sim (\beta/\alpha)^2$ gives rise to a lifting of the Pauli spin blockade even for weakly coupled quantum dots. Accordingly, $\xi \to 0$ one finds that $N_T \to 3/5$, see Fig. 10.3 where N_T (bold) and N_1 (faint) are plotted for varying degree of asymmetry γ between the double quantum dot and the leads. The plots clearly shows that N_T decays as γ increases, that is, when the coupling to the right lead grows stronger than the coupling to the left. In particular for $\gamma \sim (\beta/\alpha)^2$, the plot shows that N_T is close to $3/5$. The population numbers for the spin singlets have a negligible dependence on γ. Hence, the increasing remainder $1 - N_T$ is accumulated on the one-electron states $|1, n\rangle$, $n = 1, 2$, which is clear since $N_1 \propto \gamma N_T$, see (10.5a), (10.5b). This tendency is also verified in Fig. 10.3 for N_1.

The absolute magnitude of the current is, however, not increased by the lifting of the Pauli spin blockade through the asymmetric couplings to the leads. In the limit $\gamma = x(\beta/\alpha)^2$ and $\xi \ll 1$, where $x > 0$ is an arbitrary constant, the current in the bias voltage regime $eV \in [0.1, 1]U$ becomes

$$I_L = \frac{4e}{\hbar}\frac{x}{3+2x}\Gamma^L = \frac{8\pi e}{h}\Gamma\frac{1}{1+\gamma}\frac{x}{3+2x}, \tag{10.11}$$

where we have used that $\Gamma = \Gamma^L + \Gamma^R = (1+\gamma)\Gamma^L$. The displayed relation shows that the current remains small, and decreasing, for increasing asymmetries γ of the couplings to the leads.

10.1.2 Reverse Bias Regime

In the experiments [1], it was suggested that the unit occupation of the spin triplet was lifted for reverse bias voltages, which is a natural conjecture when considering the set-up of the system. Here, we cannot be satisfied, however, with a conjecture but we would rather like to see if this can be justified, if not proven, on a theoretical basis.

Hence, we let $eV \in -[1, 0.1]U$ and $\xi \ll 1$. Then, by the same arguments as in the previous discussion only the population numbers N_1, N_T, N_{24}, and N_{25}, are finite. It turns out that we obtain the following relations between those population numbers [2], that is,

$$N_1 = 2\frac{\Gamma^L}{\Gamma^R}\left(\frac{L_4}{R_4}\right)^2 N_{24}, \tag{10.12a}$$

$$N_T = \frac{3}{2}\left(\frac{\alpha}{\beta}\cdot\frac{L_4}{R_4}\right)^2 N_{24}, \tag{10.12b}$$

$$N_{25} = \left(\frac{L_4}{R_4}\cdot\frac{R_5}{L_5}\right)^2 N_{24}, \tag{10.12c}$$

$$N_{24} = \left\{1 + \left(\frac{L_4}{R_4}\right)^2\left[2\frac{\Gamma^L}{\Gamma^R} + \frac{3}{2}\left(\frac{\alpha}{\beta}\right)^2 + \left(\frac{R_5}{L_5}\right)^2\right]\right\}^{-1}. \tag{10.12d}$$

The expression in (10.12c) tell us that $N_T \approx 0$ for small ξ, while none of the other population numbers has a strong dependence on ξ. Thus, all population numbers in (10.12a)–(10.12d) but N_T are finite. Reverse bias voltages correspond to the negative voltage regime in Fig. 10.1, in which panel (a) shows that the spin triplet state hardly is populated for small ξ, while the populations of the singlet states, N_{24} and N_{25}, are finite. The sum of those states is less that one, however, which leads to that remaining population is (equally) distributed among N_{11} and N_{12}. The finiteness of N_1, N_{24}, and N_{25}, means that the electron population in the double quantum dot varies between 1 and 2 electrons, which is the same as to say that there is current running through the system. This current is given by [2]

$$I_L = \frac{2e}{\hbar}\Gamma^L L_4^2\left[1+\left(\frac{R_5}{R_4}\right)^2\right]N_{24}, \tag{10.13}$$

and is expected to be substantially larger than in the forward direction, which is also seen in Fig. 10.1(b).

10.1.3 Linear Regime

In the linear regime, i.e. $eV \in (-0.1, 0.1)U$, we can employ the equilibrium results

$$N_1 = 2e^{\beta(\Delta_{24,11}-\mu)}N_{24}, \tag{10.14a}$$

$$N_T = 3e^{\beta(\Delta_{24,11}-\Delta_{21,11})}N_{24}, \tag{10.14b}$$

$$N_{25} = e^{\beta(\Delta_{24,11}-\Delta_{25,11})}N_{24}, \tag{10.14c}$$

$$N_{24} = \left\{1+2e^{\beta(\Delta_{24,11}-\mu)}+3e^{\beta(\Delta_{24,11}-\Delta_{21,11})}+e^{\beta(\Delta_{24,11}-\Delta_{25,11})}\right\}^{-1}. \tag{10.14d}$$

Using that $\Delta_{24,11} < \mu = 0$, cf. Fig. 2.1, we see from those relations that $N_{24} \approx 1$ whenever $|\Delta_{24,11}| \gg 1/\beta$ which is satisfied for $\xi \gtrsim 0.1$ and $1/\beta \sim U/10$. Accordingly, the population numbers in (10.14a)–(10.14d) will be exponentially suppressed by the same condition on ξ, since the gap between the energies between the lowest singlet state, $|2,4\rangle$, and the other two-electron states increase roughly by ξ^2. Hence, for $\xi \gtrsim 0.1$, the current will be blockaded by the lowest spin singlet in a finite range around zero bias voltage, which is also illustrated in Fig. 10.1. The plots for N_{24} (dotted) in panel (a) shows the tendency of increasing N_{24} around equilibrium for increasing ξ, which gives rise to the current blockade shown in panel (b) for $\xi = 0.5$ (dotted). It is, moreover, clear that $N_{24} \to 1/7$ and $\xi \to 0$, since the energetic distance between $\Delta_{24,11}$ and μ, and $\Delta_{2n,11}$, $n = 1, 2, 3, 5$, approaches zero. In the limit of weakly coupled quantum dots, the equilibrium population numbers satisfy $N_1 = 2/7$, $N_T = 3/7$, and $N_{24} = N_{25} = 1/7$. The conductance of the system in the linear regime is therefore substantially larger than in the Pauli spin blockade regime, cf. Fig. 10.1(b).

Since the first observation of the Pauli spin blockade phenomenon in vertical coupled quantum dots [1], the effect has been recorded in other types of configurations, such as lateral semi-conducting quantum dots [5–7], and in carbon nanotubes [8]. Further discussions about the Pauli spin blockade phenomenon can be found in [9]. In the following sections we will focus on a few other theoretical aspects of the Pauli spin blockade regime.

10.2 Formation of Pure Spin One State in Double Quantum Dots

We now manipulate the external conditions and investigate how the double quantum dot react to those changes. In particular, we tune the system into the Pauli spin blockade regime as in the previous discussions, but replace one of the leads by a ferromagnetic lead keeping the other lead non-magnetic. The results discussed in this section can also be found in [3].

The extension to ferromagnetic leads is done be introducing the parameters $p_{L/R} = \Gamma_\uparrow^{L/R} - \Gamma_\downarrow^{L/R}$ for the spin-polarization, such that $\Gamma_\sigma^\chi = \Gamma_0(1+\sigma p_\chi)/2$, where $\Gamma_0 = \Gamma_\uparrow^\chi + \Gamma_\downarrow^\chi$, since we are considering symmetric couplings to the leads. A non-magnetic state in the lead corresponds to $p_\chi = 0$, whereas half-metallicity is obtained for $p_\chi = \pm 1$, corresponding to $\uparrow$ and $\downarrow$, respectively.

Changing the external condition does not change the general condition of the double quantum dot. Hence, in the bias voltage regime $eV \in (-1, 1)U$, only the population numbers N_{1n}, $n = 1, 2$, and N_{2n}, $n = 1, \ldots, 5$, are finite. However, since the spin-degenerate conditions are broken we must acknowledge that $N_{11} \neq N_{12}$, and $N_{21} \neq N_{22} \neq N_{23}$, in general.

First, we consider forward bias voltages $eV \in (0.1, 1)U$, where we have the relations

$$N_{1n} = \frac{\sum_{\sigma,n'=1}^{5} \Gamma_\sigma^R |(d_{B\sigma})_{12}^{nn'}|^2 N_{2n'}}{\sum_{\sigma,n'=1}^{5} \Gamma_\sigma^L |(d_{A\sigma})_{12}^{nn'}|^2}, \quad n = 1, 2, \tag{10.15a}$$

$$N_{2n'} = \frac{\sum_{\sigma,n=1}^{2} \Gamma_\sigma^L |(d_{A\sigma})_{12}^{nn'}|^2 N_{1n}}{\sum_{\sigma,n'=1}^{2} \Gamma_\sigma^R |(d_{B\sigma})_{12}^{nn'}|^2}, \quad n = 1, \ldots, 5, \tag{10.15b}$$

which, by using the matrix elements in Table 2.1, can be further simplified to read

$$N_{11} = \frac{1+p_R}{1+p_L}\left(\frac{\alpha}{\beta}\right)^2 N_{21}, \qquad N_{12} = \frac{1-p_L}{1-p_R}\left(\frac{1+p_R}{1+p_L}\right)^2\left(\frac{\alpha}{\beta}\right)^2 N_{21}, \tag{10.16a}$$

$$N_{22} = \left(\frac{1-p_L}{1+p_L}\frac{1+p_R}{1-p_R}\right)^2 N_{21}, \qquad N_{23} = \frac{1-p_L}{1+p_L}\frac{1+p_R}{1-p_R} N_{21}, \tag{10.16b}$$

$$N_{2n} = \frac{1-p_L}{1+p_L}\frac{1+p_R}{1-p_R}\left(\frac{\alpha}{\beta}\frac{L_n}{R_n}\right)^2, \quad n = 4, 5. \tag{10.16c}$$

Conservation of charge finally gives

$$N_{21} = \left\{1 + \frac{1+p_R}{1+p_L}\left[\frac{1-p_L}{1-p_R} + \frac{1+p_R}{1+p_L}\left(\frac{1-p_L}{1-p_R}\right)^2 + \left(\frac{\alpha}{\beta}\right)^2\left(1 + \frac{1-p_L}{1-p_R}\frac{1+p_R}{1+p_L} + \frac{1-p_L}{1-p_R}\sum_{n=4,5}\left(\frac{L_n}{R_n}\right)^2\right)\right]\right\}^{-1}. \tag{10.17}$$

Consider the case $p_L = 1$ and $p_R = 0$, which corresponds to letting the left be half-metallic with spin $\uparrow$ and the right lead being non-magnetic. Then, the population number $N_{21} = 1/[1 + (\alpha/\beta)/2] = 2(1 + \xi^2)/(2 + 3\xi^2) \to 1$ as $\xi \to 0$. Physically, this result means that the probability to populate the spin triplet state $|\uparrow\rangle|\uparrow\rangle$ is unity and that the double quantum dot acquires the definite spin moment $m_z = 1$. This result is consistent with (10.16a)–(10.16c), since all population numbers but N_{11} vanish for $p_L = 1$ and $p_R = 0$, while $N_{11} \propto (\alpha/\beta)^2 \to 0$ as $\xi \to 0$. The result is also consistent with the spin-degenerate Pauli spin blockade, since the population number of the spin triplet must be unity. While a definite spin moment is obtained in the double quantum dot, this result is not so surprising since there are only spin $\uparrow$ electrons available in the left lead. Hence, the accumulated charge in the double quantum dot must eventually become strongly spin-polarized.

The similar result is obtained by letting the left lead be non-magnetic ($p_L = 0$) and the right lead half-metallic with spin $\downarrow$ ($p_R = -1$). Then, $N_{21} = 1$ independently of the interdot tunneling rate t and detuning $\Delta\varepsilon$. This result is, of course, consistent with (10.16a)–(10.16c), since all population numbers but N_{21} are identically zero.

Neither of those results are very surprising. In the former case, for instance, only spin $\uparrow$ electrons are present in the left lead and are, thus, the only electrons that can flow from the left to the right. Accumulation of electrons in the double quantum dot necessarily have to be of spin $\uparrow$. In the latter example, both spin projections are available in the left lead, however, only spin $\downarrow$ electrons are permitted to enter the right lead. Despite the tunneling rate between the quantum dots and the right lead is small for the spin $\downarrow$ electrons, it is vanishing for the spin $\uparrow$ electrons. Hence, the accumulated electron density in the double quantum dot will eventually consist of spin $\uparrow$ solely.

The reverse regime, i.e. $eV \in (-1, -0.1)U$, is, in this respect, more intriguing. Using the same arguments are before, we find that the population numbers are related through

$$N_{11} = \frac{1+p_R}{1+p_L}\left(\frac{1-p_L}{1-p_R}\right)^2\left(\frac{\beta}{\alpha}\right)^2 N_{22}, \qquad N_{12} = \frac{1-p_L}{1-p_R}\left(\frac{\beta}{\alpha}\right)^2 N_{22}, \tag{10.18a}$$

$$N_{21} = \left(\frac{1-p_L}{1+p_L}\frac{1+p_R}{1-p_R}\right)^2 N_{22}, \qquad N_{23} = \frac{1-p_L}{1+p_L}\frac{1+p_R}{1-p_R} N_{22}, \tag{10.18b}$$

$$N_{2n} = \frac{1-p_L}{1+p_L}\frac{1+p_R}{1-p_R}\left(\frac{\beta}{\alpha}\cdot\frac{R_n}{L_n}\right)^2 N_{22}, \quad n = 4, 5. \tag{10.18c}$$

Conservation of charge gives

$$N_{22} = \left\{1 + \frac{1-p_L}{1-p_R}\left[\frac{1+p_R}{1+p_L} + \frac{1-p_L}{1-p_R}\left(\frac{1+p_R}{1+p_L}\right)^2 \right.\right.$$
$$\left.\left. + \left(\frac{\beta}{\alpha}\right)^2\left(1 + \frac{1-p_L}{1-p_R}\frac{1+p_R}{1+p_L} + \frac{1+p_R}{1+p_L}\sum_{n=4,5}\left[\frac{R_n}{L_n}\right]^2\right)\right]\right\}^{-1}. \tag{10.19}$$

From (10.19) it follows that $N_{22} = 1$ for $p_L = 1$, i.e. when the left lead is half-metallic spin $\uparrow$. We also note that this results irrespective of the spin-polarization in the right lead and of the ratio ξ, since all terms in the square bracket in (10.19) are multiplied by $1 - p_L = 0$. Simultaneously, all other population numbers vanish, see (10.18a)–(10.18c). Physically, this means that the state $|\downarrow\rangle_A |\downarrow\rangle_B$ becomes fully populated for low temperatures whenever the left lead is a spin $\uparrow$ half-metal.

Having the discussion in Sect. 10.1 in mind, this result is quite remarkable, since without the spin-polarized leads all triplet states would have small population numbers in this regime, $eV \in (-1, -0.1)U$. In fact, in the spin-degenerate case, the electron density is rather uniformly distributed among the spin singlet states $|2, n\rangle$, $n = 4, 5$, and the one-electron states $|1, n\rangle$, $n = 1, 2$. The spin-polarization does not change the transition matrix elements, see Table 2.1, but from those we, nevertheless, find the explanation for the remarkable accumulation of electron density in the spin triplet. Unless, the right lead is a spin $\uparrow$ half metal, it provides the double quantum dot with electrons of both spin projections, and from the discussion in Sect. 10.1 we know that the population numbers of the spin triplet states are finite and, hence, are part of the conduction of electrons from the right to the left lead. The probability amplitudes for the transitions $|1, 1\rangle\langle 2, 2|$ and $|1, 2\rangle\langle 2, 2|$ are 0 and β (recall that $|1, 1\rangle = \alpha|\uparrow\rangle|0\rangle + \beta|0\rangle|\uparrow\rangle$ and $|1, 2\rangle = \alpha|\downarrow\rangle|0\rangle + \beta|0\rangle|\downarrow\rangle$). Due to the absence of spin $\downarrow$ electron states in the left lead, the condition $p_L = 1$ implies that the latter transition is unavailable. Hence, while both spin $\uparrow$ and $\downarrow$ electrons can tunnel into the double quantum dot from the right, only the spin $\uparrow$ electrons can continue to the left lead. Then, since the spin triplet state $|2, 2\rangle$ is involved in the conduction, this inevitably leads to an accumulation of electron density in this state.

If we instead assume that the right lead is half-metallic spin $\downarrow$, i.e. $p_R = -1$, we find from (10.19) that $N_{22} = 2\xi^2/(1 + 3\xi^2) \to 0$ as $\xi \to 0$, for any spin-polarization in the left lead. Simultaneously, all population numbers but N_{12} vanish identically, whereas

$$N_{12} = \frac{1}{2}\left(\frac{\beta}{\alpha}\right)^2 N_{22} = \frac{1 + \xi^2}{1 + 3\xi^2} \to 1, \quad \text{as } \xi \to 0, \tag{10.20}$$

that is, the one-electron state $\alpha|\downarrow\rangle|0\rangle + \beta|0\rangle|\downarrow\rangle$ acquires an almost full population for weakly coupled quantum dots. Hence, the double quantum dot ends up in a spin-polarized configuration, however, carrying only one electron instead of two.

10.3 Non-Equilibrium Triplet Blockade in T-Shaped Double Quantum Dot

Before leaving the discussion of the Pauli spin blockade in double quantum dots, we ask ourselves whether it would be possible to obtain a similar state in parallel, or T-shaped, double quantum dot systems, see Fig. 10.4(a). We find that this is indeed the case if there is a ferromagnetic exchange interaction between the quantum dots. The results discussed in this section can also be found in [4].

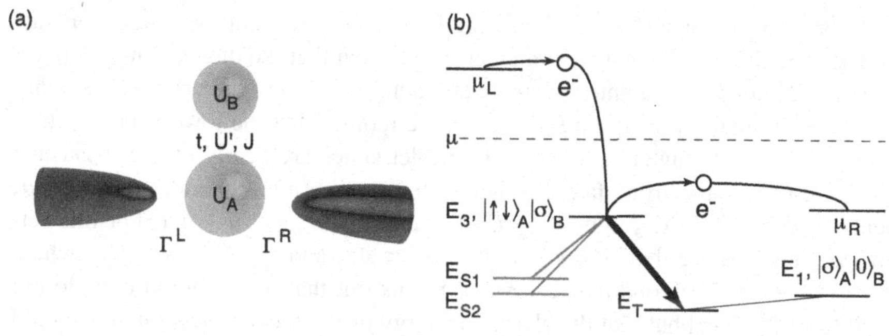

Fig. 10.4 (**a**) Schematic view of the T-shaped double quantum dot system, where only one of the quantum dots is coupled to the leads. (**b**) Processes leading to the non-equilibrium triplet, Pauli spin blockade, blockade. Faint and bold lines signify small and large transition probabilities, respectively. See text for notation

As we shall see below, the quantities that are important for obtaining the Pauli spin blockade comprise the conditions that the quantum dots are coupled through Coulomb and exchange interactions, and weakly through tunneling. In absence of the spin-spin exchange interaction, there may be regimes of normal Coulomb blockade in a finite range of bias voltages around equilibrium.

Existence of a sufficiently large ferromagnetic exchange coupling leads to that the spin triplet states $|2, n\rangle$, $n = 1, 2, 3$, acquire a lower energy than the lowest spin singlet $|2, 4\rangle$. Hence, the spin triplet naturally becomes the ground state with a unit probability of being populated, provided that the spin triplet energy is lower than the energies of all other states. The spin triplet remains fully occupied for bias voltages smaller than the energy separation between the triplet and singlet states, although transitions between the one-electron and the singlet states may open for conduction. For larger bias voltages, however, this low bias triplet blockade is lifted as the transitions between triplet and one-electron states become resonant with the chemical potentials of the leads. The current is then mediated by transitions between the two-electron singlet and one-electron states.

The proper non-equilibrium Pauli spin blockade regime is entered at bias voltages such that transitions between the three-electron states and, at least, one of the singlet states become resonant, see Fig. 10.4(b). At those conditions, an electron enters the double quantum dot from the lead with the higher (chemical) potential under a transition from a two-electron singlet to a three-electron state. Transitions from the spin triplet to the three-electron states are suppressed since the bias voltage does not support tunneling through the energy barrier between those states. Next, an electron tunnels out from the double quantum dot under a transition from the three-electron state to the spin triplet, since the transition matrix element for this process is about unity whereas the matrix elements for transitions between the three-electron states and the spin singlets are at most one half. Eventually, charge accumulate in the spin triplet since the transitions between those states and the one-electron states occur with a negligible rate.

In order to be quantitative, assume the two quantum dots to be spin-degenerate and assume that the exchange parameter $J \geq 0$ such that the direct spin-spin interaction does not favor an anti-ferromagnetic configuration of the spins in the quantum dots. In analogy with the serial case, we require that the lowest one-electron states, the lowest singlet states, and the triplet states are nearly aligned, and that the lowest three-electron states lie below the equilibrium chemical potential μ, hence, $E_T \approx E_{24} \approx E_{25} \approx \min_{n=1}^4\{E_{1n}\} < \min_{n=1}^4\{E_{3n}\} < \mu < E_4$. For this setup we, thus, require that $\mu - \varepsilon_B \approx \Delta\varepsilon$, $U' \approx \Delta\varepsilon$, and $U_A \approx 2\Delta\varepsilon \leq U_B$, where $\Delta\varepsilon = \varepsilon_B - \varepsilon_A$. The condition $U_A \leq U_B$ points out that the quantum dots do not need to be identical but that the charging energy of the second quantum dot should be bounded below by the charging energy of the first. The presence of the second quantum dot is, however, essential in order to obtain the Pauli spin blockade regime in the T-shaped system. Finally, weakly coupled quantum dots, $\xi \ll 1$, implies that the energies for the lowest one- and three-electron states acquire their main weights on quantum dot A. This condition yields a low (large) probability for the transitions between the triplet and the lowest lying one-electron (three-electron) states.

The discussion is here focused to the non-equilibrium Pauli spin blockade since the low bias blockade can be obtained for weakly coupled quantum dots whenever $J > 0$ is sufficiently large. As mentioned, the Pauli spin blockade is generated by opening transitions between the two- and three-electron states. It turns out that the restrictions $eV < 7\Delta\varepsilon/4$, $k_BT < U_A/10$, and $\xi < 1/5$ are sufficient for these purposes, which give that only the population numbers P_{1n}, $n = 1, 2$, $N_T = P_{2n}/3$, $n = 1, 2, 3$, P_{24}, P_{25}, and P_{3n}, $n = 1, 2$, are non-negligible. The population numbers for all other states are negligible since any transition involving those states are non-resonant. As in Sect. 10.1, we consider spin-degenerate conditions, for which it is required that $P_{1n} = N_1/2$, and $P_{3n} = N_3/2$, and which reduces the system to five equations for the population numbers, given by e.g. (10.2a)–(10.2c).

The Pauli spin blockade regime enters when, at least one of, the transition energies $\Delta_{3n,24}$, $\Delta_{3n,25}$ lie within the bias voltage window $eV = \mu_L - \mu_R$, whereas the transition energy between the spin triplet and the three-electrons state, i.e. $\Delta_{3n,T}$, is non-resonant, where the subscript T denotes the spin triplet states. In addition, in order to acquire an accumulation of charge in the spin triplet states under those conditions, and in order to prevent a large leakage of charge from the spin triplet states, it is necessary that transitions between the three-electron (one-electron) states and the spin triplet states occur with large (small) probability. In particular, the transitions $|3, n\rangle\langle T|$ must be very much more likely to occur than the transitions $|3, n\rangle\langle 2, 4|$ and $|3, n\rangle\langle 2, 5|$, since otherwise the charge would not tend to accumulate in the spin triplet.

We obtain the desired conditions by tuning the bias voltage such that e.g. $\min_{nm}\{\Delta_{3m,2n}\} < \mu_L < \max_{nm}\{\Delta_{3m,2n}\}$, $n = 1, \ldots, 5$, and $m = 1, 2$ (here $eV > 0$ whereas the case $eV < 0$ follow by symmetry). In this regime $f_L(\Delta_{2n,1m}) = f_R(-\Delta_{2n,1m}) = 1$, $n = 1, \ldots, 5$, $m = 1, 2$, and $f_R(\Delta_{3m,2n}) = 0$, $n = 1, \ldots, 5$, $m = 1, 2$. From the above discussion it is clear that the charge accumulation in the spin triplet will be lifted for bias voltages that support transitions between the spin triplet states and the three-electron states. We, therefore, require the bias voltage to be such that $f_{L/R}(\Delta_{3m,T})$, which is obtained for $\Delta_{3m,T} = E_{3m} - E_T > \mu_L + k_B T$, $m = 1, 2$. The equations for the population numbers can thus be written

$$N_1 = \frac{1}{p} N_3 = \frac{2/3}{1 + 2p(\kappa/\beta)^2} N_T, \qquad P_{2n} = \frac{p}{2} \frac{L_n^2/p + \Lambda_n^2 \sum_\chi f_\chi(-\Delta_{31,2n})}{L_n^2 + \Lambda_n^2 f_L(\Delta_{31,2n})} N_1, \tag{10.21a}$$

$$p = \frac{\sum_{n=4,5} L_n^2 \frac{\Lambda_n^2 f_L(\Delta_{31,2n})}{L_n^2 + \Lambda_n^2 f_L(\Delta_{31,2n})}}{3\kappa^2 + \sum_{\chi, n=4,5} \Lambda_n^2 f_\chi(-\Delta_{31,2n})[1 - \frac{\Lambda_n^2 f_L(\Delta_{31,2n})}{L_n^2 + \Lambda_n^2 f_L(\Delta_{31,2n})}]}, \tag{10.21b}$$

where $n = 4, 5$, in the expression for P_{2n}. Here, also $\beta^2 \equiv \sum_\sigma |(d_{A\sigma})_{12}^{m1}|^2 = \xi^2/[(1 + \sqrt{1 + \xi^2})^2 + \xi^2]$ and $L_n^2 \equiv \sum_\sigma |(d_{A\sigma})_{12}^{mn}|^2$, $m = 1, 2$, $n = 4, 5$, whereas $\kappa^2 \equiv \sum_\sigma |(d_{A\sigma})_{23}^{1m}|^2 = (1 + \xi^2)/[(1 + \sqrt{1 + \xi^2})^2 + \xi^2]$ and $\Lambda_n^2 \equiv \sum_\sigma |(d_{A\sigma})_{12}^{mn}|^2$, $m = 1, 2$, $n = 4, 5$, are the matrix elements for the involved transitions. The above equation follow from spin-degeneracy since $\Delta_{2n,11} = \Delta_{2n,12}$ and $\Delta_{31,2n} = \Delta_{32,2n}$, $n = 1, \ldots, 5$. Combining (10.21a)–(10.21b) with charge conservation ($N_1 + N_T + \sum_{n=4}^{5} P_{2n} + N_3 = 1$), it follows that

$$N_T = \frac{1}{1 + \frac{2/3}{1+2p(\kappa/\beta)^2}\left(1 + p + \frac{p}{2}\sum_{n=4,5} \frac{L_n^2/p + \Lambda_n^2 \sum_\chi f_\chi(-\Delta_{31,2n})}{L_n^2 + \Lambda_n^2 f_L(\Delta_{31,2n})}\right)}. \tag{10.22}$$

The matrix elements L_n and Λ_n, $n = 4, 5$, are finite and bounded, and we have the limits $L_4^2 \to 1$, $L_5^2 \to 0$, $\Lambda_4^2 \to 0$, and $\Lambda_5^2 \to 1/2$, as $\xi \to 0$. Those limits also lead to that $p \to 0$ as $\xi \to 0$ in the considered regime of bias voltages. The last term in the parentheses in the denominator of (10.22) is, hence, at most 1/2 for weakly coupled quantum dots. The ratio $2p(\kappa/\beta)^2$ is, however, finite for all ξ and $J > 0$, while it diverges as $\xi \to 0$ for $J = 0$, see the main panel in Fig. 10.5. For weakly coupled quantum dots one, thus, finds that $N_T \approx 1/(1 + [1 + 2p(\kappa/\beta)^2]^{-1}) \approx 1$, whenever the ratio $2p(\kappa/\beta)^2 \gg 1$. The inset in Fig. 10.5 illustrates a subset in (t, J)-space where this ratio is larger/smaller than 100, where the boundary between the regions is approximately given by $J(t) = J_0 - 15t^2[1 + 100t^2]$.

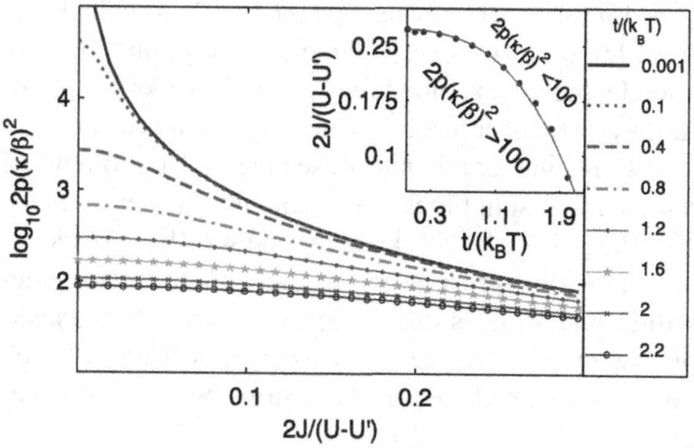

Fig. 10.5 Variation of the ratio $2p(\kappa/\beta)^2$ as function of the exchange parameter J for different tunneling rates t at constant $\Delta\varepsilon$, U', and $U_{A/B}$. The inset shows the region in (t, J)-space where $2p(\kappa/\beta)^2 > 10$

The current in the considered regime is given by

$$I = \frac{e\Gamma_0}{6\hbar}\Bigg[3(\beta^2-\kappa^2) + \sum_{n=4,5}\Bigg\{\big(L_n^2 - \Lambda_n^2 f_L(-\Delta_{31,2n})\big) + 2\Lambda_n^2 f_L(\Delta_{31,2n})\frac{L_n^2 + p\Lambda_n^2\sum_\chi f_\chi(-\Delta_{31,2n})}{L_n^2+\Lambda_n^2 f_L(\Delta_{31,2n})}\Bigg\}\Bigg]\frac{N_T}{1+2p(\kappa/\beta)^2}, \qquad (10.23)$$

which clearly shows that a large value of the ratio $2p(\kappa/\beta)^2$ generates a suppression of the current at the formation of the unit population probability of the spin triplet. Furthermore, at bias voltages such that $\mu_L < \min_{nm}\{\Delta_{3m,2n}\}$ it follows that $f_L(\Delta_{3m,2n}) \approx 0$ which, in turn, leads to that $p \approx 0$. Under those conditions, the triplet occupation is lifted resulting in an about $2p(\kappa/\beta)^2$ larger current than in the blockade regime.

In the T-shaped system, the Pauli spin blockade depends on the interplay between J and t, where a reduced t leads to a strong localization of the odd number states in either of the quantum dots. Especially for $\Delta\varepsilon > 0$, this leads to that the lowest odd number states are strongly localized in quantum dot A. In this case, the probability for transitions between the spin triplet and one-/three-electron states is small/large, since $\beta \to 0/\kappa \to 1$, as $\xi \to 0$.

As we have seen before, are the singlet states expanded in terms of the Fock states $\{|\Phi^{AB}\rangle, |\Phi^A\rangle, |\Phi^B\rangle\}$, see Table 9.1, with weights that are slowly varying functions of t but strongly dependent on J. While the two lowest spin singlet states, $|2,4\rangle$ and $|2,5\rangle$, are almost equally weighted on the Fock states $|\Phi^{AB}\rangle$ and $|\Phi^A\rangle$ at negligible J, for increasing $J > 0$ their weights are redistributed such that $|2,4\rangle$ and $|2,5\rangle$ acquires an increasing weight on $|\Phi^A\rangle$ and $|\Phi^{AB}\rangle$, respectively. For a finite $J > 0$ and $t \to 0$ this redistribution, thus, leads to an enhanced (reduced) probability for the transitions $|2,4\rangle\langle 1,n|$ ($|2,5\rangle\langle 1,n|$), $n = 1,2$ and a reduced (enhanced) probability for the transitions $|2,4\rangle\langle 3,n|$ ($|2,5\rangle\langle 3,n|$), $n = 1,2$ to occur. This is equivalently stated through the limits $L_4^2 \to 1$ ($L_5^2 \to 0$), and $\Lambda_4^2 \to 0$ ($\Lambda_5^2 \to 1/2$), which implies that $p \to 0$ as $t \to 0$, while the ratio $p(\kappa/\beta)^2$ remains almost constant. This ratio $p(\kappa/\beta)^2$, however, increases (decreases) for smaller (larger) J, which can be seen in Fig. 10.5.

The typical variation of the spin triplet population number N_T for $0 < J < J_0 - 15t^2[1 + 100t^2]$ and $t/(k_BT) < 2$, as calculated from (10.22), is plotted in Fig. 10.6(a), as function of the bias voltage and the chemical potential. The dark diamond which is extended by two *arms* marks the region in which the spin triplet is (almost) fully occupied, in which region the current flowing between the leads in almost vanishing, or blockaded, see panel (b). From the plot of the current it is legible that the Pauli spin blockade regime is a subset of a larger domain of current blockade. The part of the blockade regime which is not caused by the Pauli spin blockade i.e. the region corresponding to the white triangle centered around zero bias voltage with in panel (a), is due to normal Coulomb blockade. The two diamonds within the low current regime are caused by a lifting of the Pauli spin blockade, where the current is mediated by transitions between the one-electron states and the spin singlets.

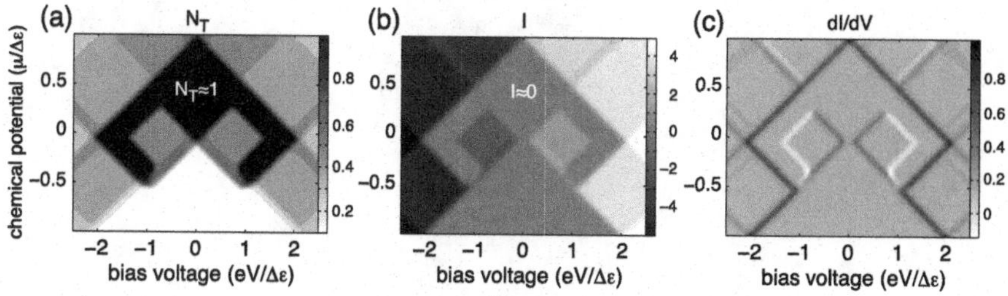

Fig. 10.6 (**a**) Variation of the spin triplet population number probability N_T, (**b**) the current I (arb. units), and (**c**) the differential conductance dI/dV (arb. units), as function of the bias voltage and the chemical potential. Here, $\xi = 0.01$, $k_B T = 0.01 U_A = 4t$, and $J = 0.2(U_A - U')/2$

It is clear from the plots that shifting the chemical potential μ in the range $\varepsilon_B + (\Delta\varepsilon - J, 2\Delta\varepsilon)$ extends the low bias triplet regime since the transitions between the spin triplet and the one-electron states become resonant at higher bias voltages. The Pauli spin blockade is shifted, on the other hand, to lower bias voltages since the chemical potential is closer to the transitions between the spin singlets and the three-electron states. The two blockade regimes merge into a single one as $|\mu - \Delta_{3m,2n}| < |\mu - \Delta_{T,1m}|$, i.e. for $\mu - \varepsilon_B \in (3\Delta\varepsilon/2, 2\Delta\varepsilon)$, see Fig. 10.6. Furthermore, shifting the chemical potential in the interval $\varepsilon_B + (\Delta\varepsilon/2, \Delta\varepsilon - J)$ removes the low bias Pauli spin blockade since the one-electron states become the equilibrium ground state. Thus, shifted towards lower bias voltages, the non-equilibrium Pauli spin blockade is here caused by transitions between the one- and two-electron states which tend to accumulate charge in the spin triplet.

Finally, we notice that while we here have only considered the case $\Delta\varepsilon > 0$, the non-equilibrium Pauli spin blockade can also be found in the opposite case, i.e. $\Delta\varepsilon < 0$ and $\mu - \varepsilon_A \approx \Delta\varepsilon$. The system has to be gated such that only the four-electron state lies above the equilibrium chemical potential, whereas the charge accumulation in the spin triplet state is meditated by the same processes as described here.

References

1. Ono, K., Austing, D.G., Tokura, Y., Taurcha, S.: Science **297**, 1313–1317 (2002)
2. Fransson, J., Råsander, M.: Phys. Rev. B **73**, 205333 (2006)
3. Fransson, J.: Nanotechnology **17**, 5344 (2006)
4. Fransson, J.: New J. Phys. **8**, 114 (2006)
5. Johnson, A.C., Petta, J.R., Taylor, J.M., Yacoby, A., Lukin, M.D., Marcus, C.M., Hanson, M.P., Gossard, A.C.: Nature **435**, 925 (2005)
6. Koppens, F.H.L., Folk, J.A., Elzerman, J.M., Hanson, R., van Beveren, L.H.W., Vink, I.T., Tranitz, H.O., Wegscheider, W., Kouwenhoven, L.P., Vandersypen, L.M.K.: Science **309**, 1346 (2005)
7. Liu, H.W., Fujisawa, T., Hayashi, T., Hirayama, Y.: Phys. Rev. B **72**, 161305(R) (2005)
8. Buitelaar, M.R., Fransson, J., Cantone, A.L., Smith, C.G., Anderson, D., Jones, G.A.C., Ardavan, A., Khlobystov, A.N., Watt, A.A.R., Porfyrakis, K., Briggs, G.A.D.: Phys. Rev. B **77**, 245439 (2008)
9. Hanson, R., Kouwenhoven, L.P., Petta, J.R., Taraucha, S., Vandersypen, L.M.K.: Rev. Mod. Phys. **79**, 001217 (2007)

Chapter 11
Detection of Exchange Interaction Through Fano-Like Interference Effects

Abstract We study Fano-like interference effects in connection with scanning tunneling microscopy experimental set-up of e.g. two-level systems. The physics of the systems suggests a mechanism that allows detection of the two-electron singlet-triplet exchange splitting in diatomic molecular systems. Different pathways for the tunneling electrons lead to interference effects and generate kinks in the differential conductance at the energies for the spin singlet and triplet states.

11.1 Introduction

Fano resonances can be realized in a variety of system, ranging from systems with interactions between continuum states and a localized state, as originally formulated by Fano [1], to systems comprising clusters of atoms, e.g. diatomic molecules, which leads to branching of the tunneling electron wave functions. Here reformulated for the purpose of STM and nanoscale systems, interference occur between waves going through the different tunneling paths in real space, where one path leads through the sample to the substrate whereas the other path goes directly into the substrate.

Typically, in STM measurements of a sample located on a substrate surface, it is desired that the tunneling current flows through the sample. While this, of course, occurs as the STM tip is sufficiently close to the sample, there is still a part of the current that flows directly between the STM tip. The tunneling electron wave functions are, thus, branched between different pathways. This branching of the wave functions gives rise to interference effects when the partial waves reassemble into one in the tip or the substrate [2]. The interference leads to a suppressed transmission probability for the tunneling electrons at certain energies, and the suppressed transmission is a fingerprint of Fano resonances [1], which generally appear in systems where tunneling electrons are branched between different pathways. Recently, these ideas have been exploited in double and triple quantum dots systems [3–7], where the quantum dots in parallel constitute the different pathways for the tunneling electrons.

J. Fransson, *Non-Equilibrium Nano-Physics*,
Lecture Notes in Physics 809,
DOI 10.1007/978-90-481-9210-6_11,

Here, we address the Fano interference effects that arise due to different pathways in phase space. In two-level systems, tunneling paths such as $|N=2,n\rangle \to |N=1,m\rangle \to |N=2,n'\rangle \to |N=1,m'\rangle \to |N=2,n\rangle$ give rise to phase space interference. Here, $N=1,2$, denote the number of electrons in the state, whereas n, m are states indices [8, 9]. Although such tunneling paths are of second order, they provide significant contributions to the transmission coefficient, and hence to the conductance, thereby causing detectable signatures. The interference effects can be described in mean-field approximation of the sample correlation functions, hence, we do not discuss any fluctuations caused by electronic correlations or by the couplings to the tip and substrate.

11.2 Tunneling Current

Before we approach the two-level system itself, we begin by setting up a framework which can be used to understand the STM current and (differential) conductance. Generally, the STM system, that we have in mind, can be described by the Hamiltonian

$$\mathcal{H} = \mathcal{H}_{\text{tip}} + \mathcal{H}_{\text{sub}} + \mathcal{H}_{\text{sample}} + \mathcal{H}_T, \tag{11.1}$$

where the first and second terms describe the electronic structure in the tip and substrate, respectively. Here, we assume flat band free-electron like models for the states in the tip and substrate, and define $\mathcal{H}_{\text{tip}} = \sum_{\mathbf{p}\sigma\in\text{tip}} \varepsilon_{\mathbf{p}\sigma} c^\dagger_{\mathbf{p}\sigma} c_{\mathbf{p}\sigma}$ and $\mathcal{H}_{\text{sub}} = \sum_{\mathbf{k}\sigma} \varepsilon_{\mathbf{k}\sigma} c^\dagger_{\mathbf{k}\sigma} c_{\mathbf{k}\sigma}$, for the tip and substrate, respectively, and we let the momentum $\mathbf{p}$ ($\mathbf{k}$) belong to the tip (substrate). The third term describes the electronic structure of the sample, which is to be studied in the STM experiment. The last term includes the tunneling interaction between the sample and the tip and substrate, and can be written as (using obvious notation)

$$\mathcal{H}_T = \sum_{\mathbf{p}n\sigma} v_{\mathbf{p}n\sigma} c^\dagger_{\mathbf{p}\sigma} d_{n\sigma} + \sum_{\mathbf{k}n\sigma} v_{\mathbf{k}n\sigma} c^\dagger_{\mathbf{k}\sigma} d_{n\sigma} + H.c. \tag{11.2}$$

Using the expression for the current given by (4.4), we can here formulate it as

$$I = \frac{e}{h} \int T(\omega)[f_{\text{tip}}(\omega) - f_{\text{sub}}(\omega)]d\omega, \tag{11.3}$$

where the transmission coefficient

$$T(\omega) = \operatorname{tr} \Gamma^{\text{tip}} \mathbf{G}^r(\omega) \Gamma^{\text{sub}} \mathbf{G}^a(\omega), \tag{11.4}$$

whereas $\Gamma^{\text{tip}} = \{\Gamma^{\text{tip}}_{nm\sigma}(\omega)\}_{nm} = \{2\pi \sum_{\mathbf{p}} v^*_{\mathbf{p}n\sigma} v_{\mathbf{p}m\sigma} \delta(\omega - \varepsilon_{\mathbf{p}\sigma})\}_{nm}$ defines the coupling matrix between the tip and sample, while the matrix for the couplings between the sample and substrate is analogously defined. For later reference, we also introduce the chemical potential $\mu_{\text{tip}} = \mu + eV$ ($\mu_{\text{sub}} = \mu$) for the tip (substrate) such that $eV = \mu_{\rm tip} - \mu_{\rm sub}$. The tunneling through the sample is, thus, conveniently described within the retarded and advanced Green functions for the electronic structure of the sample.

As it can be noticed, we neglect the possibility for the electrons to tunnel directly between the tip and substrate. This choice is made for simplicity, since we first of all want to see whether there is any effect arising due to the various pathways through the different atoms in the molecule. When this has been established, one can proceed and also include the effects from the tunneling between the tip and substrate. This is, however, beyond the scope of the present analysis.

11.3 Probing the Two-Level System

We begin by considering a diatomic molecule comprising two identical atoms, which allows one to think of the atoms in terms of quantum dots and the molecular structure as a double quantum dot. Further, we assume that the atoms are coupled through Coulomb and exchange interactions with their respective strengths given in terms of the parameters U and J. The system in schematically illustrated in Fig. 11.1, displaying the molecular structure, comprising the two atoms, adsorbed onto the substrate, and the STM tip above the molecule. The whole system is connected to a bias voltage source.

For simplicity, we assume infinite intralevel Coulomb interactions in order to avoid the possibility of ending up with two electrons in one of the atoms. We also assume that the tunneling between the atoms is negligible. While this set of assumptions is not crucial for the effect that we discuss here, it merely permits a convenient framework for heuristic and qualitative studies of the approach. The approach can be straightforwardly generalized, something that we omit to do, however, in the present discussion for the benefit of focusing on the more interesting details.

We model the molecule, or sample, by the Hamiltonian

$$\mathcal{H}_{\rm sample} = \sum_{n=1,2;\sigma} \varepsilon_0 d^\dagger_{n\sigma} d_{n\sigma} - 2J\mathbf{S}_1 \cdot \mathbf{S}_2 + \left(U - \frac{J}{2}\right)(n_{1\uparrow} + n_{1\downarrow})(n_{2\uparrow} + n_{2\downarrow}), \tag{11.5}$$

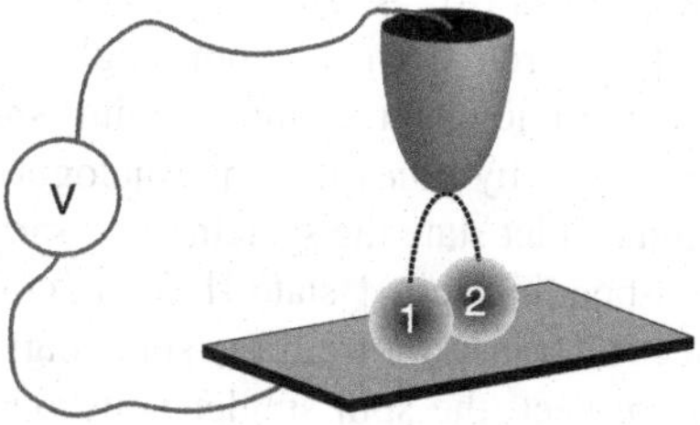

Fig. 11.1 Cartoon of the diatomic "1"+"2" molecule coupled to the STM tip and substrate. The localized electrons in the atoms are subject to an interatomic Coulomb and exchange interactions, U and J, respectively

requiring that there may be at most one electron per level n. Here also, the spin operator $\mathbf{S}_n$ is defined by $\mathbf{S}_n = \sum_{\sigma\sigma'} d^\dagger_{n\sigma}\boldsymbol{\sigma}_{\sigma\sigma'}d_{n\sigma'}$. We may also use the model given in (1.5) for a more general description of the molecular structure.

As we have discussed previously in this book, the two-level system may, in general, be described in terms of 16 eigenstates. By forbidding the double occupancy in the atoms, the system is reduced to a set of 11 states. For completeness we provide them here within the simplified model. Expressing the eigenstates in terms of the empty state $|0\rangle$, the one-electron states

$$|1,1(2)\rangle = \frac{d^\dagger_{1\uparrow(\downarrow)} - d^\dagger_{2\uparrow(\downarrow)}}{\sqrt{2}}|0\rangle, \qquad |1,3(4)\rangle = \frac{d^\dagger_{1\uparrow(\downarrow)} + d^\dagger_{2\uparrow(\downarrow)}}{\sqrt{2}}|0\rangle, \tag{11.6}$$

describe the bonding and anti-bonding configurations, respectively, with the corresponding eigenenergies $E_{1n} = \varepsilon_0$. In the same way we write the molecular two-electron states as the spin triplet configurations

$$|2,1\rangle = d^\dagger_{2\uparrow}d^\dagger_{1\uparrow}|0\rangle, \qquad |2,2\rangle = d^\dagger_{2\downarrow}d^\dagger_{1\downarrow}|0\rangle, \qquad |2,3\rangle = \frac{d^\dagger_{2\downarrow}d^\dagger_{1\uparrow} + d^\dagger_{2\uparrow}d^\dagger_{1\downarrow}}{\sqrt{2}}|0\rangle, \tag{11.7}$$

with energies $E_{2n} = 2\varepsilon_0 + U + J/2$, $n = 1, 2, 3$, or as the spin singlet state

$$|2,4\rangle = \frac{d^\dagger_{2\downarrow}d^\dagger_{1\uparrow} - d^\dagger_{2\uparrow}d^\dagger_{1\downarrow}}{\sqrt{2}}|0\rangle, \tag{11.8}$$

with energy $E_{24} = 2\varepsilon_0 + U - J/2$.

The effect that we point out using this model, and which may be used for measurements of the singlet-triplet splitting J, is caused by phase space interference between wave functions traversing through the singlet and triplet states. This interference results in that some states become more or less isolated from the surrounding environment which, in turn, generate conductance suppressions at biases that correspond to the spin singlet and triplet state energies. These conductance suppressions are direct responses to that there exist states that are only very weakly coupled to the surrounding electron bath(s), states in which electron density can accumulate and thereby cause the decreased conductance.

Heuristically, we can understand the decoupling of some certain states from the electron baths by the following consideration. Assume that the spin singlet state $|2,4\rangle$ is initially occupied. A spin $\uparrow$ electron can be removed from either atom through the process $(d_{1\uparrow} + d_{2\uparrow})|2,4\rangle = -(d^\dagger_{2\downarrow} + d^\dagger_{1\downarrow})|0\rangle/\sqrt{2} = -|1,4\rangle$. Removing a spin $\uparrow$ electron from the spin singlet state is, thus, always governed by a transition to the one-electron anti-bonding spin $\downarrow$ state. This state is orthogonal to all other states, by construction. Analogously, by removing a spin $\downarrow$ electron from the spin singlet state the system necessarily undergoes a transition to the one-electron anti-bonding spin $\uparrow$ state $|1,3\rangle$. In conclusion, the system cannot undergo a single-electron transition from the spin singlet to the one-electron bonding states and, in this respect, the spin singlet is decoupled from the one-electron bonding states. An analogous analysis shows that while the spin triplet states are decoupled from the one-electron anti-bonding states, they couple to the one-electron bonding states, since e.g. $(d_{1\uparrow}+d_{2\uparrow})|2,1\rangle=(-d^\dagger_{2\downarrow}+d^\dagger_{1\downarrow})|0\rangle=\sqrt{2}|1,2\rangle$.

We proceed by making use that our sample is given in terms of its eigenstates, for which it is beneficial to write the sample Hamiltonian in diagonal form, i.e. $\mathcal{H}_{\text{sample}}=\sum_{Nn}E_{Nn}h^n_N$. The tunneling Hamiltonian $\mathcal{H}_T$ is then rewritten according to

$$\mathcal{H}_T=\sum_{Nnm}\left(\sum_{\mathbf{p}\sigma}v_{\mathbf{p}\sigma Nnm}c^\dagger_{\mathbf{p}\sigma}+\sum_{\mathbf{k}\sigma}v_{\mathbf{k}\sigma Nnm}c^\dagger_{\mathbf{k}\sigma}\right)X^{nm}_{NN+1}+H.c., \tag{11.9}$$

where the tunneling rates

$$v_{\mathbf{p}(\mathbf{k})\sigma Nnm}=\langle N,n|[v_{\mathbf{p}(\mathbf{k})1\sigma}d_{1\sigma}+v_{\mathbf{p}(\mathbf{k})2\sigma}d_{2\sigma}]|N+1,m\rangle, \tag{11.10}$$

also include the matrix elements for single electron transitions in the sample. This form of the tunneling rates reflects the branching of the tunneling electron wave functions in real space, as the electrons tunnel between the tip, sample, and substrate.

Now, suppose that the molecule is prepared in the spin singlet state. The tunneling current between the tip and substrate is then mediated by the sequence e.g. $|2,4\rangle\rightarrow|1,3\rangle\rightarrow|2,n\rangle$, where $n=1,2,3,4$, is random. The second arrow in the sequence may be directed to the spin singlet, of course, but also to the spin triplet since e.g. $d^\dagger_{1\downarrow}|1,3\rangle=-[|2,3\rangle-|2,4\rangle]/2$ and $d^\dagger_{2\downarrow}|1,3\rangle=[|2,3\rangle+|2,4\rangle]/2$. Simply adding those processes, which corresponds to equal coupling strengths between the molecular atoms and tip, result in that the transitions to the spin triplet, $|2,3\rangle$, cancel. Taking into account that these couplings, in general, are different one can realize that there will be a non-vanishing contribution from the transitions to the spin triplet, as will be discussed in more technical terms below. Quantitatively, we have

$$|\langle 1,3|[v_{\mathbf{k}1\downarrow}d_{1\downarrow}+v_{\mathbf{k}2\downarrow}d_{2\downarrow}|2,4\rangle|=\frac{1}{2}|v_{\mathbf{k}1\downarrow}+v_{\mathbf{k}2\downarrow}|, \tag{11.11a}$$

$$|\langle 1,3|[v_{\mathbf{k}1\downarrow}d_{1\downarrow}+v_{\mathbf{k}2\downarrow}d_{2\downarrow}|2,3\rangle|=\frac{1}{2}|v_{\mathbf{k}1\downarrow}-v_{\mathbf{k}2\downarrow}|. \tag{11.11b}$$

Hence, depending on the relation between the tunneling rates $v_{\mathbf{k}1\sigma}$ and $v_{\mathbf{k}2\sigma}$ there will be a larger or smaller degree of cancellation of the coupling between anti-bonding one-electron states and the spin triplet. Following the argument made in the heuristic discussion, the spin triplet state $|2,3\rangle$ becomes completely decoupled from the anti-bonding states $|1,3\rangle$ when the tunneling rates $v_{\mathbf{k}1\sigma}=v_{\mathbf{k}2\sigma}$. In this case, the spin singlet has a unit coupling to the anti-bonding states.

The full set of possible transitions is given by diagram

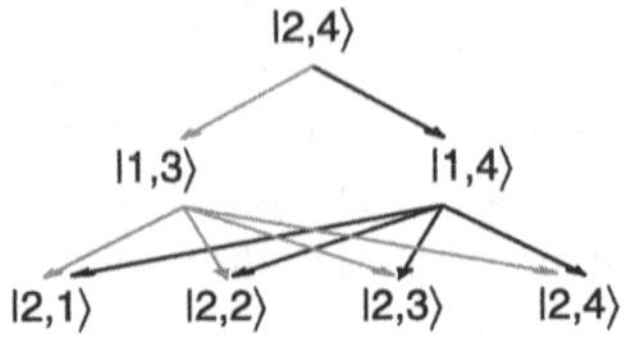

which illustrates that there is a coupling of the spin singlet and triplet states through the tunneling process, that is, there occur indirect transitions between the spin singlet and triplet states. These indirect singlet-triplet transitions generate interference effects between the tunneling electron waves traversing the different transport channels which, moreover, implies that multiple-scattering events have to be included into the picture.

Before we go in to the specific set-ups, we introduce the framework for our succeeding discussions. Since we are only interested in probing the two-electron states, the spin singlet and spin triplets, we restrict ourselves to the transitions between only the one- and two-electron states.[1] The sample Green function, thus, becomes a 20×20 matrix which, in general, is not diagonal due to higher order transitions that may contribute significantly to the electronic structure of the molecular system. We simplify our system, however, by assuming that only diagonal processes $|2,n\rangle \to |1,m\rangle \to |2,n\rangle$ and off-diagonal processes like $|2,n\rangle \to |1,m\rangle \to |2,n'\rangle \to |1,m\rangle \to |2,n\rangle$ contribute to the tunneling. The introduced simplification reduces the system to a set of 2×2 matrix equation, each of which may be analytically solved.

We introduce the Green functions

$$G_{mnm'}(t,t') = (-i)\langle \mathrm{T} X_{12}^{nm}(t) X_{21}^{m'n}(t')\rangle. \tag{11.12}$$

In the simplified system, we obtain the equations of motion

$$\begin{aligned}(i\partial_t - \Delta_{mn})&G_{mnm'}(t,t')\\ &= \delta(t-t')P_{mnm'}(t) + \sum_{\mu\mu'}\int_C P_{mn\mu}(t)V_{n\mu\mu' n}(t,t'')G_{\mu' nm'}(t'',t')dt'',\end{aligned} \tag{11.13}$$

in the Hubbard-I-approximation. Here, the interaction

$$V_{n\mu\mu' n}(t,t') = \sum_{\mathbf{p}\sigma} v^*_{\mathbf{p}\sigma 2n\mu} v_{\mathbf{p}\sigma 2n\mu'} g_{\mathbf{p}\sigma}(t,t') + \sum_{\mathbf{k}\sigma} v^*_{\mathbf{k}\sigma 2n\mu} v_{\mathbf{k}\sigma 2n\mu'} g_{\mathbf{k}\sigma}(t,t'), \tag{11.14}$$

where $g_{\mathbf{p}(\mathbf{k})\sigma}(t,t')$ is the (free-electron-like) Green function for the electrons in the tip (substrate).

---

[1] Recall that the three- and four-electron states are discarded already in the formulation of the Hamiltonian, since the intra-dot Coulomb interaction strength is infinite.

As we know from previous chapters, the equations for the Green functions are to be self-consistently solved, with respect to the occupation numbers N_{Nn}, for each point in the vast parameters space of the Hamiltonian, e.g. bias voltage, temperature etc. Through such calculations we would obtain a valid non-equilibrium description of the Fano-like interference effects. In order to focus on the physical mechanism rather than on the quantitative variations, we omit such a treatment here. The qualitative features of the system remain the same and, therefore, we discuss the physics without the self-consistency condition by further simplifying the equations of motion. We assume, without loss of phase space interference effects, that the end-factors $P_{mnm'} = \delta_{mm'}$, i.e. we assume that the off-diagonal occupation numbers are negligible. In absence of spin-flip transitions in the system, the equations for transitions between the one-electron spin $\uparrow$ and $\downarrow$ states are equal. Thus, we omit any reference to the spin degree of freedom.

We note that transitions between different spin triplet configurations do not generate the interference effects we discuss here and it is, therefore, reasonable to consider only transitions between the spin-singlet and one of spin-triplet states, say $|2,3\rangle$. The other processes, that is, the couplings of the spin-singlet to the spin triplet states $|2,1\rangle$ and $|2,2\rangle$, merely renormalizes the coefficients in the final expression for the transmission.

Calculation of the Fourier transformed retarded Green function for transitions between the states $|2,3\rangle$ and $|2,4\rangle$ through the one-electron state $|1,n\rangle$ results in

$$G^r_{3n4}(\omega) = \frac{1}{(\omega-\omega_{n+})(\omega-\omega_{n-})}\begin{pmatrix} \omega - \Delta_{4n} + \frac{i}{2}\Gamma_{4n4} & -\frac{i}{2}\Gamma_{3n4} \\ -\frac{i}{2}\Gamma_{4n3} & \omega - \Delta_{3n} + \frac{i}{2}\Gamma_{3n3} \end{pmatrix}, \tag{11.15}$$

where $\Gamma_{mnm'}(\omega) = -2\,\mathrm{Im}\,V^r_{nmm'n}(\omega)$ defines the total coupling between the sample and the tip and substrate. We define the coupling between tip and sample by $\Gamma^{\mathrm{tip}}_{mnm'}(\omega) = 2\pi\sum_{\mathbf{p}\sigma} v^*_{\mathbf{p}\sigma 2nm} v_{\mathbf{p}\sigma 2nm'}\delta(\omega - \varepsilon_{\mathbf{p}\sigma})$ and analogously for the coupling between the substrate and sample, such that $\Gamma_{mnm'} = \Gamma^{\mathrm{tip}}_{mnm'} + \Gamma^{\mathrm{sub}}_{mnm'}$. Finally, we define the denominator $C_n(\omega) = (\omega-\omega_{n+})(\omega-\omega_{n-})$, where the poles $\omega_\pm$ are given by

$$\omega_{n\pm} = \frac{1}{2}\Bigg\{\Delta_{3n} + \Delta_{4n} - \frac{i}{2}[\Gamma_{3n3} + \Gamma_{4n4}] \pm\sqrt{\left(\Delta_{3n} - \Delta_{4n} - \frac{i}{2}[\Gamma_{3n3} - \Gamma_{4n4}]\right)^2 - 4\Gamma_{3n4}\Gamma_{4n3}}\Bigg\}. \tag{11.16}$$

11.3.1 Symmetric Coupling to the Substrate

Assume that the atoms in the diatomic molecule couple equally strong to the substrate. Quantitatively, this means that $v_{\mathbf{k}\sigma 1} = v_{\mathbf{k}\sigma 2} = v_{\mathbf{k}\sigma}$, which results in the coupling matrices

$$\Gamma^{\text{sub}}_{|1,1\rangle} = \Gamma^{\text{sub}}_0 \begin{pmatrix} 1 & 0 \\ 0 & 0 \end{pmatrix}, \qquad \Gamma^{\text{sub}}_{|1,3\rangle} = \Gamma^{\text{sub}}_0 \begin{pmatrix} 0 & 0 \\ 0 & 1 \end{pmatrix}, \tag{11.17}$$

where $\Gamma_0^{\text{sub}} = 2\pi \sum_{\mathbf{k}\sigma} |v_{\mathbf{k}\sigma}|^2 \delta(\omega - \varepsilon_{\mathbf{k}\sigma})$. Further, we assume that the tunneling rate between the tip and sample can be parametrized by $v_{\mathbf{p}\sigma n} = \gamma_n v_{\mathbf{p}\sigma}$, where $\gamma_n \in [0, 1]$. Then, the coupling matrices for the transmission between the tip and sample can be written

$$\Gamma^{\text{tip}}_{|1,1\rangle} = \frac{\Gamma^{\text{tip}}_0}{4} \begin{pmatrix} (\gamma_1+\gamma_2)^2 & -\gamma_1^2+\gamma_2^2 \\ -\gamma_1^2+\gamma_2^2 & (\gamma_1-\gamma_2)^2 \end{pmatrix}, \tag{11.18a}$$

$$\Gamma^{\text{tip}}_{|1,3\rangle} = \frac{\Gamma^{\text{tip}}_0}{4} \begin{pmatrix} (\gamma_1-\gamma_2)^2 & -\gamma_1^2+\gamma_2^2 \\ -\gamma_1^2+\gamma_2^2 & (\gamma_1+\gamma_2)^2 \end{pmatrix}. \tag{11.18b}$$

We, finally, assume that $\Gamma_0^{\text{tip}} = \lambda \Gamma_0/2$ and $\Gamma_0^{\text{sub}} = \Gamma_0/2$, where $\lambda \ll 1$, such that the broadening of the localized states $\Gamma_{|2,m\rangle} \simeq \Gamma^{\text{sub}}_{|2,m\rangle}$. The final assumption is performed in order to simplify the analytical treatment. In particular for the poles of the Green functions, this assumption leads to

$$\omega_{1\pm} = \varepsilon_0 + U \pm \frac{J}{2} - \frac{i}{8}\Gamma_0(1 \pm 1), \tag{11.19a}$$

$$\omega_{3\pm} = \varepsilon_0 + U \pm \frac{J}{2} - \frac{i}{8}\Gamma_0(1 \mp 1), \tag{11.19b}$$

for the transitions through the bonding and anti-bonding one-electron states, respectively, since $\Delta_{m1} = \Delta_{m3}$, $m = 1, \ldots, 4$, and since the transition energies $\Delta_{31} = E_{23} - E_{11} = 2\varepsilon_0 + U + J/2 - \varepsilon_0 = \varepsilon_0 + U + J/2$ and $\Delta_{41} = E_{24} - E_{11} = 2\varepsilon_0 + U - J/2 - \varepsilon_0 = \varepsilon_0 + U - J/2$. We see that the poles of the Green function are located at the singlet and triplet energies. We also see that the poles ω_{1-} and ω_{3+} acquire vanishing widths, which correspond to the energies where the transitions occur with zero probability. Inclusion of the widths that are due to the tunneling between the tip and the sample, these expressions show that there are states which couple very weakly to the surrounding delocalized electrons. We, thus, expect both sharp and broad peaks, centered around more or less at the same energy, in the spectrum for the density of electrons states in the molecular system, and that those peaks are associated with the singlet and triplet states.

The expected sharp peaks in the local density of electron states in the sample is a direct result of the interference between the waves, of the tunneling electrons, that are branched in phase space. This destructive interference manifests itself in that the widths of the states associated with the energies ω_{1-} and ω_{3+} become small, which is the same as to say that electrons occupying those states have very little interactions with the surrounding delocalized electrons in the baths. The states associated with the energies ω_{1+} and ω_{3-} have large widths, something which is provided by constructive interference between the branched tunneling electron waves.

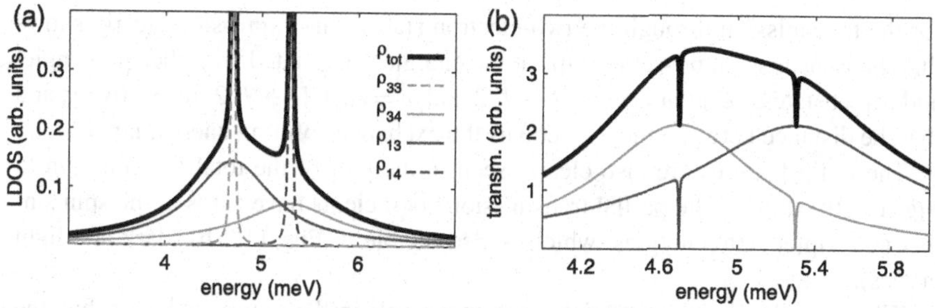

Fig. 11.2 (**a**) Typical density of electron states (*bold*) around the spin singlet and triplet states in the sample. The density of electron states associated with the pose $\omega_{1\pm}$ (*dark*) and $\omega_{3\pm}$ (*light*) are plotted separately. (**b**) Transmission coefficient for the system, with the total (*bold*) and partial transmissions through the channels which couple the two-electron states to the bonding (*dark*) and anti-bonding (*light*) one-electron states. Here, we have used $\varepsilon_0 = 2$, $U = 3$, $J = 0.6$, $\Gamma_0 = 2$ (units: meV), $\lambda = 0.005$, $\gamma_1 = 0$ and $\gamma_2 = 1$

We calculate the total local density of electron states ρ_{tot} through

$$\rho_{\text{tot}}(\omega) = \sum_{nm} \rho_{nm}(\omega) = -\frac{1}{\pi}\sum_{nm} \text{Im}\, G^r_{mnm}(\omega), \tag{11.20}$$

see Fig. 11.2(a), where we have plotted ρ_{tot} (bold) along with its components ρ_{nm}. It is readily seen from the partial densities of states that there are two sharp peaks and two wide ones, which in the total density of states is expressed as a single broad peak with two sharp features on top of it. The example illustrated in the plots, is reasonable from the point of view that the single-triplet splitting may be smaller than the broadening due to the coupling between the sample and the substrate and tip. In the calculations of the poles, a finite contributions from coupling to the electrons in the tip have been included in addition to the one generated by the coupling to the substrate. Hence, without the additional features due to the interference, it would be very hard to realize whether the broad peak arises due to one or two, or more, states. The interference, thus, enables a possibility to actually detect the location of the spin singlet and triplet states, or, at least, the energy splitting between those states.

The narrow, or sharply peaked, states in the sample give rise to dips in the conductance of the system. This can be directly seen in the transmission coefficient $\mathcal{T}(\omega)$. For simplicity, we use the above assumption that the widths of the states in the sample can be written $\Gamma_{m1m'} = \Gamma^{\text{tip}}_{m1m'} + \delta_{mm'}\delta_{m3}$ and $\Gamma_{m3m'} = \Gamma^{\text{tip}}_{m3m'} + \delta_{mm'}\delta_{m4}$, we find

$$\begin{aligned}\mathcal{T}(\omega) &= \Gamma^{\text{tip}}_{|1,1\rangle}\mathbf{G}^r_{|1,1\rangle}\Gamma^{\text{sub}}_{|1,1\rangle}\mathbf{G}^a_{|1,1\rangle} + \Gamma^{\text{tip}}_{|1,3\rangle}\mathbf{G}^r_{|1,3\rangle}\Gamma^{\text{sub}}_{|1,3\rangle}\mathbf{G}^a_{|1,3\rangle} \\ &= \lambda\left(\frac{\Gamma_0}{4}\right)^2(\gamma_1+\gamma_2)^2\left(\left|\frac{\omega-\varepsilon_0-U+J/2}{C_1(\omega)}\right|^2 + \left|\frac{\omega-\varepsilon_0-U-J/2}{C_3(\omega)}\right|^2\right),\end{aligned} \tag{11.21}$$

for the transmission through the two-electron states. The expression clearly shows that the transmission possesses dips at the energies associated with the spin singlet and triplet states, i.e. at $\omega = \varepsilon_0 + U - J/2$ and $\omega = \varepsilon_0 + U + J/2$, respectively, and that the distance between the dips equals the exchange splitting energy, i.e. J.

These dip features are also clearly seen in the plot of the total transmission in Fig. 11.2(b) (bold). The partial transmission coefficients have zeros at the spin singlet and triplet state energies, which is also legible in Fig. 11.2(b) (dark and light curves).

Whenever the broadening of the quantum levels in the sample is larger than the singlet-triplet splitting J there will not be two distinct peaks associated with the singlet and triplet state in the transmission, nor in the differential conductance. Visibility of the individual dips, moreover, requires low temperatures since the transmission coefficient is convolved with the thermal distribution function. Therefore it would be preferable, under those circumstances, to resolve the spectrum through measurements of the second current derivative with respect to the bias voltage, i.e. d^2I/dV^2, rather than the differential conductance. Such measurements would provide further information about the long-lived states in the sample in terms of the very sharp features at the energies corresponding to those states.

The plots in Fig. 11.3 show the differential conductance (a) and d^2I/dV^2 (b) corresponding to the density of electron states and transmission in Fig. 11.2. As we already have mentioned, the broad electron densities gives rise to wide conductance peaks which significantly overlap such that the individual peaks cannot be distinguished. On the other hand, the dips occurring in the conductance plot, originating from the narrow peaks in the density of electron states, can be used to pinpoint the positions of the spin singlet and triplet states. However, the large level broadening makes an unambiguous read-out of the singlet and triplet states difficult. The plot of the d^2I/dV^2 does resolve this issue, since the dips in the conductance display themselves as large and narrow variations from the, almost, vanishing underlying d^2I/dV^2.

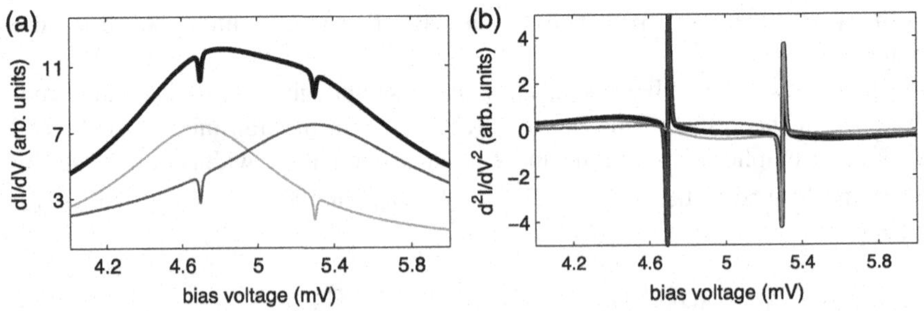

Fig. 11.3 (**a**) Total differential conductance (*bold*) and the partial conductances through the channel which couple the two-electron states through the bonding (*dark*) and anti-bonding (*light*) states. (**b**) Total and partial voltage derivatives d^2I/dV^2. Here, $T = 0.05$ K, while other parameters are as in Fig. 11.2

11.3.2 Asymmetric Coupling to the Substrate

In the above discussion, we approached the interference effects from simplest point of view and assumed that the atoms in the molecular structure couple equally strong to the substrate. Moreover, we assumed that the molecular structure comprised two atoms. Neither of these two assumptions are necessary and here we make contact with the more general situation, that is, either the atoms in the molecule couple asymmetrically to the substrate or the molecule consists of a single entity with two levels. In both situations, we can parametrize the tunneling rate between the sample and substrate according to $v_{\mathbf{k}n\sigma} = \kappa_n v_{\mathbf{k}\sigma}$, with $\kappa_n \in [0, 1]$.

Under those circumstances we have to rederive the equations for the poles of the Green function and, hence, the transmission coefficient using the additional parametrization. It is quite obvious that the coupling matrices $\Gamma^{\text{sub}}_{|1,n\rangle}$ now take on the forms

$$\Gamma^{\text{sub}}_{|1,1\rangle} = \frac{\Gamma^{\text{sub}}_0}{4}\begin{pmatrix} (\kappa_1+\kappa_2)^2 & -\kappa_1^2+\kappa_2^2 \\ -\kappa_1^2+\kappa_2^2 & (\kappa_1-\kappa_2)^2 \end{pmatrix}, \tag{11.22a}$$

$$\Gamma^{\text{sub}}_{|1,3\rangle} = \frac{\Gamma^{\text{sub}}_0}{4}\begin{pmatrix} (\kappa_1-\kappa_2)^2 & -\kappa_1^2+\kappa_2^2 \\ -\kappa_1^2+\kappa_2^2 & (\kappa_1+\kappa_2)^2 \end{pmatrix}, \tag{11.22b}$$

in analogy to the couplings between the sample and the tip. By still assuming that the tunneling between atoms, or levels, in the molecule is vanishingly small, and that tunneling between the tip and sample has a negligible effect on the widths of the molecular levels, we find that the localized states can be written as

$$\omega_{1\pm} = \varepsilon_0 + U - \frac{i}{16}\Gamma_0(\kappa_1^2+\kappa_2^2) \pm \frac{1}{2}\sqrt{\left(J - \frac{i}{4}\kappa_1\kappa_2\Gamma_0\right)^2 - (\kappa_1^2-\kappa_2^2)^2\left(\frac{\Gamma_0}{4}\right)^2}, \tag{11.23a}$$

$$\omega_{3\pm} = \varepsilon_0 + U - \frac{i}{16}\Gamma_0(\kappa_1^2+\kappa_2^2) \pm \frac{1}{2}\sqrt{\left(J + \frac{i}{4}\kappa_1\kappa_2\Gamma_0\right)^2 - (\kappa_1^2-\kappa_2^2)^2\left(\frac{\Gamma_0}{4}\right)^2}. \tag{11.23b}$$

These expressions show that both poles $\omega_{n\pm}$, $n = 1, 3$, acquire a significant width whenever the couplings between the atoms, or levels, and the substrate are unequal. A large width of a state implies that electrons residing in this state interacts strongly with the electron bath in the substrate, for which reason the conductance dips due to interference cease to exist. This is, hence, and indication that asymmetric couplings between the levels in the sample and the substrate is detrimental to the possibility for observing the sharp features in the dI/dV or d^2I/dV^2.

We can conclude that the presence of sharp features in the conductance that are generated by the interference depends on the asymmetry of the couplings between the levels in the sample and the de-localized electrons in the substrate. In general, a weakly coupled state is obtained by requiring fairly symmetric tunneling rates between the sample levels and either the tip or the substrate. The interference effects are, of course, amplified whenever the tunneling rates are symmetric to both the tip and substrate.

The transmission coefficient for the system is in the present set-up given by

$$\mathcal{T}(\omega) = \lambda\left(\frac{\Gamma_0}{4}\right)^2 (\gamma_1\kappa_1 + \gamma_2\kappa_2)^2 \left(\left|\frac{\omega - \varepsilon - U + qJ/2}{C_1(\omega)}\right|^2 + \left|\frac{\omega - \varepsilon - U - qJ/2}{C_3(\omega)}\right|^2\right), \tag{11.24}$$

where the asymmetry of the coupling introduces the factor

$$q = \frac{\gamma_1\kappa_2 + \gamma_2\kappa_1}{\gamma_1\kappa_1 + \gamma_2\kappa_2}. \tag{11.25}$$

Except from the change in the overall amplitude of the transmission, the asymmetry shifts the positions of the transmission dips. Quantitatively we obtain an about 6% shift of the transmission dips for coupling asymmetries $\gamma_2/\gamma_1 \gtrsim 0.7$ and $\kappa_2/\kappa_1 \gtrsim 0.7$. Hence, even for a rather high degree of asymmetry in the couplings, this indicates that there is a reasonably good chance to measure the exchange parameter J.

11.3.3 Nonresonant Levels

Finally, we approach the case where the levels in the atoms are non-resonant, in which we access molecular structures that are comprised of non-equivalent atoms. We can also think of spin-split levels, in which case the first term in the sample Hamiltonian has to be modified according to

$$\sum_{n\sigma} \varepsilon_{n\sigma} d^\dagger_{n\sigma} d_{n\sigma} = \sum_{n\sigma}\left[\varepsilon_n - \sigma\frac{\Delta_n}{2}\right] d^\dagger_{n\sigma} d_{n\sigma}, \tag{11.26}$$

where Δ_n is the spin-split of the nth level. The spin-split may be imposed by external and/or internal magnetic fields.

The two-electron states remain the same as in the previous discussion, and we keep the nomenclature of spin-singlet and spin-triplet states, although referring to the spin-triplet is not correct in a strict sense since the energies associated with those states are non-degenerate in the spin-dependent case. Nonetheless, the triplet state energies become $E_{21} = E_{23} + (\Delta_1 + \Delta_2)/2$, $E_{22} = E_{23} - (\Delta_1 + \Delta_2)/2$, and $E_{23} = \varepsilon_1 + \varepsilon_2 + U + J/2$, whereas the singlet state energy becomes $E_{24} = \varepsilon_1 + \varepsilon_2 + U - J/2$.

It is preferable to write the one-electron states as

$$|1,n\rangle = \{\delta_{n1}d_{1\uparrow}^{\dagger} + \delta_{n2}d_{1\downarrow}^{\dagger} + \delta_{n3}d_{2\uparrow}^{\dagger} + \delta_{n4}d_{2\downarrow}^{\dagger}\}|0\rangle, \tag{11.27}$$

with the corresponding energies $E_{1n} = (\delta_{1n} + \delta_{3n})\varepsilon_{n\uparrow} + (\delta_{2n} + \delta_{4n})\varepsilon_{n\downarrow}$.

In the remainder of this discussion we assume that the tunneling rates can be parametrized as in the previous section, i.e. according to $v_{\mathbf{p}n\sigma} = \gamma_n v_{\mathbf{p}\sigma}$ and $v_{\mathbf{k}n\sigma} = \kappa_n v_{\mathbf{k}\sigma}$. The coupling matrices between the tip and the sample then become

$$\Gamma_{|1,1\rangle}^{\text{tip}} = \gamma_2^2 \frac{\Gamma_0^{\text{tip}}}{2} \begin{pmatrix} 1 & 1 \\ 1 & 1 \end{pmatrix}, \qquad \Gamma_{|1,3\rangle}^{\text{tip}} = \gamma_1^2 \frac{\Gamma_0^{\text{tip}}}{2} \begin{pmatrix} 1 & -1 \\ -1 & 1 \end{pmatrix}, \tag{11.28}$$

while the couplings between the sample and the substrate are given by

$$\Gamma_{|1,1\rangle}^{\text{sub}} = \kappa_2^2 \frac{\Gamma_0^{\text{sub}}}{2} \begin{pmatrix} 1 & 1 \\ 1 & 1 \end{pmatrix}, \qquad \Gamma_{|1,3\rangle}^{\text{sub}} = \kappa_1^2 \frac{\Gamma_0^{\text{sub}}}{2} \begin{pmatrix} 1 & -1 \\ -1 & 1 \end{pmatrix}. \tag{11.29}$$

Here, we have only considered the coupling matrices for the triplet state $|2,3\rangle$, whereas the coupling matrices to the other triplet configurations are the same.

Assuming that the level broadening due to the tunneling between the tip and the sample is negligible, we obtain the following poles of the Green function, involving $|2,3\rangle$ and $|2,4\rangle$,

$$\omega_{1\pm} = \frac{1}{2}\left[\varepsilon_{1\uparrow} + \varepsilon_{2\uparrow} + 2U - \frac{i}{2}\kappa_2^2\Gamma_0 \pm \sqrt{J^2 - \left(\frac{\kappa_2^2\Gamma_0}{2}\right)^2}\,\right], \tag{11.30a}$$

$$\omega_{3\pm} = \frac{1}{2}\left[\varepsilon_{1\uparrow} + \varepsilon_{2\uparrow} + 2U - \frac{i}{2}\kappa_1^2\Gamma_0 \pm \sqrt{J^2 - \left(\frac{\kappa_1^2\Gamma_0}{2}\right)^2}\,\right]. \tag{11.30b}$$

The poles $\omega_{2\pm}$ and $\omega_{4\pm}$ associated with the spin $\downarrow$ channel are obtained by letting $\uparrow \rightarrow \downarrow$ in the above equations.

The form of the poles reveal for the spin-degenerate case, i.e. $\Delta_n = 0$, that there may be sharp localized states in the molecule only for $J \ll \kappa_n\Gamma_0/2$. In this case, we have e.g.

$$\begin{aligned}\omega_{1\pm} &\approx \frac{1}{2}\left[\varepsilon_{1\uparrow} + \varepsilon_{2\uparrow} + 2U - \frac{i}{2}\kappa_2^2\Gamma_0 \pm i\frac{\kappa_2^2\Gamma_0}{2}\left\{1 - \frac{1}{2}\left(\frac{2J}{\kappa_2^2\Gamma_0}\right)^2\right\}\right] \\ &\approx \begin{cases} \frac{1}{2}[\varepsilon_{1\uparrow} + \varepsilon_{2\uparrow} + 2U - i\frac{J^2}{\kappa_2^2\Gamma_0}], & +, \\ \frac{1}{2}[\varepsilon_{1\uparrow} + \varepsilon_{2\uparrow} + 2U - i\kappa_2^2\Gamma_0], & -. \end{cases}\end{aligned} \tag{11.31}$$

In the symmetric case it should, thus, be possible to measure J in cases where the singlet-triplet splitting is small. Then, however, the theory presented previously applies, and therefore we proceed to the discussion for the spin-dependent system.

The above discussion revealed that non-resonant levels will not give rise to any sharp features, generated by interference, in the conductance. Therefore, we again assume that the levels $\varepsilon_1 = \varepsilon_2 = \varepsilon_0$ while the spin-split is uniform, i.e. $\Delta_1 = \Delta_2 = \Delta_0$. For simplicity, we also assume that both levels in the sample couple equally strong to the substrate. We can, thus, describe the one-electron states using (11.6) and the coupling matrices between the sample and substrate by (11.17).

A reason to use spin-dependent transport for the types of measurements corresponding to the situation discussed here, is that the spin-dependence of the current will enable studies of the spin-splitting of the two-electron triplet states. The singlet state couples to all triplet configurations trough the one-electron states. From the previous discussion we know that these couplings generate sharp localized states in the sample at energies which correspond to the triplet and singlet states. By introducing the spin-splitting of the levels, we expect there to appear more sharp states and, hence, more sharp features in the conductance. We, moreover, expect that some of those features should be located at energies which correspond to the different spin configurations of the triplet states.

Quantitatively, we have to consider all Green functions G^r_{mn4}, $m = 1, 2, 3$ for the triplet configurations and $n = 1, \ldots, 4$ for the one-electron states. The poles of G^r_{3n4} are given by

$$\omega^{(0)}_{1\pm} = \varepsilon_0 + \frac{\Delta_0}{2} + U \pm \frac{J}{2} - i\frac{\Gamma_0}{8}(1 \pm 1), \qquad \omega^{(0)}_{2\pm} = \omega^{(0)}_{1\pm} - \Delta_0, \tag{11.32a}$$

$$\omega^{(0)}_{3\pm} = \varepsilon_0 + \frac{\Delta_0}{2} + U \pm \frac{J}{2} - i\frac{\Gamma_0}{8}(1 \mp 1), \qquad \omega^{(0)}_{4\pm} = \omega^{(0)}_{3\pm} - \Delta_0, \tag{11.32b}$$

where the superscript refers to that the poles are associated with the triplet state $|2,3\rangle = |S = 1, m_z = 0\rangle$. Likewise we have for G^r_{1n4} and G^r_{2n4} the poles

$$\omega^{(1)}_{1\pm} = \omega^{(0)}_{1\pm} - \Delta_0(1 \pm 1), \qquad \omega^{(-1)}_{2\pm} = \omega^{(0)}_{2\pm} + \Delta_0(1 \pm 1), \tag{11.33a}$$

$$\omega^{(1)}_{3\pm} = \omega^{(0)}_{3\pm} - \Delta_0(1 \pm 1), \qquad \omega^{(-1)}_{4\pm} = \omega^{(0)}_{4\pm} + \Delta_0(1 \pm 1). \tag{11.33b}$$

We notice that the spin-dependence gives rise to four sets of poles, since there is now a different coupling between the singlet and the three triplet configurations. We also notice that the triplet configuration e.g. $|2,1\rangle = |S = 1, m_z = 1\rangle = d^\dagger_{2\uparrow} d^\dagger_{1\uparrow}|0\rangle$ does not couple to the spin $\downarrow$ states $|1,2\rangle$ and $|1,4\rangle$. Likewise, the triplet state $|2,2\rangle = |S = 1, m_z = -1\rangle = d^\dagger_{2\downarrow} d^\dagger_{1\downarrow}|0\rangle$ does not couple to the spin $\uparrow$ states $|1,1\rangle$ and $|1,3\rangle$. This is, of course, obvious in absence of spin-flip processes. The observations lead, however, to that there will not appear any sharp features in the conductance which are associated with transitions between, say, $|2,1\rangle$ and $|1,2\rangle$. We notice, further the equalities

$$\omega^{(1)}_{n-} = \omega^{(0)}_{n-}, \quad n = 1, 3, \qquad \omega^{(-1)}_{n-} = \omega^{(0)}_{n-}, \quad n = 2, 4,$$

$$\omega^{(1)}_{1+} = \omega^{(0)}_{2+}, \qquad \omega^{(-1)}_{2+} = \omega^{(0)}_{1+},$$

$$\omega^{(1)}_{3+} = \omega^{(0)}_{4+}, \qquad \omega^{(-1)}_{4+} = \omega^{(0)}_{3+}.$$

Hence, there are only four distinct energies and we should, therefore, only expect to find four sharp dips in the conductance.

References

1. Fano, U.: Phys. Rev. **124**, 1866 (1961)
2. Madhavan, V., Chen, W., Jamneala, T., Crommie, M.F., Wingreen, N.S.: Phys. Rev. B **64**, 165412 (2001)
3. Kubala, B., König, J.: Phys. Rev. B **65**, 245301 (2002)
4. Ladrón de Guevara, M.L., Claro, F., Orellana, P.A.: Phys. Rev. B **67**, 195335 (2003)
5. Orellana, P.A., Ladrón de Guevara, M.L., Claro, F.: Phys. Rev. B **70**, 233315 (2004)
6. Ladrón de Guevara, M.L., Orellana, P.A.: Phys. Rev. B **73**, 205303 (2006)
7. Lu, H.Z., Lu, R., Zhu, B.F.: Phys. Rev. B **71**, 235320 (2005)
8. Fransson, J., Balatsky, A.V.: Phys. Rev. B **75**, 153309 (2007)
9. Fransson, J.: Phys. Rev. B **76**, 045416 (2007)

Chapter 12
Spin Systems in Non-Equilibrium

Abstract A quantum Langevin equation for the dynamics of a local spin embedded in a tunnel junction is derived. Under non-equilibrium conditions the dynamics is considered on the Keldysh contour. Several issues regarding the properties of the leads and their influence on the local spin dynamics are discussed.

12.1 Introduction

In previous chapters we have encountered questions about how the current can become spin-polarized by manipulations of the quantum dot, see e.g. Chaps. 6 and 10. Such questions are highly relevant in the context of how one should be able to e.g. pick up particular signatures in the current that is related to one or another set-up of the local spins. Having those questions in mind, it also becomes natural to ask whether the local spin would be affected by the electronic and magnetic properties in the surroundings, e.g. the leads. In other words, how would we perform a certain manipulation of the local spin through applications of external gate and bias voltages, or magnetic fields. The answers to those questions very much rely on the electronic and magnetic properties of the leads.

In this chapter we address the dynamics of a single spin embedded in the tunnel junction between metallic leads [1, 2]. The leads may be ferromagnetic. Taking the non-equilibrium conditions into account, we derive a quantum Langevin equation for the spin dynamics. Through the parameters comprised in this equation, we are able to analyze the effects on the local spin dynamics of various electronic and magnetic properties of the leads.

We proceed to study a few particular cases for observing the spin dynamics through e.g. magnetic resonance force mircoscopy (MRFM) [1], and tunneling measurements [3, 4].

The present chapter might seem a bit off the target when it comes to the previously discussed X-operator formalism. However, although we do not employ the X-operators, we nonetheless utilize many-body operators as we are considering spin operators, which also have non-Fermionic algebraic properties. It is, hence, pertinent in this text on many-body approach to physics.

J. Fransson, *Non-Equilibrium Nano-Physics*,
Lecture Notes in Physics 809,
DOI 10.1007/978-90-481-9210-6_12,

12.2 Equation for Local Spin

Consider two metallic leads coupled to one another through a tunnel junction in which a spin **S** is embedded, see Fig. 12.1 for a schematic view. The tunnel junction may be thought of as e.g. a point contact between the leads, so that possible magnetic fields generated at the tips of the leads (if those are magnetic) can be neglected. The Hamiltonian of the system may be written

$$\mathcal{H} = \mathcal{H}_L + \mathcal{H}_R + \mathcal{H}_S + \mathcal{H}_T, \tag{12.1}$$

where $\mathcal{H}_\chi = \sum_{\mathbf{k}\sigma} \varepsilon_{\mathbf{k}\sigma} c^\dagger_{\mathbf{k}\sigma} c_{\mathbf{k}\sigma}$, $\mathbf{k} \in \chi = L, R$. In the following, we assign the subscripts $\mathbf{p}$ ($\mathbf{q}$) to electrons in the left (right) lead. The local spin may be exposed to an external magnetic field $\mathbf{B}_{\text{ext}}(t)$, and we model the free spin according to $\mathcal{H}_S = -g\mu_B \mathbf{B}_{\text{ext}} \cdot \mathbf{S}$, where g and μ_B are the gyromagnetic ratio (g-factor) and Bohr magneton, respectively. The leads are weakly coupled through the tunneling Hamiltonian

$$\mathcal{H}_T = \sum_{\mathbf{p}\mathbf{q}\sigma\sigma'} c^\dagger_{\mathbf{p}\sigma} \hat{T}_{\sigma\sigma'} c_{\mathbf{q}\sigma'} + H.c., \tag{12.2}$$

where the matrix element $\hat{T}_{\sigma\sigma'}$ defines the rate of the electron transfer through the magnetically active tunnel barrier. Here, a spin is embedded in the barrier which leads to that the matrix element becomes a spin operator [5, 6]. For the present discussions, we will restrict ourselves to

$$\hat{T}_{\sigma\sigma'} = T_0 \delta_{\sigma\sigma'} + T_1 \sigma_{\sigma\sigma'} \cdot \mathbf{S} + T_2 \sigma^x_{\sigma\sigma'}. \tag{12.3}$$

The first term, T_0, describes the spin-independent tunneling rate, which is the same as the tunneling electrons would experience also without the presence of the local spin. The second term, $T_1\sigma_{\sigma\sigma'} \cdot \mathbf{S}$, accounts for the tunneling rate which is affected by the local spin through spin-spin, or exchange, interaction. The last term, $T_2\sigma^x_{\sigma\sigma'}$, denotes the rate of spin-flip transition inside the barrier generated by e.g. spin-orbit interaction. This term may also be present even in absence of the local spin. In principle, we can lump the first and last term into a single spin-dependent tunneling rate, i.e. $T_{\sigma\sigma'} = T_0\delta_{\sigma\sigma'} + T_2\sigma^x_{\sigma\sigma'}$. For more details on the tunneling rates, see [5–7].

We assume that the system is biased with the time-dependent voltage $V(t) = V_{\text{dc}} + V_{\text{ac}}(t)$, where V_{dc} and V_{ac} are dc and ac components. Under the non-equilibrium conditions, there forms a dipole around the barrier region due to the

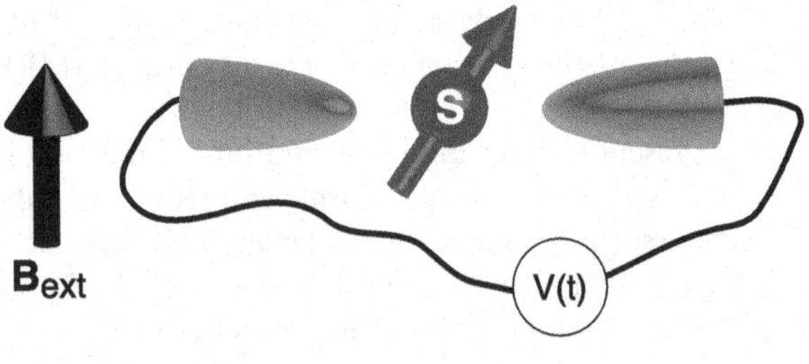

Fig. 12.1 Local spin embedded in the tunnel junction between two metallic, possibly ferromagnetic, leads

accumulation or depletion of electron charge at the interfaces. As a consequence of this process we can talk about the single-electron energies $E_{\mathbf{k}\sigma} = \varepsilon_{\mathbf{k}\sigma} + W_L(t)$, which are constrained by the relation $W_L(t) - W_R(t) = V_{\mathrm{ac}}(t)$. Disregarding pathological time-dependences of the bias voltage, such as monotonically and indefinitely increasing voltages, the occupation of the states in the respective contact remain determined by the distribution established before the time-dependence is turned on. Hence, the chemical potentials in the left and right leads differ by the dc component only, or in other words, $\mu_L - \mu_R = eV_{\mathrm{dc}}$.

It turns out to be preferable to apply the gauge transformation

$$\hat{U} = e^{-i\int_{t_0}^{t}[\mu_L + W_L(t')]N_L dt'} e^{-i\int_{t_0}^{t}[\mu_R + W_R(t')]N_R dt'}, \tag{12.4}$$

to the model, where $N_\chi = \sum_{\mathbf{k}\sigma\in\chi} c^\dagger_{\mathbf{k}\sigma} c_{\mathbf{k}\sigma}$. The model $\mathcal{H}$ is then turned into $\mathcal{K} = \mathcal{K}_L + \mathcal{K}_R + \mathcal{K}_R + \mathcal{K}_T$. Here, $\mathcal{K}_\chi = \sum_{\mathbf{k}\sigma\in\chi} \xi_{\mathbf{k}\sigma} c^\dagger_{\mathbf{k}\sigma} c_{\mathbf{k}\sigma}$, $\xi_{\mathbf{k}\sigma} = \varepsilon_{\mathbf{k}\sigma} - \mu_\chi$, and $\mathcal{K}_S = \mathcal{H}_S$, whereas

$$\mathcal{K}_T = \sum_{\mathbf{pq}\sigma\sigma'} c^\dagger_{\mathbf{p}\sigma} \hat{T}_{\sigma\sigma'} c_{\mathbf{q}\sigma'} e^{i\phi(t)} + H.c., \quad \phi(t) = e\int_{t_0}^{t} V(t')dt'. \tag{12.5}$$

We now derive an effective action for the local spin using the Keldysh technique [8]. Considering all external fields to be the same on both the forward and backward branches of the Keldysh contour C, the partition function $\mathcal{Z}$ has to satisfy the condition

$$\mathcal{Z} = \mathrm{tr}\,\mathrm{T}_C e^{-i\oint_C \mathcal{K}_L(t)dt} = 1, \tag{12.6}$$

where the trace runs over both electronic and spin degrees of freedom. Here, T_C denotes time-ordering on the Keldysh contour. In order to obtain an action for the spin, we take the partial trace over the electronic degrees of freedom only. The tunneling contribution to the spin action, thus, reads

$$i\delta\mathcal{S}_T = -\frac{1}{2}\oint_C \langle \mathrm{T}_C \mathcal{K}_T(\mathbf{S}(t), t)\mathcal{K}_T(\mathbf{S}(t'), t')\rangle dt dt'. \tag{12.7}$$

Notice that this action is given by the first non-trivial contribution in the expansion of the exponential operator $\mathrm{T}_C e^{-i\oint_C \mathcal{K}_L(t)dt}$. The traces over the electronic degrees of freedom of the first and second contributions, $\oint_C \langle \mathrm{T}_C 1\rangle dt$ and $\oint_C \langle \mathrm{T}_C \mathcal{K}_T(\mathbf{S}(t), t)\rangle dt$, respectively, vanish, for which reason (12.7) provides the first non-trivial contribution to the spin-action.

We introduce the tunneling operator $\mathcal{A}_{\sigma\sigma'} = \sum_{\mathbf{pq}} c^\dagger_{\mathbf{p}\sigma} c_{\mathbf{q}\sigma'}$ and the propagator $\mathcal{D}_{\sigma\sigma'}(t, t') = (-i)\langle \mathrm{T}_C \mathcal{A}_{\sigma\sigma'}(t)\mathcal{A}^\dagger_{\sigma\sigma'}(t')\rangle$, in terms of which the tunneling Hamiltonian reads

$$\mathcal{K}_T = \sum_{\sigma\sigma'} \hat{T}_{\sigma\sigma'} \mathcal{A}_{\sigma\sigma'} e^{i\phi(t)} + H.c., \tag{12.8}$$

whereas the effective action is written

$$i\delta\mathcal{S}_T = (-i)\oint_C \sum_{\sigma\sigma'} \hat{T}_{\sigma\sigma'}(t)\mathcal{D}_{\sigma\sigma'}(t,t')\hat{T}_{\sigma'\sigma}(t')e^{i\delta\phi(t,t')}dtdt', \tag{12.9}$$

where $\delta\phi(t,t') = \int_{t'}^{t} V(t')dt'$.

The next step is to expand the two contour integrals in terms of the upper and lower branches of the Keldysh contour. We use the superscripts u and l, in order to distinguish between quantities on the upper and lower branches, $(-\infty,\infty)$ and $(\infty,-\infty)$, respectively. Moreover, the propagator $\mathcal{D}_{\sigma\sigma'}(t,t')$ has four different behaviors depending on which branch the time variables t and t' lie. We have

$$\mathcal{D}^T_{\sigma\sigma'}(t,t') = (-i)\langle \mathrm{T}\mathcal{A}_{\sigma\sigma'}(t)\mathcal{A}^\dagger_{\sigma\sigma'}(t)\rangle, \quad t,t' \in (-\infty,\infty), \tag{12.10a}$$

$$\mathcal{D}^<_{\sigma\sigma'}(t,t') = (-i)\langle \mathcal{A}^\dagger_{\sigma\sigma'}(t)\mathcal{A}_{\sigma\sigma'}(t)\rangle, \quad t \in (-\infty,\infty),\ t' \in (\infty,-\infty), \tag{12.10b}$$

$$\mathcal{D}^>_{\sigma\sigma'}(t,t') = (-i)\langle \mathcal{A}_{\sigma\sigma'}(t)\mathcal{A}^\dagger_{\sigma\sigma'}(t)\rangle, \quad t \in (\infty,-\infty),\ t' \in (-\infty,\infty), \tag{12.10c}$$

$$\mathcal{D}^{\bar{T}}_{\sigma\sigma'}(t,t') = (-i)\langle \bar{\mathrm{T}}\mathcal{A}_{\sigma\sigma'}(t)\mathcal{A}^\dagger_{\sigma\sigma'}(t)\rangle, \quad t,t' \in (\infty,-\infty), \tag{12.10d}$$

where T and $\bar{T}$ is the time ordering and anti-time ordering operator, respectively. Recall that $\mathcal{A}_{\sigma\sigma'}$ is a Bose-like operator. In principle we could have been satisfied by working with only the lesser and greater propagators, only requiring knowledge of whether t is less than or greater than t' in the contour sense. It is, however, convenient to make use of the time and anti-time ordered propagators whenever t and t' lie on the same branch of the contour, which will become apparent below.

In terms of the propagators in (12.10a)–(12.10d), we write the action as

$$\begin{aligned}
i\delta\mathcal{S}_T &= (-i)\sum_{\sigma\sigma'}\int_{-\infty}^{\infty} \big(\hat{T}^u_{\sigma\sigma'}(t)\mathcal{D}^T_{\sigma\sigma'}(t,t')\hat{T}^u_{\sigma'\sigma}(t') - \hat{T}^u_{\sigma\sigma'}(t)\mathcal{D}^<_{\sigma\sigma'}(t,t')\hat{T}^l_{\sigma'\sigma}(t') \\
&\quad - \hat{T}^l_{\sigma\sigma'}(t)\mathcal{D}^>_{\sigma\sigma'}(t,t')\hat{T}^u_{\sigma'\sigma}(t') + \hat{T}^l_{\sigma\sigma'}(t)\mathcal{D}^{\bar{T}}_{\sigma\sigma'}(t,t')\hat{T}^l_{\sigma'\sigma}(t')\big)e^{i\delta\phi(t,t')}dtdt' \\
&= (-i)\sum_{\sigma\sigma'}\int_{-\infty}^{\infty} e^{i\delta\phi(t,t')} \\
&\quad \times \begin{pmatrix}\hat{T}^u_{\sigma\sigma'}(t) & -\hat{T}^u_{\sigma\sigma'}(t)\end{pmatrix}\begin{pmatrix}\mathcal{D}^T_{\sigma\sigma'}(t,t') & \mathcal{D}^<_{\sigma\sigma'}(t,t') \\ \mathcal{D}^>_{\sigma\sigma'}(t,t') & \mathcal{D}^{\bar{T}}_{\sigma\sigma'}(t,t')\end{pmatrix}\begin{pmatrix}\hat{T}^u_{\sigma'\sigma}(t') \\ -\hat{T}^u_{\sigma'\sigma}(t')\end{pmatrix} dtdt'.
\end{aligned} \tag{12.11}$$

The four propagators in (12.10a)–(12.10d) are related to the retarded, advanced, and Keldysh forms of the propagators, defined by

$$\mathcal{D}^{r/a}_{\sigma\sigma'}(t,t') = (\mp i)\theta(\pm t \mp t')\langle[\mathcal{A}_{\sigma\sigma'}(t), \mathcal{A}^\dagger_{\sigma\sigma'}(t')]\rangle, \tag{12.12a}$$

$$\mathcal{D}^K_{\sigma\sigma'}(t,t') = (-i)\langle\{\mathcal{A}_{\sigma\sigma'}(t), \mathcal{A}^\dagger_{\sigma\sigma'}(t')\}\rangle, \tag{12.12b}$$

through (3.48) along with the equations

$$\mathcal{D}^{T}_{\sigma\sigma'}(t,t') = \theta(t-t')\mathcal{D}^{>}_{\sigma\sigma'}(t,t') + \theta(t'-t)\mathcal{D}^{<}_{\sigma\sigma'}(t,t'), \tag{12.13a}$$

$$\mathcal{D}^{\bar{T}}_{\sigma\sigma'}(t,t') = \theta(t'-t)\mathcal{D}^{>}_{\sigma\sigma'}(t,t') + \theta(t-t')\mathcal{D}^{<}_{\sigma\sigma'}(t,t'), \tag{12.13b}$$

$$\mathcal{D}^{K}_{\sigma\sigma'}(t,t') = \mathcal{D}^{>}_{\sigma\sigma'}(t,t') + \mathcal{D}^{<}_{\sigma\sigma'}(t,t'). \tag{12.13c}$$

Therefore, we introduce the *fast* and *slow* variables $\hat{T}^c_{\sigma\sigma'} = (\hat{T}^u_{\sigma\sigma'} + \hat{T}^l_{\sigma\sigma'})/2$ and $\hat{T}^q_{\sigma\sigma'} = \hat{T}^u_{\sigma\sigma'} - \hat{T}^l_{\sigma\sigma'}$, such that $\hat{T}^q_{\sigma\sigma'} \cdot \hat{T}^c_{\sigma\sigma'} = 0$. We, furthermore, apply the rotation

$$\mathcal{R} = \frac{1}{\sqrt{2}}\begin{pmatrix} 1 & 1 \\ -1 & 1 \end{pmatrix}, \qquad \mathcal{R}^{-1} = \mathcal{R}^{\dagger}, \tag{12.14}$$

to the matrix with propagators in (12.11). This leads to that we finally can write the effective action according to

$$\begin{aligned}
i\delta\mathcal{S}_T &= -\frac{i}{2}\sum_{\sigma\sigma'}\int_{-\infty}^{\infty} e^{i\delta\phi(t,t')} \\
&\quad \times \begin{pmatrix} 2\hat{T}^c_{\sigma\sigma'}(t) & \hat{T}^q_{\sigma\sigma'}(t) \end{pmatrix} \begin{pmatrix} 0 & \mathcal{D}^a_{\sigma\sigma'}(t,t') \\ \mathcal{D}^r_{\sigma\sigma'}(t,t') & \mathcal{D}^K_{\sigma\sigma'}(t,t') \end{pmatrix} \begin{pmatrix} 2\hat{T}^c_{\sigma'\sigma}(t') \\ \hat{T}^q_{\sigma'\sigma}(t') \end{pmatrix} dtdt' \\
&= (-i)\sum_{\sigma\sigma'}\int_{-\infty}^{\infty} \Big(\hat{T}^c_{\sigma\sigma'}(t)\mathcal{D}^a_{\sigma\sigma'}(t,t')\hat{T}^q_{\sigma'\sigma}(t') + \hat{T}^q_{\sigma\sigma'}(t)\mathcal{D}^r_{\sigma\sigma'}(t,t')\hat{T}^c_{\sigma'\sigma}(t') \\
&\quad + \frac{1}{2}\hat{T}^q_{\sigma\sigma'}(t)\mathcal{D}^K_{\sigma\sigma'}(t,t')\hat{T}^q_{\sigma'\sigma}(t') \Big) e^{i\delta\phi(t,t')} dtdt'.
\end{aligned} \tag{12.15}$$

Now, we can begin to analyze the components of the action due to tunneling. Recall that $\hat{T}_{\sigma\sigma} = T_{\sigma\sigma} + T_1\sigma_{\sigma\sigma'} \cdot \mathbf{S}$, where only the second term has a time-dependence. It is then straightforward to see that $\hat{T}^c = T_{\sigma\sigma'} + T_1\sigma_{\sigma\sigma'} \cdot \mathbf{S}^c$, whereas $\hat{T}^q = T_1\sigma_{\sigma\sigma'} \cdot \mathbf{S}^q$. Those observations lead to that the action can be partitioned into $\delta\mathcal{S}_T = \delta\mathcal{S}_c + \delta\mathcal{S}_q$, where the action of the slow and fast variables can be written

$$\delta\mathcal{S}_c = \frac{1}{e}\int j^{(1)}_m(t) S^q_m(t) dt + \frac{1}{e}\int S^q_m(t) j^{(2)}_{mn}(t,t') S^c_n(t') dtdt', \tag{12.16a}$$

$$\delta\mathcal{S}_q = \frac{1}{e}\int S^q_m(t) j^{(3)}_{mn}(t,t') S^q_n(t') dtdt', \tag{12.16b}$$

$m,n = x,y,z$, respectively. The current densities $j^{(k)}_{mn}$, $k = 1,2,3$, $j^{(1)}_m \equiv j^{(1)}_{mm}$, introduced here are defined by

$$j_m^{(1)}(t) = -eT_1 \sum_{\sigma\sigma'} \sigma_{\sigma\sigma'}^m T_{\sigma'\sigma} \int \left(\mathcal{D}_{\sigma\sigma'}^r(t,t')e^{i\delta\phi(t,t')} + \mathcal{D}_{\sigma'\sigma}^a(t',t)e^{-i\delta\phi(t,t')}\right)dt', \tag{12.17a}$$

$$j_{mn}^{(2)}(t,t) = -eT_1^2 \sum_{\sigma\sigma'} \sigma_{\sigma\sigma'}^m \left(\mathcal{D}_{\sigma\sigma'}^r(t,t')e^{i\delta\phi(t,t')} + \mathcal{D}_{\sigma'\sigma}^a(t',t)e^{-i\delta\phi(t,t')}\right)\sigma_{\sigma'\sigma}^n, \tag{12.17b}$$

$$j_{mn}^{(3)}(t,t) = -\frac{e}{2}T_1^2 \sum_{\sigma\sigma'} \sigma_{\sigma\sigma'}^m \mathcal{D}_{\sigma\sigma'}^K(t,t')e^{i\delta\phi(t,t')}\sigma_{\sigma'\sigma}^n. \tag{12.17c}$$

The first current density, $j_m^{(1)}$ arises due to the spin-imbalance in the leads along with the non-equilibrium conditions present in the system. It is readily seen in (12.17a) that this current vanishes if the leads are non-magnetic and in the absence of spin-flip transitions inside the barrier, but is finite if at least either of the mechanisms is present. The second and third current densities, $j_{mn}^{(2)}$, $j_{mn}^{(3)}$, are both tensors of second order (matrices), and arise due to the interactions between the local spin in the tunnel barrier and the de-localized spins constituted by the tunneling electrons.

We now make the model complete by adding the action that stems from the interaction between the local spin and tunneling electrons to the actions arising from the interaction between the external magnetic field, $\delta\mathcal{S}_{\text{ext}} = g\mu_B \oint_C \mathbf{B}_{\text{ext}} \cdot \mathbf{S}dt = g\mu_B \int \mathbf{B}_{\text{ext}} \cdot \mathbf{S}^q dt$, and the Wess-Zuminov-Witten-Novikov (WZWN) action which describes the Berry phase accumulated by the spin as a result of its motion on the sphere. On the Keldysh contour, the latter term reads [9]

$$\mathcal{S}_{WSWN} = \frac{1}{S^2}\int \mathbf{S}^q \cdot (\partial_t \mathbf{S}^c \times \mathbf{S}^c)dt. \tag{12.18}$$

The total action can be summarized by

$$\begin{aligned}\mathcal{S} = &\int \left(\frac{1}{S^2}\mathbf{S}^q \cdot (\partial_t \mathbf{S}^c \times \mathbf{S}^c) + g\mu_B \mathbf{B}_{\text{ext}} \cdot \mathbf{S}^q + \frac{1}{e}\mathbf{j}^{(1)} \cdot \mathbf{S}^q\right)dt \\ &+ \frac{1}{e}\int \left(\mathbf{S}^q \cdot \mathbf{j}^{(2)} \cdot \mathbf{S}^c + \mathbf{S}^q \cdot \mathbf{j}^{(3)} \cdot \mathbf{S}^q\right)dtdt'.\end{aligned} \tag{12.19}$$

Effective equations for the dynamics of the local spin are obtained by effecting a functional differentiation on the action with respect to either of the spin fields $\mathbf{S}^c$ or $\mathbf{S}^q$ and putting the resulting expression to zero. Here, we will consider the slow dynamics, which means that we will differentiate out the field $\mathbf{S}^q$. A problem will arise in the last term since it contains the fast spin variable $\mathbf{S}^q$ twice. Hence, we will not get rid of the fast variable completely through the functional differentiation. By application of the Hubbard-Stratonovich transformation to the action $\int \mathbf{S}^q \cdot \mathbf{j}^{(3)} \cdot \mathbf{S}^q dtdt'$, the fast spin variable can be removed, however, at the cost of introducing an auxiliary random field ξ, i.e.

$$e^{i\delta S_q} = e^{i\int S^q_m j^{(3)}_{mn} S^q_n dt dt'/e} = e^{-(i/2)\int S^q_m[-2j^{(3)}_{mn}/e]S^q_n dt dt'}$$
$$= \int e^{(i/2)\int \xi^q_m[-2j^{(3)}_{mn}/e]^{-1}\xi^q_n dt dt'} e^{-i\int \xi_m S^q_m dt}\mathcal{D}\xi, \tag{12.20a}$$

$$\mathcal{D}\xi = \lim_{\varepsilon\to 0}\prod\sqrt{\det\varepsilon[-2\mathbf{j}^{(3)}/e]^{-1}}\frac{d\xi_m}{\sqrt{i2\pi}}. \tag{12.20b}$$

The random fields can be related to the current tensor $j^{(3)}_{mn}$ through the following observation. Assume that there is a random magnetic field ξ coupled to the local spin $\mathbf{S}^q$ via $\mathcal{H}_\xi = -g\mu_B\xi\cdot\mathbf{S}$. The partition function with respect to the random fields then assumes the form

$$\mathcal{Z}[\xi] = \mathrm{tr}_\xi e^{-\int\mathcal{H}_\xi dt} \approx e^{-\int\langle\mathcal{H}_\xi(t)\mathcal{H}_\xi(t')\rangle dt dt'/2} = e^{-(g\mu_B)^2\int\langle\xi_m(t)S^q_m(t)\xi_n(t')S^q_n(t')\rangle dt dt'/2}$$
$$= e^{-(g\mu_B)^2\int\langle\xi_m(t)\xi_n(t')\rangle S^q_m(t)S^q_n(t')dt dt'/2}. \tag{12.21}$$

Inspection of (12.20a), (12.20b) and (12.21) indicates that the correlation function of the random fields should satisfy the relation $(g\mu_B)^2\langle\xi_m(t)\xi_n(t')\rangle = -i2j^{(3)}_{mn}(t,t')/e$.

After this transformation, the total action can be written

$$\mathcal{S} = \int\mathbf{S}^q\cdot\left(\frac{1}{S^2}\partial_t\mathbf{S}^c\times\mathbf{S}^c + g\mu_B[\mathbf{B}_{\mathrm{ext}}+\xi] + \frac{1}{e}\mathbf{j}^{(1)} + \frac{1}{e}\int\mathbf{j}^{(2)}\cdot\mathbf{S}^c dt'\right)dt, \tag{12.22}$$

for which we obtain the functional derivative

$$0 = \frac{\delta\mathcal{S}}{\delta\mathbf{S}^q(t)} = \frac{1}{S^2}\partial_t\mathbf{S}^c\times\mathbf{S}^c + g\mu_B[\mathbf{B}_{\mathrm{ext}}+\xi] + \frac{1}{e}\mathbf{j}^{(1)} + \frac{1}{e}\int\mathbf{j}^{(2)}\cdot\mathbf{S}^c dt'. \tag{12.23}$$

We obtain an appealing equation of motion for $\mathbf{S}^c$ by cross-multiplying this equation with $\mathbf{S}^c$ under the requirement that $|\mathbf{S}^c| = S$ is constant, since then (dropping the superscript c)

$$\partial_t\mathbf{S}(t) = g\mu_B\mathbf{S}(t)\times[\mathbf{B}_{\mathrm{ext}}(t)+\xi(t)] + \frac{1}{e}\mathbf{S}(t)\times\mathbf{j}^{(1)}(t)$$
$$+\frac{1}{e}\int\mathbf{S}(t)\times\left[\mathbf{j}^{(2)}(t,t')\cdot\mathbf{S}(t')\right]dt'. \tag{12.24}$$

Here, it has been tacitly used that the slow spin variable can be considered as being semi-classical to a good approximation. We notice, however, that quantum effects have indeed been included in this semi-classical equation through the presence of action δS_q given solely in terms of the fast, or quantum, spin variable $\mathbf{S}^q$.

In the presence of the electronic tunneling processes the semi-classical approximation can be further elaborated to simplify the integral in the last term. For this, we assume that the electrons in the leads are sufficiently described by the Green functions $g_{\mathbf{k}\sigma}(t,t') = (-i)\mathrm{T}_C\exp[-i\xi_{\mathbf{k}\sigma}(t-t')]$. Then, for low temperatures the elements of the current tensor $\mathbf{j}^{(2)}$ acquire the time-dependence

$$j^{(2)}_{mn}(t,t') \sim \theta(t-t')\frac{\sin D(t-t')}{(t-t')^2}\big(1-\cos D(t-t')\big), \tag{12.25}$$

where D is half the band width of the conduction band in the leads. The time-dependence of this current tensor suggests a peaked onset at $t-t'=0$, however, oscillating with a quadratic decay for larger times. The characteristic time-scale is set by $1/D$, which essentially provides the width of the peak around $t-t'=0$. Hence, for spin-precession (Larmor) frequencies $\omega_L \ll D$, we are in the regime where the spin dynamics is much slower than the processes of the conduction electrons. The slowness of the spin dynamics, thus, allows the local, or Born-Oppenheimer, approximation $\mathbf{S}(t') \approx \mathbf{S}(t) - (t-t')\partial_t \mathbf{S}(t)$, analogous to the approximation introduced in Chap. 8. The equation of motion for the spin, then assumes the simpler form

$$\partial_t \mathbf{S} = \alpha(t)\partial_t \mathbf{S}\cdot\mathbf{S} + g\mu_B \mathbf{S}\times\left[\mathbf{B}_{\text{ext}} + \xi + \frac{1}{eg\mu_B}\mathbf{j}^{(1)} + \frac{1}{eg\mu_B}\int \mathbf{j}^{(2)}(t,t')dt'\cdot\mathbf{S}\right], \tag{12.26}$$

where $\alpha(t) = \int (t-t')\mathbf{j}^{(2)}(t,t')dt'/e$ defines a damping parameter. This equation for $\mathbf{S}$ captures the main dynamics of the local spin under the influence of the tunneling electrons, and is reminiscent of the Landau-Lifshitz-Gilbert (LLG) equation [10, 11]. From here on, we shall proceed with an approximative discussion of the spin dynamics.

The random, or Langevin, term ξ arises due to the exchange interaction between the tunneling electrons and the local spin. In the present set-up, the energy scale for the exchange interaction is given by the parameter T_1, here in the order of 0.1–1 meV. The precession frequency ω_L is set by the external magnetic field and the internal currents. Assuming that the effective magnetic field acting on the local spin is less than 0.5 T, the Larmor frequency $\omega_L < 60$ μeV. Under those conditions it is reasonable to expect that the random processes are suppressed, and we can neglect the random term in the following.

The damping parameter α is also generated by the exchange interaction. For low temperatures and dc voltages, it is straightforward to show that it is negligible as well. To see this, consider the contribution $\mathcal{D}_{\sigma\sigma'}(t,t')e^{i\delta\phi(t,t')}$ in the current tensor $\mathbf{j}^{(2)}$. We then have the time-integration

$$\begin{aligned}
&\int_{-\infty}^{t} (t-t')\mathcal{D}^r_{\sigma\sigma'}(t,t')e^{i\delta\phi(t,t')}dt' \\
&\quad\sim (-i)\int_{-D}^{D}\int_{-\infty}^{t}(t-t')\big(f(E)-f(E')\big)e^{i(E-E'+eV_{\text{dc}})(t-t')}dt'dEdE' \\
&\quad= i\int_{-D}^{D}\frac{f(E)-f(E')}{(E-E'+eV_{\text{dc}})^2}dEdE' \\
&\quad= i\int_{-D}^{D} f(E)\bigg(\frac{2E}{E^2-(D-eV_{\text{dc}})^2}-\frac{2E}{E^2-(D+eV_{\text{dc}})^2}\bigg)dE \\
&\quad\approx i\ln\left|\frac{1-eV_{\text{dc}}/D}{1+eV_{\text{dc}}/D}\right|.
\end{aligned} \tag{12.27}$$

It is, thus, clear that the damping parameter is vanishingly small for bias voltages such that $eV_{dc} \ll D$.

12.3 Dynamics of Local Spin

For our discussion of the local spin dynamics we take the above approximations for granted, under which the spin equation of motion reduces to

$$\partial_t \mathbf{S} = g\mu_B \mathbf{S} \times \left[\mathbf{B}_{\text{ext}} + \frac{1}{eg\mu_B} \mathbf{j}^{(1)} + \frac{1}{eg\mu_B} \int \mathbf{j}^{(2)}(t,t')dt'\mathbf{S} \right]. \tag{12.28}$$

In order to make life a little easier, we assume that the electrons in the leads can be described by the simplified Green function $g_{\mathbf{k}\sigma}(t,t') = (-i)\mathrm{T}\exp[-i\xi_{\mathbf{k}\sigma}(t-t')]$. For simplicity, we also assume a simple harmonic time-dependence of the bias voltage such that $V(t) = V_{dc} + V_{ac}\cos\omega_0 t$, and that the densities of states in the leads are slowly varying with energy/momentum. Under those approximations, the current densities involved in the equation are given by

$$\begin{aligned} j_m^{(1)}(t) = {} & ieT_1 \sum_{\sigma\sigma'\mu\nu} \sigma^m_{\sigma\sigma'} T_{\sigma\sigma'} J_\mu(eV_{ac}/\omega_0) J_\nu(eV_{ac}/\omega_0) \\ & \times \int_{-\infty}^{t} \int \big[N_{L\sigma} N_{R\sigma'} e^{i(E-E'+eV_{dc})(t-t')+i\omega_0(\mu t-\nu t')} \\ & - N_{L\sigma'} N_{R\sigma} e^{-i(E-E'+eV_{dc})(t-t')-i\omega_0(\mu t-\nu t')} \big] \\ & \times [f(E) - f(E')] dE dE' dt', \end{aligned} \tag{12.29a}$$

and

$$\begin{aligned} j_{mn}^{(2)}(t,t') = {} & ieT_1^2 \theta(t-t') \sum_{\sigma\sigma'\mu\nu} \sigma^m_{\sigma\sigma'} \sigma^n_{\sigma'\sigma} J_\mu(eV_{ac}/\omega_0) J_\nu(eV_{ac}/\omega_0) \\ & \times \int_{-\infty}^{t} \int \big[N_{L\sigma} N_{R\sigma'} e^{i(E-E'+eV_{dc})(t-t')+i\omega_0(\mu t-\nu t')} \\ & - N_{L\sigma'} N_{R\sigma} e^{-i(E-E'+eV_{dc})(t-t')-i\omega_0(\mu t-\nu t')} \big] \\ & \times [f(E) - f(E')] dE dE'. \end{aligned} \tag{12.29b}$$

Here, we have used that $e^{iz\sin\phi} = \sum_n J_n(z) e^{in\phi}$. From these expressions we can obtain dc conditions by letting $V_{ac} \to 0$ and $\omega_0 \to 0$, in that order, in which case only the term with $\mu = \nu = 0$ survives.

For later reference, we also rewrite the spin components of the densities of states in terms of the electronic and magnetic densities N_χ and $\mathbf{m}_\chi$, such that $N_{\chi\sigma} = \sum_{\sigma'} [1 + \sigma\sigma'] N_{\chi\sigma'}/2 = (N_\chi + \sigma m_\chi^z)/2$.

12.3.1 Non-Magnetic Leads

First, we study the interaction between the tunneling current and the local spin for non-magnetic leads. Then, the densities of states are spin-independent, so that $N_{\chi\sigma} = N_\chi/2$, which leads to the current densities

$$j_m^{(1)}(t) \approx -e\pi T_1 N_L N_R \sum_{\sigma\sigma'\mu\nu} \sigma^m_{\sigma\sigma'} T_{\sigma\sigma'} J_\mu(eV_{ac}/\omega_0) J_\nu(eV_{ac}/\omega_0) \times [eV_{dc} + \nu\omega_0] \sin\omega_0(\mu-\nu)t, \tag{12.30a}$$

$$\int j_{mn}^{(2)}(t,t')dt' \approx -e\pi T_1^2 N_L N_R \sum_{\sigma\sigma'\mu\nu} \sigma^m_{\sigma\sigma'} \sigma^n_{\sigma'\sigma} J_\mu(eV_{ac}/\omega_0) J_\nu(eV_{ac}/\omega_0) \times [eV_{dc} + \nu\omega_0] \sin\omega_0(\mu-\nu)t. \tag{12.30b}$$

Here, we first notice in (12.30b) that the only non-vanishing components are the diagonal ones, that is, $j_{mm}^{(2)}$, $m = x, y, z$. We, furthermore, notice that these diagonal components are equal. In the case of non-magnetic leads, they do not have any effect on the spin dynamics, however, since $\mathbf{S} \times \mathbf{j}^{(2)}\mathbf{S} = (S_y S_z[j_{zz}^{(2)} - j_y^{(2)}], S_x S_z[j_{xx}^{(2)} - j_{zz}^{(2)}], S_x S_y[j_{yy}^{(2)} - j_{xx}^{(2)}]) = 0$, where we have used that the spins are semi-classical.

We now study the effect of the current density in (12.30a). By recalling that $T_{\sigma\sigma'} = \delta_{\sigma\sigma'} T_0 + \sigma^x_{\sigma\sigma'} T_2$, it is clear that the only non-vanishing component in (12.30a) is $j_x^{(1)}$. Hence, the equation of motion for the spin can be written

$$\partial_t \mathbf{S} = g\mu_B \mathbf{S} \times \mathbf{B}_{ext} + \frac{1}{e}\begin{pmatrix} 0 & 0 & 0 \\ 0 & 0 & j_x^{(1)} \\ 0 & -j_x^{(1)} & 0 \end{pmatrix}. \tag{12.31}$$

In absence of external magnetic fields, the spin component S_x is a constant of motion, and the local spin acquires a rotational motion in the yz-plane about the x-axis, with frequency $\omega_L = |j_x^{(1)}|/e$.

This result suggests force microscopy experiment, MRFM, which is set-up such that there is a magnetic particle mounted on a nanomechanical cantilever above of the tunnel junction, see Fig. 12.2. The magnetic particle is coupled to the local spin through a magnetic force [12], e.g.

$$\mathbf{F}(\mathbf{r},t) = -[\mathbf{m}_e(t)\cdot\nabla]\mathbf{B}_{ext}(\mathbf{r}), \tag{12.32}$$

where $\mathbf{r}$ is the mean distance between the micromagnet and the local spin, $\mathbf{m}_e(t) = g\mu_B\mathbf{S}(t)$ is the magnetic moment of the local spin, and the total external magnetic field $\mathbf{B}(\mathbf{r})$ comprise the external field $\mathbf{B}_{ext}$ and the magnetic field $\mathbf{B}_t(\mathbf{r})$ generated by the cantilever micromagnet. Assuming, for instance, that $|\partial B_t^z/\partial z| \sim 10^{-1}$ T/μm yields a force signal of about 10 aN.

The experimental set-up according to this scheme would provide a direct measurement of the magnetic resonance signal from a single spin, particularly in ab-

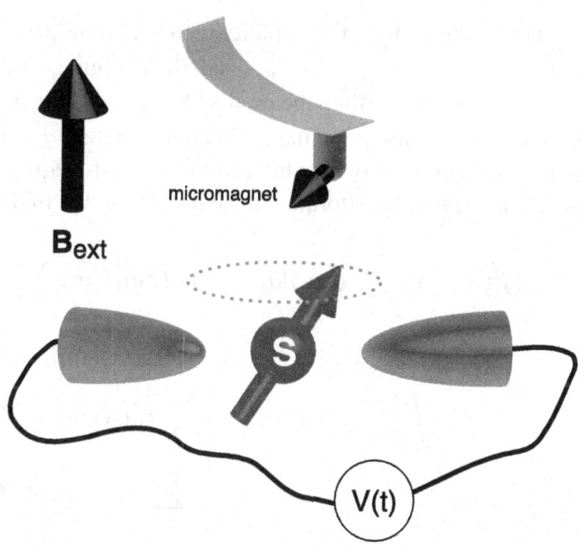

Fig. 12.2 Schematic view of the tunnel junction, in which a spin is embedded, in combination with MRFM. A static external magnetic field polarized the spin, while an ac electric field generates an effective time-dependent spin-polarized current, through spin-flip transitions in the tunnel junction, which influences the dynamics of the local spin. The dynamical variations in the local effective magnetic field is picked up by a resonator, nanomechanical cantilever, onto which a micromagnet is mounted

sence of an externally applied time-dependent field. Recording such a signal would be a very strong evidence for spin-flip transitions across the tunnel junction.

12.3.2 Ferromagnetic Leads

We now turn our attention to the case of ferromagnetic leads. In the following we put the spin-flip tunneling matrix element $T_2 = 0$, in order to purify the system. Then, the only surviving contribution from the current density in (12.29a) is the z-component, which essentially behaves as

$$\begin{aligned} j_z^{(1)}(t) &= -\pi e T_0 T_1 (N_R m_L^z + N_L m_R^z \cos\theta) \\ &\quad \times \sum_{\mu\nu} J_\mu(eV_{\mathrm{ac}}/\omega_0) J_\nu(eV_{\mathrm{ac}}/\omega_0)[eV_{\mathrm{dc}} + \nu\omega_0] \sin\omega_0(\mu - \nu)t, \end{aligned} \tag{12.33}$$

where θ defines the angle between the magnetic moments of the left and right leads. In this way, we allow for non-collinear arrangements of the magnetic moments. In the above formula we have let the global reference frame to coincide with the spin reference frame of the left lead.

This current density, (12.33) contributes a finite shift to the Larmor frequency whenever $N_R m_L^z + N_L m_R^z \neq 0$. This requirement is satisfied neither in case of non-magnetic leads (see previous discussion), nor in case in which both leads are ferromagnetic but such that $N_L m_R^z = -N_R m_L^z$, or $N_L/N_R = -m_L^z/m_R^z$. This latter case can be realized, for example, when both leads have identical electronic and magnetic structures but are aligned in an anti-ferromagnetic order.

The current density $\mathbf{j}^{(2)}$ is more interesting. We find, for instance, that the components $j_{xx}^{(2)} = j_{yy}^{(2)}$, $j_{xy}^{(2)} = -j_{yx}^{(2)}$, and $j_{zz}^{(2)}$ are finite, while the others are identically zero. Considering, further, the product $\mathbf{S}\times\mathbf{j}^{(2)}\mathbf{S}$, it can be understood that the difference $j_{xx}^{(2)} - j_{zz}^{(2)}$ lies along the z-direction, whereas $j_{xy}^{(2)}$ lies in the xy-plane. While both contributions provide shifts to the Larmor frequency, the latter is interesting since it also provides a torque on the local spin. We find that

$$\int_{-\infty}^{t} [j_{xx}^{(2)}(t,t') - j_{zz}^{(2)}(t,t')]dt' = 2\pi e T_1^2 \mathbf{m}_L \cdot \mathbf{m}_R \sum_{\mu\nu} J_\mu(eV_{\mathrm{ac}}/\omega_0) J_\nu(eV_{\mathrm{ac}}/\omega_0) \times [eV_{\mathrm{dc}} + \nu\omega_0] \sin\omega_0(\mu-\nu)t, \tag{12.34a}$$

$$\int_{-\infty}^{t} j_{xy}^{(2)}(t,t')dt' = -\frac{\pi e}{2} T_1^2 (N_R m_L^z - N_L m_R^z \cos\theta) \times \sum_{\mu\nu} J_\mu(eV_{\mathrm{ac}}/\omega_0) J_\nu(eV_{\mathrm{ac}}/\omega_0)[eV_{\mathrm{dc}} + \nu\omega_0] \times \cos\omega_0(\mu-\nu)t. \tag{12.34b}$$

The longitudinal component, (12.34a), is non-vanishing only if both leads are spin-polarized. In addition, it only contributes to the spin dynamics in presence of a time-dependent bias voltage. The transversal component, on the other hand, only requires that the difference $N_R m_L^z - N_L m_R^z \cos\theta \neq 0$, which is satisfied whenever the at least one of the leads is spin-polarized or, if they are identical, they are not in ferromagnetic arrangement ($\theta = 0$). It is, furthermore, clear that this contribution is finite for stationary biases, something that we shall discuss below.

The effects of the current densities above can be summarized in that the components in (12.33) and (12.34a) contributes to the precession of the local spin about its local direction. The component in (12.34b) is generated by spin-flips of the tunneling electrons when interacting with the local spin moments. Under those processes the local spin is subject to a torque which tends to slightly change the local direction of the local spin. For example, in case of anti-ferromagnetically arranged leads, the current density component (12.34b) acts as to line the local spin up in the direction of the magnetic moment of the source lead. An ac bias voltage causes a wobbling, or nutation, of the local spin motion, as we shall see below.

We have already taken into account that the local spin is described in a semi-classical sense, hence, here we take the full advantage of this description in that we parametrize the spin on the unit sphere according to $\mathbf{S} = S(\cos\varphi\sin\vartheta, \sin\varphi\sin\vartheta, \cos\vartheta)$, where φ and ϑ is the azimuthal and polar angles, respectively, whereas $S = |\mathbf{S}|$. The equation of motion for the local spin can then be written as

$$\dot{\varphi} = -g\mu_B B_{\mathrm{ext}}(t) + \frac{1}{e} S j_z^{(1)}(t) + \frac{1}{e} S \int [j_{xx}^{(2)}(t,t') - j_{zz}^{(2)}(t,t')]dt' \cos\vartheta, \tag{12.35a}$$

$$\dot{\vartheta} = \frac{1}{e} S \int j_{xy}^{(2)}(t,t')dt' \sin\vartheta. \tag{12.35b}$$

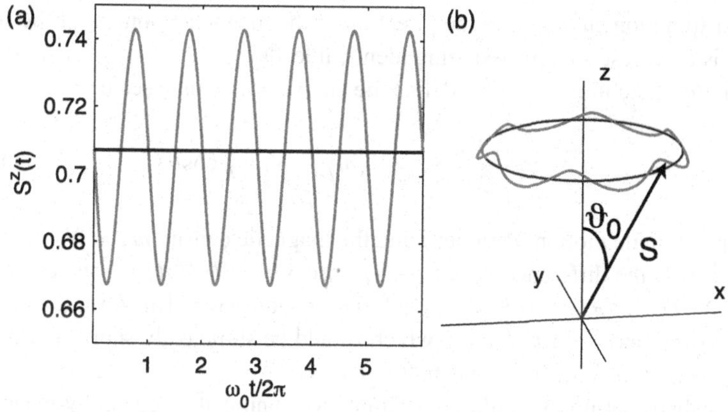

Fig. 12.3 The different qualitative behavior of the local spin in presence of an applied ac bias voltage when the magnetizations of the leads are equal and ferromagnetically (*black*) and anti-ferromagnetically (*grey*) arranged. (**a**) The polar displacement of the spin. (**b**) The resulting spin motion on the unit sphere. Here, $N_L T_1 \sim N_R T_1 \sim 0.1$, $m_\chi^z/N_\chi \sim 0.9$ $\theta = 0$ (**a**) and $\theta = \pi$ (**b**), and $\omega_0 = 3$ meV

The solution of the second equation is given by

$$\vartheta(t) = 2\arctan\Bigg(\tan\frac{\vartheta_0}{2} e^{-\pi S T_1^2 (N_R m_L^z - N_L m_R^z \cos\theta)/2} \times \prod_{\mu\nu} \exp\bigg\{ J_\mu\bigg(\frac{eV_{\rm ac}}{\omega_0}\bigg) J_\nu\bigg(\frac{eV_{\rm ac}}{\omega_0}\bigg) [eV_{\rm dc} + \nu\omega_0] \frac{\sin\omega_0(\mu-\nu)t}{\omega_0(\mu-\nu)}\bigg\}\Bigg), \quad (12.36)$$

where ϑ_0 is the polar orientation of the spin at some initial time t_0. The full motion of the local spin, where the planar precession is given by the equation for φ and the deviation from the planar motion is given by ϑ, is obviously rather complicated, even in the simplified case we are considering here, see Fig. 12.3. Especially, we see that the nutation of the spin, which is caused by its coupling to the tunneling electrons, is directly modulated by the time-dependence of the bias voltage. In case of harmonic bias voltages, it can also be seen from the expression of ϑ, that the lower the frequency of the bias, i.e. $\omega_0 \ll eV_{\rm dc}/10$, the larger the amplitude of the polar angular displacement, something that will provide a signal for detection in the tunneling current.

In static limit, $eV_{\rm ac} \to 0$, $\omega_0 \to 0$, such that $eV_{\rm ac}/\omega_0 \to 0$, the polar angle acquires the time-dependence

$$\vartheta(t) = 2\arctan\bigg(\tan\frac{\vartheta_0}{2} e^{-\pi eV_{\rm dc} S T_1^2 (N_R m_L^z - N_L m_R^z \cos\theta)(t-t_0)/2}\bigg). \quad (12.37)$$

Thus, assuming $\vartheta_0 \neq 0$ and $V_{\rm dc} > 0$, it is clear that ϑ approaches 0 (π) as time grows for positive (negative) $N_R m_L^z - N_L m_R^z \cos\theta$, and it can be understood that the local spin moment assumes the direction of the magnetic moment of the source lead. This is easiest seen by assuming identical leads.

From the static limit result we define the characteristic time scale

$$\tau_c^{-1} \simeq \frac{\pi e V_{dc}}{2} S T_1^2 (N_R m_L^z - N_L m_R^z \cos\theta) \tag{12.38}$$

for the polar angle motion. Parametrizing the magnetic density $m_\chi^z = N_\chi p_\chi$, where $-1 \leq p_\chi \leq 1$, the difference $N_R m_L^z - N_L m_R^z \cos\theta = N_L N_R (p_L - p_R \cos\theta)$. Assuming $N_L T_1 \sim N_R T_1 \sim 0.1$, $p_\chi = 1/2$, $\theta = \pi$, and $V_{dc} \sim 1$ mV, we arrive at the characteristic time-scale $\tau_c \approx 5$ ps, which should be sufficiently short to switch the spin from e.g. $\uparrow$ to $\downarrow$ within a bias pulse of 1 ns.

In a realistic set-up, it would be desirable to manipulate the spin by means of a bias pulse with time span τ_s. A sudden onset of the bias inevitably leads to transient effects in the induced current densities, which transfer to the motion of the spin. Therefore, consider the voltage pulse $V(t) = V_{dc} + V_{ac}[\theta(t-\tau_0) - \theta(t-\tau_1)]$, where $\tau_1 = \tau_0 + \tau_s$. As we are interested in the dynamics induced by the current, we assume that $\mathbf{B}_{ext} = 0$. From the above discussion we know that a stationary bias eventually will line the local spin up in the direction parallel to the magnetic moment of the source (or anti-parallel to the magnetic moment of the drain if the source is non-magnetic). We, thus, set $V_{dc} = 0$ in order to focus on the transient behavior of the spin. Under those conditions, the polar angle motion is given by

$$\begin{aligned}\dot{\vartheta}(t) = & -\frac{\pi e V_{ac}}{2} S T_1^2 (N_R m_L^z - N_L m_R^z \cos\theta)\big(\big[1 - e^{-(t-\tau_0)/\tau}\big] \\ & \times \big[\theta(t-\tau_0) - \theta(t-\tau_1)\big] \\ & + \big[e^{-(t-\tau_1)/\tau} - e^{-(t-\tau_0)/\tau}\big] \cos e V_{ac}(t-\tau_1)\theta(t-\tau_1)\big) \sin\vartheta,\end{aligned} \tag{12.39}$$

where the time-scale τ is related to the electronic tunneling processes, and is of the order of 1 fs. Those processes are much faster than both τ_c and the time-scale of the Larmor frequency ω_L for bias voltages between 1–100 mV. For e.g. a 1 ns bias pulse of 1 mV, the critical time-scale is ~100 fs. Physically, the time-scale τ implies that the induced current densities are retarded (delayed) responses of the bias voltage across the junction.

The time-dependence of the polar angle is plotted in Fig. 12.4(a) for the square pulse, and it is clearly seen that the angle goes from π to 0, i.e. spin going from $-z$ to z in the global reference frame, within the time scale of the pulse. The integrated induced current density $\int_{-\infty}^{t} j_{xy}^{(2)}(t,t')dt'$ is plotted in Fig. 12.4(b). At the onset (termination) of the pulse, the amplitude of the induced current density grows (decays) exponentially, which is expected from the equation of ϑ. As the pulse terminates, however, there are additional oscillations in the induced current density, as a reaction to that the bias is removed. These oscillations are not visible in the polar angle motion since they are exponentially suppressed.

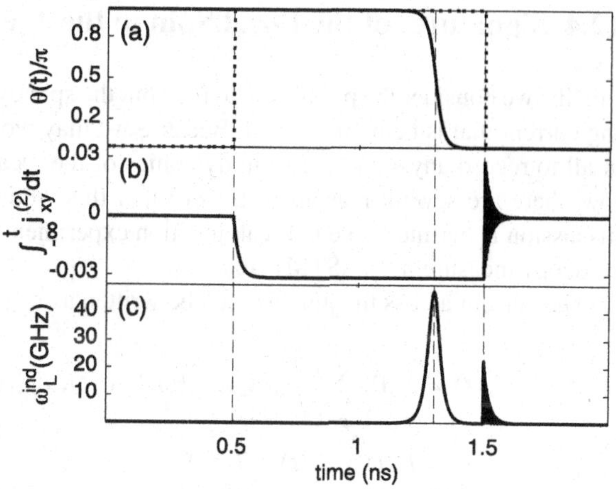

Fig. 12.4 Time-dependent variation of the polar angle (**a**), the integrated current density $\int_{-\infty}^{t} j_{xy}^{(2)}(t,t')dt'$ (**b**), and the Larmor frequency (**c**) for a square bias voltage pulse, cf. *dotted line* in (**a**). Here, $V_{ac} = 1$ mV, $N_L T_1 \sim N_R T_1 \sim 0.1$, $p_\chi = 1/2$ and $\theta = \pi$

The Larmor frequency ω_L of the precession is affected by the time-dependent variation of ϑ. In absence of external magnetic field, we find

$$\omega_L^2(t) = \frac{S^2}{e^2}\left(j_z^{(1)}(t) + \int_{-\infty}^{t} [j_{xx}^{(2)}(t,t') - j_{zz}^{(2)}(t,t')]dt'\right)^2 + \frac{S^2}{e^2}\left(\int_{-\infty}^{t} j_{xy}^{(2)}(t,t')dt'\right)^2 \sin^2\vartheta. \tag{12.40}$$

The time-dependence of the first term is the square of

$$eV_{ac}\left[e^{-(t-\tau_0)/\tau} - e^{-(t-\tau_1)/\tau}\right]\sin eV_{ac}(t-\tau_1)\theta(t-\tau_1). \tag{12.41}$$

Essentially, this field does not affect the Larmor frequency until after the bias pulse has terminated. Its response to this termination is an exponentially decaying oscillatory variation of ω_L, see Fig. 12.4(c) around 1.5 ns.

The last contribution to the Larmor frequency, vanishes for as long as $\vartheta = n\pi$, n integer, that is, as long as the local spin is directed along the z-direction in the global reference frame. However, during the time interval of the spin flips, the polar angle $0 < |\vartheta| < \pi$, leading to that $\sin^2\vartheta \neq 0$. One would, therefore, expect a sharp peak in the time-evolution of the Larmor frequency, $\omega_L(t)$. This is also depicted in Fig. 12.4(c), showing a peaked Larmor frequency simultaneously with the spin flip.

One should bear in mind here, that we have omitted any anisotropy fields that may have the effect to stabilize the local spin moment under equilibrium conditions, that is, for zero bias voltage. The purpose of the above discussion is, however, to demonstrate that the bias voltage itself may have a stabilizing effect on the local spin moment, hence, it can be regarded as an anisotropy field in its own respect.

12.4 Signatures of the Local Spin in the Transport

Finally, we consider the possibility to read out the spin dynamics through the tunneling current that is being measured. Because we may wonder—would it be possible at all to record any signature from dynamics of the local spin. As we shall see below, there are several reasons to believe that this would be the case. The present discussion is pertinent to e.g. break junction experiments or scanning tunneling microscopy measurements (STM) etc.

The current across the junction can be written as

$$\begin{aligned} I(t) &= 2e\,\mathrm{Re}\sum_{\sigma\sigma'}\int_{-\infty}^{t}\langle[\hat{T}_{\sigma\sigma'}(t)\mathcal{A}_{\sigma\sigma'}(t),\hat{T}_{\sigma'\sigma}(t')\mathcal{A}^{\dagger}_{\sigma\sigma'}(t')]\rangle dt' \\ &= I_0(t)+I_z(t)+I_{\perp}(t), \end{aligned} \tag{12.42}$$

where

$$I_0(t) = 2eT_0^2\,\mathrm{Re}\sum_{\mathbf{pq}\sigma}\int_{-\infty}^{t}\big(f(\xi_{\mathbf{p}\sigma})-f(\xi_{\mathbf{q}\sigma})\big)e^{i(\xi_{\mathbf{p}\sigma}-\xi_{\mathbf{q}\sigma})(t-t')+i\delta\phi(t,t')}dt', \tag{12.43a}$$

$$\begin{aligned} I_z(t) &= 4eT_0T_1\,\mathrm{Re}\sum_{\mathbf{pq}\sigma}\int_{-\infty}^{t}\left(\sigma^z_{\sigma\sigma}+\frac{T_0}{2T_1}\cos\vartheta(t)\right)\cos\vartheta(t) \\ &\quad\times\big(f(\xi_{\mathbf{p}\sigma})-f(\xi_{\mathbf{q}\sigma})\big)e^{i(\xi_{\mathbf{p}\sigma}-\xi_{\mathbf{q}\sigma})(t-t')+i\delta\phi(t,t')}dt', \end{aligned} \tag{12.43b}$$

$$\begin{aligned} I_{\perp}(t) &= 2eT_1^2\,\mathrm{Re}\sum_{\mathbf{pq}\sigma}\int_{-\infty}^{t}\sin^2\vartheta(t)\big(f(\xi_{\mathbf{p}\sigma})-f(\xi_{\mathbf{q}\bar{\sigma}})\big) \\ &\quad\times e^{i(\xi_{\mathbf{p}\sigma}-\xi_{\mathbf{q}\bar{\sigma}})(t-t')+i\delta\phi(t,t')}dt'. \end{aligned} \tag{12.43c}$$

These currents describe, from above, the direct tunneling between the leads without noticing the local spin moment, the tunneling in which the current becomes spin-polarized due to the presence of the local spin, and the tunneling in which the tunneling electrons undergo spin-flips generated by their interactions with the local spin moment. Note here, that we have used the semi-classical description of the local spin when obtaining the above expressions for the current. In a quantum description one would be able to obtain an analogous separation of the currents, however, in terms of the average of the local spin moments, $\langle S_z\rangle$, and the spin-spin correlation functions, $\sigma\cdot\langle\mathbf{S}(t)\mathbf{S}(t')\rangle\cdot\sigma$, see e.g. [13, 14].

It is important to notice that the tunneling current is modulated by the presence of the local spin through the variation of the polar angle $\vartheta(t)$, cf. terms proportional to $\cos\vartheta$ and $\cos^2\vartheta$ in I_z, as well as the term proportional to $\sin^2\vartheta$ in $I_{\perp}$. The quadratic dependence on the sine and cosine leads to a modulation of the current by the doubled frequency 2ϑ, as well. We can, therefore, expect that the current contains components which show a slightly deviated time-dependence compared to the fundamental time-dependence introduced through the bias voltage. The expressions of the current, furthermore, show that there is no direct influence on the current from the azimuthal motion of the local spin, i.e. there is no explicit dependence on φ.

In order to be more concrete, consider the biasing of the system according to

$$\delta\phi(t,t') = e\int_{t'}^{t}\big(V_{\mathrm{dc}} + V_{\mathrm{ac}}[\theta(\tau-\tau_0) - \theta(\tau-\tau_1)]\big)d\tau. \tag{12.44}$$

Under those conditions the equation of motion for ϑ becomes

$$\begin{aligned}\dot{\vartheta}(t) = {}& -\frac{\pi e}{2}ST_1^2(N_R m_L^z - N_L m_R^z\cos\theta)\big(V_{\mathrm{dc}}\theta(t-\tau_0) \\ & + \big[V_{\mathrm{dc}}\cos eV_{\mathrm{ac}}(t-\tau_0)e^{-(t-\tau_0)/\tau} + (V_{\mathrm{dc}}+V_{\mathrm{ac}})\big(1-e^{-(t-\tau_0)/\tau}\big)\big] \\ & \times\big[\theta(t-\tau_0)-\theta(t-\tau_1)\big] + \big[V_{\mathrm{dc}}\cos eV_{\mathrm{ac}}(\tau_1-\tau_0)e^{-(t-\tau_0)/\tau} \\ & + (V_{\mathrm{dc}}+V_{\mathrm{ac}})\big(e^{-(t-\tau_1)/\tau} - e^{-(t-\tau_0)/\tau}\big)\cos eV_{\mathrm{ac}}(t-\tau_1) \\ & + eV_{\mathrm{dc}}\big(1-e^{-(t-\tau_1)/\tau}\big)\big]\theta(t-\tau_1)\sin\vartheta(t).\end{aligned} \tag{12.45}$$

For simplicity, assume that the initial polar angle $\vartheta_0 = \pi$, so that the local spin is $\downarrow$ at $t = t_0$. Also assume that $V_{\mathrm{dc}} > 0$, and that $p_L = 1/2$, $p_R = 0$, for $N_L T_1 \sim N_R T_1 \sim 0.1$. Hence, we assume to have one ferromagnetic and one non-magnetic lead, a setup that is pertinent to e.g. STM using a spin-polarized tip. The dynamics of the local spin is plotted in Fig. 12.5, and it behaves as one would expect from the above discussion and Fig. 12.4.

It can be understood from (12.43a) that I_0 follow the time-dependence of the bias voltage only, as can be seen in Fig. 12.5(b) (grey), as well. Therefore, we leave this component. Before the spin flip event, see Fig. 12.5(a), both the current I_z (solid-dotted) are finite whereas $I_\perp = 0$ (faint), see Fig. 12.5(b). Simultaneously with the spin flip event though, both currents I_z and $I_\perp$ change their amplitude, as an effect that the polar angle varies from π to 0. Note that, I_z assumes a new steady state

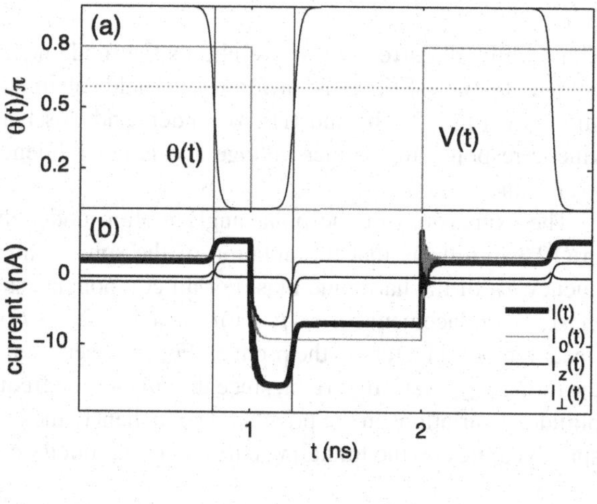

Fig. 12.5 Time-dependence of the polar angle (**a**) and the tunneling current (**b**) for the bias pulse $V(t) = eV_{\mathrm{dc}} + eV_{\mathrm{ac}}[\theta(t-\tau_0) - \theta(t-\tau_1)]$, $(\tau_0,\tau_1) = (1,2)$ ns (**a**) (faint—zero bias level is indicated by the *dotted line*). In (**b**) the plotted currents are the total $I(t)$ (*bold*), $I_0(t)$ (*grey*), $I_z(t)$ (*solid-dotted*), and $I_\perp(t)$ (*faint*)

after the spin flip event, whereas $I_\perp$ merely peaks at the event and thereafter returns back to its previous steady state.

The bias voltage abruptly changes at the time τ_0, which here occurs after the spin flip, which leads to a change in I_z. A finite time span after the onset of the bias pulse, the spin flips, which again leads to that I_z changes and assumes a new steady state, whereas $I_\perp$ peaks.

The tiny peaks in the current component $I_\perp$ at the spin flip events are too small to be recorded by themselves, since they are occurring simultaneously as the component I_z changes from one state to another, which leaves a greater impression on the current. The total current (bold), nevertheless, varies due to the sum of the changes in I_z and $I_\perp$. The changes in the total current due to the onset (τ_0) and termination (τ_1) of the bias pulse are expected and be can controlled. The changes due to the spin flip event can, however be distinguished from the other variations and, thus, be taken as signatures of that the spin flips have occurred.

We replace the bias voltage pulse by a harmonic bias, e.g.

$$\delta\phi(t,t') = e\int_{t'}^{t}\Big(V_{\text{dc}} + V_{\text{ac}}\cos\omega_0\tau\Big)d\tau. \tag{12.46}$$

For this type of biasing, it will become evident below that the current displays higher order harmonics due to the interactions between the tunneling electrons and the local spin. It will also become evident that those additional features are present in the current only when the current density $j_{xy}^{(2)}$ acting on the local spin is non-vanishing.

Under the harmonically modulated bias voltage, the induced current density $\int_{-\infty}^{t} j_{xy}(2)(t,t')dt'$, as well as the current components I_0, I_z, and $I_\perp$, all have the time-dependence

$$\begin{aligned}
&\int_{-\infty}^{t} e^{i(x+eV_{\text{dc}})(t-t')+ieV_{\text{ac}}(\sin\omega_0 t-\sin\omega_0 t')/\omega_0}dt' \\
&\quad\sim\sum_{\mu\nu} J_\mu(eV_{\text{ac}}/\omega)J_\nu(eV_{\text{ac}}/\omega)\delta(x+eV_{\text{dc}}+\nu\omega_0)e^{i\omega_0(\mu-\nu)t}.
\end{aligned} \tag{12.47}$$

Thus, while all current densities inherits the fundamental frequency ω_0 of the bias voltage, the net of the bias on I_0 merely yields a single, or simple, harmonic variation, see Fig. 12.6(b) and (f). We understand this behavior since I_0 is a simple kinetic response to the bias voltage and is not influenced by the dynamics of the local spin.

The expression for the polar angle motion under the harmonic biasing, given in (12.36), reflects that the nutation of the spin is directly modulated by the frequency ω_0 of the harmonic bias. For all components satisfying $\mu-\nu=0$, we can make the replacement $\sin\omega_0(\mu-\nu)t/[\omega_0(\mu-\nu)]\rightarrow t$, which, thus, contribute with linear temporal factors in the product. For all other components, it is intelligible that the ratio eV_{dc}/ω_0 is important, since the smaller the frequency ω_0, the larger the amplitude of variations in the polar angle and, hence, the greater the impact of the local spin dynamics on the tunneling current. The explicit occurrence of the fundamental

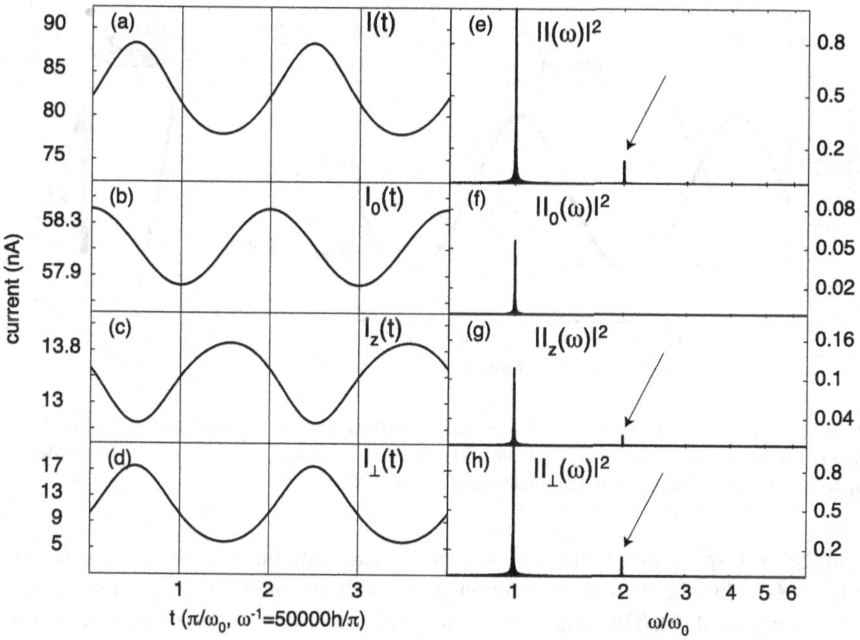

Fig. 12.6 Time-dependence and the its Fourier transforms of the total current I (**a**) and (**e**), respectively, and the corresponding current components I_0 (**b**) and (**f**), I_z (**c**) and (**g**), and $I_\perp$ (**d**) and (**h**). Here, $V(t) = eV_{dc} + eV_{ac} \cos \omega_0 t$ with $\omega_0^{-1} = 5h \times 10^4/\pi$, $p_L = p_R = 0.9$, $\theta = \pi$, $T_1/T_0 = 1/2$, $N_L T1 \sim N_R T_1 \sim 0.1$, $eV_{dc} \sim 1000\omega_0$, and $eV_{ac} \sim 5\omega_0$

frequency ω_0 in the time-dependence of ϑ suggests that we should expect to see higher order harmonics, both in ϑ but most importantly in the resulting tunneling current. The absence of such signatures would indicate that there is no (exchange) interaction between the tunneling current and the local spin.

The currents plotted in Fig. 12.6 are provided for $eV_{dc} \sim 1000\omega_0$ and $eV_{ac} \sim 5\omega_0$. Under those conditions, along with the other parameters given in the figure caption, the amplitude of the polar angle $\max(\vartheta - \min \vartheta)/\pi \sim 0.05$, see Fig. 12.7(a) (bold), which should be sufficiently large to generate an observable contribution from the currents I_z and $I_\perp$ through the components $\cos \vartheta$ and $\sin \vartheta$, see Fig. 12.7(a) (faint).

The time-dependent current components I_z and $I_\perp$ and their corresponding Fourier transforms are plotted in Fig. 12.6(c) and (d), and (g) and (h), respectively. From the plots, its clearly seen that the time-dependences of these current components are largely modified compared to the input time-dependence, cf. the harmonic behavior of I_0 in Fig. 12.6(b). The vertical lines in the plots are inserted as guides to the eye. Moreover, although the largest contribution to the current is carried by I_0, the time-dependence of the total current is significantly influenced by the contributions from I_z and $I_\perp$, cf. Fig. 12.6(a).

By studying the Fourier transforms of the currents, we also find that there is a second contribution to the total current (marked with an arrow at $2\omega_0$), see Fig. 12.6(e). Its presence has to originate from the interaction between the tunneling electrons

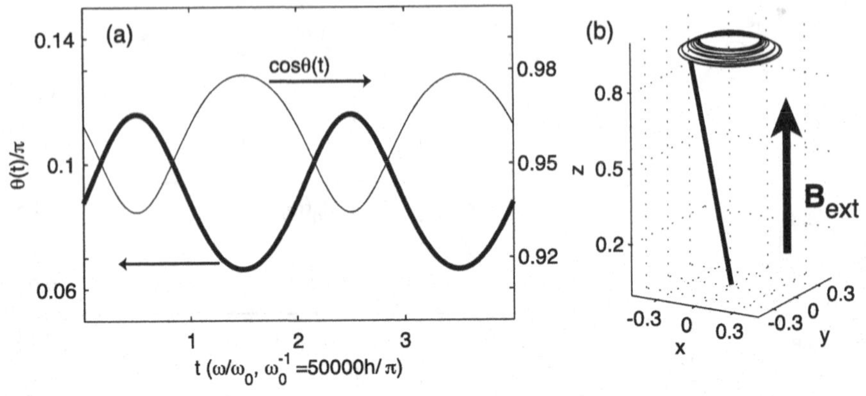

Fig. 12.7 (**a**) Polar angle motion (*bold*) and $\cos\vartheta(t)$ (*faint*) for the harmonic bias given Fig. 12.6, and (**b**) spin motion. In panel (**b**) we have added the static magnetic field $\mathbf{B}_{\text{ext}} = 1\hat{\mathbf{z}}$ T in order to speed up the azimuthal motion. Parameters are as in Fig. 12.6

and the local spin, since there is no second Fourier component in the current I_0, see Fig. 12.6(f), but there are such components in both the contributions I_z and $I_\perp$, see Fig. 12.6(g) and (h). This doubled frequency is, according to the above discussion, a direct fingerprint of the exchange interaction between the spin-polarized tunneling electrons and the local spin. Although even higher order harmonics are expected in the resulting current, the amplitudes of those are too small to provide visible signatures on the present scale.

The parameter regime in which the tunneling experiment is proposed should be within the realms of present state-of-the-art nanotechnology. Experiments aiming towards measurements of the spin dynamics would be extremely intriguing and useful for further advances within basic science of nanoscale physics.

References

1. Zhu, J.-X., Fransson, J.: J. Phys.: Condens. Matter **18**, 9929 (2006)
2. Fransson, J., Zhu, J.-X.: New J. Phys. **10**, 013017 (2008)
3. Fransson, J.: Phys. Rev. B **77**, 205316 (2008)
4. Fransson, J.: Nanotech. **19**, 285714 (2008)
5. Balatsky, A.V., Manassen, Y., Salem, R.: Phys. Rev. B **66**, 195416 (2002)
6. Zhu, J.-X., Balatsky, A.V.: Phys. Rev. Lett. **89**, 286802 (2002)
7. Zhu, J.-X., Balatsky, A.V.: Phys. Rev. B **67**, 174505 (2003)
8. Keldysh, L.V.: Sov. Phys. JETP **20**, 1018 (1965)
9. Zhu, J.-X., Nussinov, Z., Shnirman, A., Balatsky, A.V.: Phys. Rev. Lett. **92**, 107001 (2004)
10. Landau, L., Lifshitz, E.M.: Phys. Z. Sowjetunion **8**, 153 (1935)
11. Gilbert, T.L.: Phys. Rev. **100**, 1243 (1955) [Abstract only; full report, Armor Research Foundation Project No. A059, Supplementary Report, May 1, 1956] (unpublished); IEEE Trans. Magn. **40**, 34443 (2004)
12. Zhang, Z., Hammel, P.C., Wigen, P.E.: Appl. Phys. Lett. **68**, 2005 (1996)
13. Fransson, J., Eriksson, O., Balatsky, A.V.: arXiv:0812.4956v2
14. Fransson, J.: Nano Lett. **9**, 2414 (2009)

Index

J. Fransson, *Non-Equilibrium Nano-Physics*,
Lecture Notes in Physics 809,
DOI 10.1007/978-90-481-9210-6, © Springer Science+Business Media B.V. 2010